Activated Sludge Separation Problems

Activated Sludge Separation Problems

Theory, Control Measures, Practical Experiences

Second Edition

Edited by
Simona Rossetti, Valter Tandoi and Jiri Wanner

IWA PUBLISHING

Published by **IWA Publishing**
Alliance House
12 Caxton Street
London SW1H 0QS, UK
Telephone: +44 (0)20 7654 5500
Fax: +44 (0)20 7654 5555
Email: publications@iwap.co.uk
Web: www.iwapublishing.com

First published 2017

British Library Cataloguing in Publication Data
A CIP catalogue record for this book is available from the British Library

ISBN: 9781780408637 (Paperback)
ISBN: 9781780408644 (eBook)

Cover image: istockphoto

“To Flavio, Enza, Nadia and Filip”

Contents

Abbreviations

ADWF	Average Dry Weather Flow
Anammox®	Anaerobic ammonia oxydation
AOB	Ammonia Oxidizing Bacteria
A_2O®	Aerobic/Aerobic process
AS	Activated sludge
ASM	Activated Sludge Model
ATP	adenosine triphosphate
BCFS®	Biological/Chemical Phosphorus and Nitrogen removal
BD	Biodenitro
BOD	Biochemical Oxygen Demand
BNR	Biological Nutrients Removal
CE	Capillary Electrophoresis
CECs	contaminants of emerging concern
COD	Chemical Oxygen Demand
CMRs	completely mixed reactors coagulants
CPFRs	continuous plug-flow reactors
CTAB	Cetyltrimethyl ammonium bromide
CSTR	Continuous Stirred Tank Reactor
DAPI	4',6-diamidino-2-phenylindole
DCBT	Divalent Cation Bridging Theory
DNA	Deoxyribonucleic Acid
DO	Dissolved Oxygen
DSVI	Diluted SVI
EA	Extended Aeration
EBPR	Enhanced Biological Phosphorus Removal
EBNR	Enhanced Biological Nitrogen Removal
EBI	Eikelboom Index
ECP	extracellular polymer

EPS	Extracellular Polymeric Substance
FI	Filament Index
FISH	Fluorescence In Situ Hybridization
F/M ratio	food to microorganism ratio
GAO	Glycogen Accumulating Organisms
GALO	*Gordonia amarae* like organisms
GPCDR	Gram positive cocci/diphteroid Rods
HAB	Heterotrophic bacteria
HRT	Hydraulic Retention Time
IST	Individual Septic Tanks
LCFA	Long Chain Fatty Acids
MAR	microradiography
MATH	Microbial Adherence To Hydrocarbons
MBR	membrane bioreactors
MCRT	Mean Cell Retention Time (equal to sludge age)
MLSS	Mixed Liquor Suspended Solids
NAD+	Nicotinamide adenine dinucleotide
NALO	*Nocardia amarae* like organisms
N/DN	Nitrification/denitrification
NGS	Next Generation Sequencing
OD	Oxidation Ditch
OFR	overflow rate
OTU	Operational Taxonomic Units
OUR	Oxygen Uptake Rate
PAC	polyaluminium chloride
PAO	Polyphosphate Accumulating Organisms
poly-P	polyphosphate
PCR	Polymerase Chain Reaction
PE	Population Equivalent
PP	Polyphosphate
PHA	Poly-hydroxyalkanoate
PHB	Poly-hydroxybutyrate
PHV	Poly-hydroxyvalerate
q-PCR	quantitative PCR
q-FISH	quantitative-FISH
16S rDNA	16S ribosomial RNA gene (present in DNA)
RAS	Return Activated Sludge
RBC	Rotating Biological Contactors
RBCOD	Readily Biodegradable COD
RNA	ribonucleic acid
rRNA	ribosomal ribonucleic acid
SBR	Sequencing Batch Reactor
SI	Scum Index

SCFA	Short chain fatty acids
SFN	Specific filament index
SRT	Sludge Retention Time
SS	See TSS
SVI	Sludge Volume Index
TEFL	Total Extended Filament Length
THM	Triahalomethanes
TKN	Total Kjeldahl Nitrogen (the sum of organic and ammonia nitrogen)
TSS	Total Suspended Solids
TP	Total Phosphorus
UCT	University of Cape Town
UFR	underflow rate
VFA	Volatile Fatty Acids
VSS	Volatile Suspended Solids
WAS	Waste Activated Sludge
WWTP	Wastewater treatment plant
ZSV	Zone Settling Velocity

List of Contributors

A. Andreadakis
National Technical University of Athens, Athens, Greece
andre1@central.ntua.gr

S. Badoer
Gruppo Veritas, Venice, Italy
s.badoer@gruppoveritas.it

F. Bux
Durban University of Technology, Durban, South Africa
faizalb@dut.ac.za

M. Christie
Griffith University, Nathan (Qld), Australia
m.christie@griffith.edu.au

G. Daigger
University of Michigan-CEE, Ann Arbor (MI), USA
gdaigger@umich.edu

C. Davoli
Gruppo IREN, Reggio Emilia, Italy
claudia.davoli@gruppoiren.it

N. Deepnarain
Durban University of Technology, Durban, South Africa
ndeepnarain@gmail.com

J. Deurinck
Aquafin, Aartselaar, Belgium
jeroen.deurinck@aquafin.be

N. Durban
Irstea, Antony, France
nadege.durban@gmail.com

F. S. Estévez
Metropolitan Water Company of Seville (EMASESA), Seville, Spain
festevez@emasesa.com

Y. Fayolle
Irstea, Antony, France
yannick.fayolle@irstea.fr

A. Fenu
Aquafin, Aartselaar, Belgium
alessio.fenu@aquafin.be

S. Gillot
Irstea, Lyon-Villeurbanne, France
sylvie.gillot@irstea.fr

P. Griffiths
pH Water Consultants Pty Ltd, Brisbane (Qld), Australia
peter@phwater.com.au

D. Jenkins
David Jenkins and Associates Inc., Kensington (CA), USA
flocdr@yahoo.com

L. Juzan
Irstea, Antony, France

N. Kreuzinger
Technische Universität Wien, Wien, Austria
norbkreu@iwag.tuwien.ac.at

O. Krhutkova
LentiKat's a.s., Prague, Czech Republic
krhutkova@lentikats.eu

S. Kumari
Durban University of Technology, Durban, South Africa
sheenas@dut.ac.za

C. Levantesi
CNR-IRSA, Monterotondo (RM), Italy
levantesi@irsa.cnr.it

M. Majone
Sapienza University, Rome, Italy
mauro.majone@uniroma1.it

D. Mamais
National Technical University of Athens, Athens, Greece
mamais@central.ntua.gr

N. Matsché
Technische Universität Wien, Wien, Austria
nmatsche@iwag.tuwien.ac.at

D. Mosca Angelucci
CNR-IRSA, Monterotondo (RM), Italy
mosca@irsa.cnr.it

K. Muda
Universiti Teknologi Malaysia (UTM), Johor, Malaysia
khalida@utm.my

P. H. Nielsen
AAU, Aalborg, Denmark
phn@bio.aau.dk

M. Nierychlo
AAU, Aalborg, Denmark
mni@bio.aau.dk

C. Noutsopoulos
National Technical University of Athens, Athens, Greece
cnoutso@central.ntua.gr

R. Qi
RCEES, CAS, Beijing, China
qirong@rcees.ac.cn

E. Reina
Bioindication Group Seville (GBS), Seville, Spain
gbs@asociaciongbs.com

E. Rodríguez
Bioindication Group Seville (GBS), Seville, Spain
gbs@asociaciongbs.com

S. Rossetti
CNR-IRSA, Monterotondo (RM), Italy
rossetti@irsa.cnr.it

I. Ruzickova
University of Chemistry and Technology, Prague, Czech Republic
iveta.ruzickova@vscht.cz

R. Seviour
La Trobe University Bundoora, Melbourne (Victoria), Australia
R.Seviour@latrobe.edu.au

M. Sillanpää
Lappenranta University of Technology (LUT), Mikkeli, Finland
Mika.sillanpaa@lut.fi

J. Sohaili
Universiti Teknologi Malaysia (UTM), Johor, Malaysia
johansohaili@utm.my

H. Stratton
Griffith University, Nathan (Qld), Australia
h.stratton@griffith.edu.au

V. Tandoi
CNR-IRSA, Monterotondo (RM), Italy
tandoi@irsa.cnr.it

M. C. Tomei
CNR-IRSA, Monterotondo (RM), Italy
tomei@irsa.cnr.it

M. Torregrossa
University of Palermo, Palermo, Italy
michele.torregrossa@unipa.it

S. Van Damme
Aquafin, Aartselaar, Belgium
Sandra.vandamme@aquafin.be

J. Wang
RCEES, CAS, Beijing, China
kenni125@qq.com

J. Wanner
University of Chemistry and Technology, Prague, Czech Republic
jiri.wanner@vscht.cz

M. Yang
RCEES, CAS, Beijing, China
yangmin@rcees.ac.cn

N. S. Zaidi
Universiti Teknologi Malaysia (UTM), Johor, Malaysia
nursyamimi@utm.my

Preface

The Activated Sludge (AS) Process is a traditional technology more than 100 years old (invented in 1914) but is still widely adopted worldwide for its convenience and simplicity: an impressive number (many hundreds of thousands) of this kind of system are in operation.

Occasionally, problems such as bulking and foaming occur, causing regulation violations and large investments are often required immediately controlling them. For this reason, an intense research effort has been made during the last few decades to face these problems, and this book details the work undertaken during the last ten years, when the first Edition appeared. A large contribution comes from research centres, universities, water companies, consultants and from the work done in the IWA Specialist Groups, such as Microbial Ecology and Wastewater Engineering (the former "Activated Sludge Population Dynamics") and Design, Operation and Costs of Large Wastewater Treatment Plants.

This book describes first the approach for facing the problem of characterising a complex microbial community, such as the activated sludge, its main components, their distinctive properties and interconnected metabolisms. Particular emphasis is reserved for the storage processes, concerning mechanisms, kinetics, role; these fundamental data are still scarce for many filamentous and floc former bacteria. This new Edition, keeping a traditional and pragmatic practical approach, reports also the potentialities of biomolecular tools available today (such as FISH and q-PCR), to proper apply knowledge-driven remedial actions. The new set of data on strategies applied in thirteen different countries all around the world to face bulking and foaming problems are presented. It is possible to evaluate that both

doi: 10.2166/9781780408637_017

advanced and consolidated biomolecular tools are now applied in many of them with the purpose of a precise and deep understanding of the basic mechanisms that determine the activated sludge composition and eventually the wastewater treatment plant performances. The main reasons supposed to favour the different filamentous microorganisms growth in activated sludge, the efficacy of the biomolecular tools available today, together with the methods to avoid the proliferation of these organisms, are critically presented.

The book also provides an explanation of basic activated sludge process principles and of parameters necessary for process control and operation. The theory of secondary clarifiers is described in the extent necessary for understanding the construction and operation of secondary clarifiers. The activated sludge reactor and secondary clarifiers are treated as one system and the interactions are explained. The wide range of experiences around the world is documented and the methods to avoid the proliferation of these organisms are presented and critically reviewed.

The book *Activated Sludge Separation Problems: Theory, Control Measures, Practical Experiences* consists of six chapters, presenting up-to-date technical and scientific aspects of these processes. Particularly detailed is Chapter 6, "Experiences in different Countries", written by 36 authors reporting a picture of the work done at their places, located in Africa, America, Asia, Australia, and Europe.

The first Edition of the book proved to be a valuable help for students of environmental engineering and useful handbook for wastewater specialists, plant operators and designers of activated sludge plants. The Editors of the book believe that the second Edition will find also new readers from among specialists in wastewater operation laboratories, especially from those studying activated sludge separation properties. The second Edition will give hopefully faster and more comprehensive answers to questions asked by people from activated sludge plants because they can compare their problems and solutions with the experience collected from so many countries around the globe.

The Editors have been actively involved for many years in organizing international workshops and conferences under the umbrella of IWA. Therefore they know well how important it is for the career of young professional people to take part in such kinds of information exchange. Therefore five per cent of the income originating from this book sale, which were given up from the Editors, will be reserved to young researchers to sustain their participation at IWA Specialized Conferences on the subject of the book.

Simona Rossetti and Valter Tandoi
Institute for Water Research, National Research Council, Rome, Italy

Jiri Wanner
University of Chemistry and Technology, Prague, Czech Republic

Chapter 1

Wastewater characterization

M. C. Tomei and D. Mosca Angelucci

1.1 INTRODUCTION

At the beginning of wastewater treatment 'history', wastewater characterization data were limited to BOD_5 and COD, suspended solids and ammonia concentrations. In recent decades model development has required wastewater characterization in order to better describe complex biological process systems such as activated sludge.

The first simple models, based on Monod kinetics which considered one carbonaceous substrate and one biomass as homogeneous substances, were inadequate to explain the bacterial selection mechanisms, which determine the equilibrium conditions in the reactors and the plant response under dynamic conditions; consequently it was necessary to introduce a more detailed substrate characterization. The first model including the distinction between soluble and suspended carbonaceous substrate (COD) was the Clifft and Andrews model (1981); subsequent models (Dold *et al.* 1980; Henze *et al.* 1987) recognized the importance of evaluating the COD and nitrogen substrate fractions in the influent not only in terms of the physical state but also in terms of biodegradability. This approach, in fact, allows us to better take into account the role of the colloidal fraction that includes readily and slowly biodegradable components. The COD characterization, as a function of biodegradability, is also the most effective in relation to the bulking phenomena, the kinetic competition between floc-forming and filamentous bacteria for readily biodegradable COD being one of the most important selection

Theory, Control Measures, Practical Experiences
Simona Rossetti, Valter Tandoi and Jiri Wanner
doi: 10.2166/9781780408637_001

mechanisms causing the predominance of one of the two species in the sludge biocenosis (Kappeler & Gujer, 1994).

A detailed and structured knowledge of wastewater composition is also necessary for understanding the bulking and foaming problems and for finding proper control measures.

In the following paragraphs, the different approaches to wastewater characterization will be presented.

1.2 GROSS PARAMETERS

Sewage is a complex mixture of inorganic and organic substances. The chemical and physical nature of municipal wastewater varies significantly both in terms of place and time, furthermore it can depend on variable industrial contributions.

Average wastewater characteristics are difficult to determine because of their temporal variability. The simplest method of characterizing municipal wastewater is the determination of 'gross parameters' such as BOD, COD, TSS, TKN and TP concentrations. An overview of the values of these parameters is reported in Tables 1.1–1.3. Table 1.1 shows annual pollution loads for different countries, Table 1.2 the pollutant concentration data for settled and raw municipal wastewater, while the typical composition of raw domestic/municipal wastewater is given in Table 1.3 for different concentration/dilution conditions.

Table 1.1 Pollution load (referred to persons) in various countries (from Henze *et al.* 2002).

Country	Parameter (kg $pers^{-1} \cdot yr^{-1}$)				
	BOD	TSS	N_T	P_T	Detergents
Brazil	20–25	20–25	3–5	0.6–1	0.5–1
Denmark	20–25	30–35	5–7	1.5–2	0.8–1.2
Egypt	10–13	15–25	3–5	0.4–0.6	0.3–0.5
Germany	20–25	30–35	4–6	1.2–0.6	0.7–1
India	10–15	–	–	–	–
Italy	18–22	20–30	3–5	0.6–1	0.5–1
Sweden	25–30	30–35	4–6	0.8–1.2	0.7–1
Turkey	10–15	15–25	3–5	0.4–0.6	0.3–0.5
Uganda	20–25	15–20	3–5	0.4–0.6	–
USA	30–35	30–35	5–7	1.5–2	0.8–1.2

Table 1.2 Typical pollutant concentration of settled and raw (for USA and UK) municipal wastewater in different countries (Al-Salem, 1987; Henze *et al.* 1987; Ekama & Marais, 1984; Gray, 2004). Data for Italy are referring to Regione Emilia Romagna (Regione Emilia Romagna ARPA E.R., 2002).

Parameter	**Country**							
(mg L^{-1})	**Denmark**	**Italy**	**Hungary**	**USA**	**UK**	**Jordan**	**South Africa**	**Switzerland**
BOD_5	100–350	220	–	250	326	770	150–300	–
COD	210–740	620	350	500	650	1830	300–600	220
SS	120–450	300	–	220	127	900	150–300	–
TKN	20–80	–	58	40[1]	66[1]	150[1]	30–70	27
Ammonia N	12–50	42	30	25	47	–	24–56	10
Total P	4–23	7	–	12		25	6–15	–

[1]Total Nitrogen.

Table 1.3 Typical composition of raw municipal wastewater: concentrated wastewater (high) represents cases with low water consumption and/or infiltration; diluted wastewater (low) represents high water consumption and/or infiltration (data from Henze *et al.* 2008).

Parameter (mg L^{-1})	High	Medium	Low
COD total	1200	750	500
COD soluble	480	300	200
COD suspended	720	450	300
BOD	560	350	230
VFA (as acetate)	80	30	10
N-total	100	60	30
Ammonia-N	75	45	20
Nitrite + Nitrate-N	0.5	0.2	0.1
Organic-N	25	15	10
TKN	100	60	30
P total	25	15	6
Ortho-P	15	10	4
Organic-P	10	5	2
TSS	600	400	250
VSS	480	320	200

1.3 PHYSICAL PROPERTIES

The physical properties of sewage are odour and colour, temperature, pH, conductivity and solids content and distribution.

The distinctive musty smell of sewage is produced by the two main groups of nitrogen and sulfur compounds and, to a minor extent, by chlorine and phenol compounds. In the first group amines, ammonia, diamines and skatole are included, while the second is comprised of hydrogen sulfide, mercaptans, organic sulfide and sulfur dioxide. Fresh sewage has a lightly soapy and cloudy appearance on the basis of its composition/concentration. With time, the sewage becomes stale and darkened in colour due the natural decay of organic materials, with an increased and pronounced smell.

Wastewater temperature is generally higher than that of tap water and it depends on geographic location and season. Typical temperature values range from 8–10°C during the winter in Northern Europe up to 28–30°C in the summer in Asiatic countries (Gray, 2004). The pH value is in the range of 6.7–8.2, and its value can be closely linked with the tap water hardness. The conductivity of domestic wastewater generally ranges from 50 to 1500 μS cm^{-1} (WEF, 2008).

The total solids in a wastewater consist of the insoluble or suspended solids and the soluble compounds dissolved in water. The suspended solids fraction (size from 1 μm to distinguishable gross matter) includes a settleable portion and colloidal particles, which are not removed by gravity settling. A simple classification based on size (Painter, 1971) distinguishes solid particles as settleable (>100 μm, about 50%), supracolloidal (1–100 μm, 30–70%) and colloidal (1 nm – 1 μm) the remaining 17–20%. The dissolved solids consist of both organic and inorganic substances and ions, which are present in solution in water (Metcalf & Eddy, 2003).

A more detailed classification of the solids fractions is presented in Table 1.4.

Table 1.4 Typical solids fraction and related COD in sewage (modified from Gray, 2004).

Solid Fraction	Total Solids		COD	
	mg L^{-1}	%	mg L^{-1}	%
Settleable	74	15	120	29
Supracolloidal	57	11	87	21
Colloidal	31	6	43	10
Soluble	351	68	168	40
Total	513	–	418	–

1.4 ORGANIC MATTER

Wastewater contains thousands of different organics: as regards a wastewater of medium strength, the organic fractions related to suspended solids and filterable solids are about 75 and 40%, respectively (Gray, 2004). The main organic substances found in wastewater are proteins (40 to 60%), carbohydrates (25 to 50%), and fats and oils (10%). The residual fraction includes myriad materials such as surfactants, vitamins, antibiotics, purines, hormonal compounds, pesticides, hydrocarbons, pigments and other synthetic compounds, and other micropollutants. Furthermore, urea, the chief constituent of urine, is another important organic compound contributing to wastewater, but, due to its ready decomposition to ammonia, undecomposed urea is only found in very fresh wastewater (Metcalf & Eddy, 2003).

Sewage composition depends on sewer characteristics. Long anaerobic retention times favour conversion of carbohydrates to volatile fatty acids and the rapid transformation of urea in ammonia as well as solubilization of particulates.

Table 1.5 shows the distribution of organic carbon compound groups in solution and in suspension.

Table 1.5 Typical distribution of organic constituents in municipal wastewater (modified from Painter, 1971).

Constituent	In Solution		In Suspension	
	Concentration (mg L^{-1})	C/C Total (%)	Concentration (mg L^{-1})	C/C Total (%)
Fats	–	–	140	50
Carbohydrates	70	31.3	34	6.4
Free and bound Amino acids	18	10.7	42	10
Volatile acids	25	11.3	12.5[1]	2.3[1]
Non-volatile acids	34	15.2		
Detergents (ABS)	17	11.2	5.9	1.8
Uric acid	1	0.5	–	–
Creatine	6	3.9	–	–
Amino sugars	–	–	1.7	0.3
Amides	–	–	2.7	0.6

[1]Sum of volatile and non-volatile acids.

The carbohydrates, widely distributed in nature as sugars, starches, cellulose and wood fibres, are mainly found in wastewater as glucose, sucrose and lactose with smaller proportions of galactose, fructose, xylose and arabinose and have a typical concentration interval of 50–120 mg L^{-1}.

In wastewater characterization the word 'fats' is a general term used to indicate not only fats but also oils and waxes present in wastewater. They mainly originate from food preparation (fats and oil are the third major component of foodstuffs) and to a minor extent from excreta and lubricating and road oils. They are weakly soluble, and are found essentially in the suspended fraction of wastewater at a concentration ranging from 40 to 150 mg L^{-1}. Fats are among the more stable of organic compounds and are characterized by low biodegradation kinetics because they need to be hydrolyzed prior to being metabolized by microorganisms. Hydrolysis products are fatty acids that increase the content of free fatty acids present in the sewage. A fraction of 70–90% of total fatty acids is constituted by palmitic, stearic and oleic acids.

Among volatile acids detected in sewage, the most important is acetic acid, whose concentration is in the range of 10–80 mg L^{-1} (Henze *et al.* 2008). It plays an important role in the storage phenomena and consequently in microbial selection. Other volatile acids present in significant quantities are propionic, butyric and valeric that, together with acetic, constitute 75–90% of the total acidity present in municipal wastewater. Non-volatile acids such as glutaric, glycolic, lactic, citric, and benzoic are present to a minor extent with concentrations ranging from 0.1 to 1 mg L^{-1}.

Proteins are also a source of carbon but their contribution is most significant for organic nitrogen, which constitutes about 16% of their structure. The breakdown of proteins by bacterial action gives rise to the formation of free amino acids, fatty acids, nitrogenous compounds, phosphates and sulfides. Concentrations for free and bonded amino acids (peptides and proteins) are <5 mgN L^{-1} and in the range of 4–15 mgN L^{-1}, respectively.

Urea is the major source of nitrogen (2–16 mgN L^{-1}); depending on the retention time in the sewer system, a significant part of urea is converted to ammonia both in anaerobic and aerobic conditions.

The remaining fraction of biodegradable matter not included in the above mentioned categories are compounds resulting from the combination of carbohydrates, proteins and fats such as lipoproteins and nucleoproteins. Lignin and cellulose constitute a slightly degradable fraction, while the organic unbiodegradable fraction is 20–40% of the total organic matter.

Organisms deriving from human excreta, soil and water including viruses, bacteria, fungi and protozoans are also present in sewage. A considerable fraction of them is pathogenic and has to be removed in the wastewater treatment plant.

1.5 INORGANIC MATTER

The content of inorganic matter in sewage is determined by two main factors: the characteristics in terms of the hardness of the water supplied and the nature of the pollutants. In hard wastewater the predominant ions are calcium, sodium and chloride. Other components detected in significant amounts are potassium, magnesium, sulfur compounds (mainly sulfates), phosphates, bicarbonates and ammonia.

Nitrogen and phosphorus are essentially the most important nutrients for microorganism growth. In wastewater, nitrogen occurs in four basic forms: organic nitrogen, ammonia (both ionized and free ammonia), nitrite, and nitrate. Inorganic nitrogen, which gives the majority of the total nitrogen, is constituted by ammonia, which is found in solution as ammonium (NH_4^+) or ammonia (NH_3) depending on the pH. Raw wastewater has higher concentrations of organic nitrogen and ammonia while the oxidized nitrogen forms are present only in very small amounts. The forms of nitrogen present in a wastewater indicate the level of organic stabilization (WEF, 2008).

Phosphorus is found in three different forms, orthophosphate, polyphosphate and organic phosphate. Orthophosphate is the predominant form and is directly assimilable by organisms. Both organic phosphate and polyphosphates must be transformed to orthophosphate prior to being utilized both in the plant and in the receiving water; as a result, the orthophosphate concentration in the plant effluent is an important parameter in evaluating the expected impact on the receiving water.

The average data for nitrogen and phosphorus in municipal wastewater are shown in Tables 1.2 and 1.3.

Other elements necessary for biological growth are sulfur and metals (but only in traces), which are usually present in sufficient (or excessive) amounts in sewage. Excessive amounts of metals can exert a toxic or inhibitory effect on the biological processes and constitute a problem in sludge treatment and disposal.

In Table 1.6 the range of concentration of metals in sewage is reported.

Table 1.6 Typical concentration of metals in municipal wastewater.

Heavy Metal ($mg\ m^{-3}$)	Gray (2004)	Wastewater Type (Henze *et al.* 2002)			
		Concentrated	Moderate	Diluted	Very Diluted
Aluminium	–	1000	650	400	250
Arsenic	–	5	3	2	
Cadmium	20–50	4	2	2	1
Chromium	10–400	40	25	15	10
Cobalt	<20	2	1	1	0.5
Copper	160–880	100	70	40	30
Iron	150–1300	1500	1000	600	400
Lead	10–1780	80	65	30	25
Manganese	200	150	100	60	40
Mercury	–	3	2	1	1
Nickel	20–330	40	25	15	10
Silver	–	10	7	4	3
Zinc	50–840	300	200	130	80

1.6 MICROPOLLUTANTS

The occurrence of micropollutants in the aquatic environment has become a worldwide issue of increasing environmental concern. Micropollutants, also defined as contaminants of emerging concern (CECs), consist of a vast and expanding array of anthropogenic as well as natural substances present in municipal wastewater. They are not the direct target for treatment, but they contribute to the toxicity of the wastewater, either in relation to the biological processes in the treatment plant or to the receiving waters (Henze *et al.* 2008).

The occurrence of so-called emerging or unregulated contaminants in different aquatic systems has become an environmental problem and the research is documenting with increasing frequency that many chemical and microbial constituents, which have not historically been considered as contaminants, are present in the environment on a global scale. Many definitions have been given to describe these 'new' contaminants or contaminants of environmental concern (CECs). According to the United States Geological Survey (USGS), they can be broadly defined as '*any synthetic or naturally occurring chemical or any microorganism that is not commonly monitored in the environment but has the potential to enter the environment and cause known or suspected adverse ecological and (or) human health effects*'. Table 1.7 shows the effects related to the most diffused classes of CECs. In some cases, release of CECs to the environment probably occurred for a long time, but may not have been identified and documented until new detection methodologies were developed. In other cases, synthesis of new chemicals or changes in use and disposal of existing chemicals can create new sources of emerging contaminants. Surely, effluents from WWTPs represent the primary source of these micropollutants for loading to aquatic systems (Du *et al.* 2014; Reemtsma *et al.* 2006). There is not a standardized categorization of CECs, most generally used categories include pharmaceuticals, personal care products, plasticizers, flame retardants, and pesticides. The typical concentrations for the most frequently detected CECs in municipal wastewater are shown in Table 1.8. Although most of these compounds are present at low concentrations, many of them raise considerable toxicological concerns, particularly when present as components of complex mixtures (Schwarzenbach *et al.* 2006). Given that conventional WWTPs are designed to reduce loads of carbon, nitrogen, and phosphorus, most CECs are not efficiently removed and may be discharged with WWTP effluents into receiving waters, and then occur in surface waters.

Table 1.7 Contaminants of emerging concern found in municipal wastewater: categories and effects (data from Raghav *et al.* 2013).

Category	Effects
Antibiotics	Antibiotic resistance in disease causing bacteria complicating treatment of infections
Disinfectants	Genotoxicity, cytotoxicity, carcinogenicity
Fire retardants	Endocrine disruption, indications of increased risk for cancer
Life-style products *(Caffeine, Nicotine)*	Can cause cellular stress, negative effects on reproductive activity in animals
Non-prescription drugs	Unknown health effects
Other prescription drugs	Increased cancer rates, organ damage
Personal care products	Bacterial resistance, endocrine disruption
Reproductive hormones	Endocrine disruption
Steroids	Endocrine disruption

Table 1.8 Typical concentrations for the most frequently detected CECs in municipal wastewater.

Category	Concentration ($\mu g\ L^{-1}$)	Reference
Pharmaceuticals		
Analgesics/anti-inflammatory	0.027–51	Margot *et al.* (2013)
Diclofenac	0.105–4.11	Miége *et al.* (2009)
Paracetamol	51	Margot *et al.* (2013)
	120	Martin Ruel *et al.* (2012)
Ibuprofen	41	Margot *et al.* (2013)
	1.2–2.7	Clara *et al.* (2005)
Antibiotics	0.065–2.3	Margot *et al.* (2013)
	0.024–0.14	Clara *et al.* (2005)
Ciprofloxacin	0.18–0.57	Miége *et al.* (2009)
Tetracyclin	0.24–0.79	Miége *et al.* (2009)
Anticonvulsants	0.11–3.9	Margot *et al.* (2013)
	0.3–1.8	Clara *et al.* (2005)
	0.7–1	Martin Ruel *et al.* (2012)
Carbamazepine	0.1–1.9	Miége *et al.* (2009)
Life-style products		
Caffeine	>10	Margot *et al.* (2013)
Aspartame	>10	Margot *et al.* (2013)
Acesulfame K	12–46	Buerge *et al.* (2009)
Sucralose	2–9	Hollender *et al.* (2009)
Personal care products		
Fragrance	0.02–3.1	Clara *et al.* (2005)
	0.33	Margot *et al.* (2013)
Tonalide	0.99	Miége *et al.* (2009)
Galaxolide	2.51	Miége *et al.* (2009)
Disinfectant		
Triclosan	0.38	Miége *et al.* (2009)
Endocrine disruptors		
Bisphenol A	0.8	Margot *et al.* (2013)
	0.7–2	Clara *et al.* (2005)
Nonylphenol	1.2–4	Clara *et al.* (2005)
Hormones		
Estriol	0.31	Margot *et al.* (2013)
Estrone	0.13	Margot *et al.* (2013)
17α-Estradiol	0.0015–0.0172	Miége *et al.* (2009)
17β-Estradiol	0.014	Margot *et al.* (2013)
	0.025–0.125	Miége *et al.* (2009)

1.7 BIODEGRADABILITY

1.7.1 COD, Nitrogen and phosphorus fractionation

In relation to understanding of bulking and foaming, and control, detailed wastewater characterization in terms of compound groups can be simplified by distinguishing carbon and nitrogen fractions as a function of biodegradability. This approach has been proposed by general activated sludge models (Dold *et al.* 1980; Henze *et al.* 1987) and is reported in detail in the IAWQ Scientific and Technical Reports No. 1 (Henze *et al.* 1987) and No. 2 (Henze *et al.* 1995).

The carbonaceous substrate, expressed as COD (C_{COD}), as stated by the nomenclature proposed in the Activated sludge model No. 1, can be divided into the following fractions:

$$C_{COD} = S_S + S_I + X_S + X_I$$

where

S_S is (dissolved) readily biodegradable organic matter,
S_I is inert soluble organic matter,
X_S is (suspended) slowly biodegradable substrate,
X_I is inert suspended organic matter.

The readily biodegradable substrate (which can be considered soluble in first approximation) consists of simple molecules directly metabolizable by heterotrophic bacteria for their maintenance and growth. VFAs (Volatile Fatty Acids), lower alcohols, lower amino acids and simple carbohydrates (Henze *et al.* 1994) are the main groups included in this fraction.

Slowly biodegradable substrate (which in first approximation can be considered particulate) consists of more complex compounds that need to be hydrolyzed prior to being utilized by the microorganisms. The slow biodegradability is a consequence of the lower kinetics of the hydrolysis process compared to the utilization kinetics, therefore the hydrolysis is the rate- limiting step of the whole biodegradation process. Both readily biodegradable substrates present in wastewater and the hydrolysis products support the growth of a certain type of filamentous bacteria.

Inert suspended organic matter is constituted by the inorganic fraction of the suspended solids and part of the suspended organic matter which is not biodegradable. In this fraction, compounds very slowly biodegradable (so practically unbiodegradable under the usual operating conditions), such as lignin and cellulose, are also included.

Inert soluble organic compounds are presumably constituted by xenobiotic compounds, which are scarcely biodegradable or unbiodegradable and are found integrally in the effluent.

In the Activated Sludge Model No. 2 (Henze *et al.* 1995), including biological phosphorus removal, a further subdivision has been considered: the S_S is constituted by the sum of two components: fermentable readily biodegradable organic substances S_F and S_A, fermentation products assimilable to acetate.

As regards the nitrogenous substrate, the main distinction is between ammonia and organic components. For the latter the same approach as for COD is considered. Nitrite and nitrate are present to a minor extent.

Consequently, nitrogen (C_N) in municipal wastewater is constituted by the following fractions:

$$C_N = S_{NH4} + S_{NOX} + S_{ND} + S_{NI} + X_{NS} + X_{NI}$$

where

S_{NH4} is ammonium nitrogen,
S_{NOX} is nitrite plus nitrate nitrogen,
S_{ND} is soluble biodegradable organic nitrogen (assimilable to readily biodegradable nitrogen),
S_{NI} is soluble inert organic nitrogen,
X_{NS} is particulate biodegradable organic nitrogen (assimilable to slowly biodegradable nitrogen),
X_{NI} is particulate inert organic nitrogen.

The soluble inert compounds are usually negligible while the particulate substances are associated with the suspended carbonaceous substrate. The soluble oxidizable fraction is constituted essentially by ammonia and its salts and, to a minor extent, by soluble organic nitrogen, while the biodegradable particulate, likewise the particulate biodegradable COD, are hydrolyzed to soluble organic nitrogen which in turn is converted to ammonia.

Phosphorus (C_P) is also found in dissolved and suspended phases, it can be divided into the following fractions:

$$C_P = S_{PO4} + S_{p\text{-}P} + S_{org\text{-}P} + X_{org\text{-}P}$$

where

S_{PO4} is soluble inorganic ortophosphate,
$S_{p\text{-}P}$ is soluble inorganic polyphosphate,
$S_{org\text{-}P}$ is soluble organic phosphorus,
$X_{org\text{-}P}$ is particulate organic phosphorus.

The inorganic orthophosphate is often the prevalent fraction, the suspended fraction is constituted by organic phosphorus associated with suspended COD.

Figure 1.1 shows the carbonaceous substrate fractionation in terms of groups of compounds and related biodegradability in municipal wastewater while data available for different countries are reported in Table 1.9 for settled municipal wastewater and in Table 1.10 for raw wastewater. The inappropriate ratio between COD (S_S–X_S) and N and P concentration results in bulking by certain types of filaments.

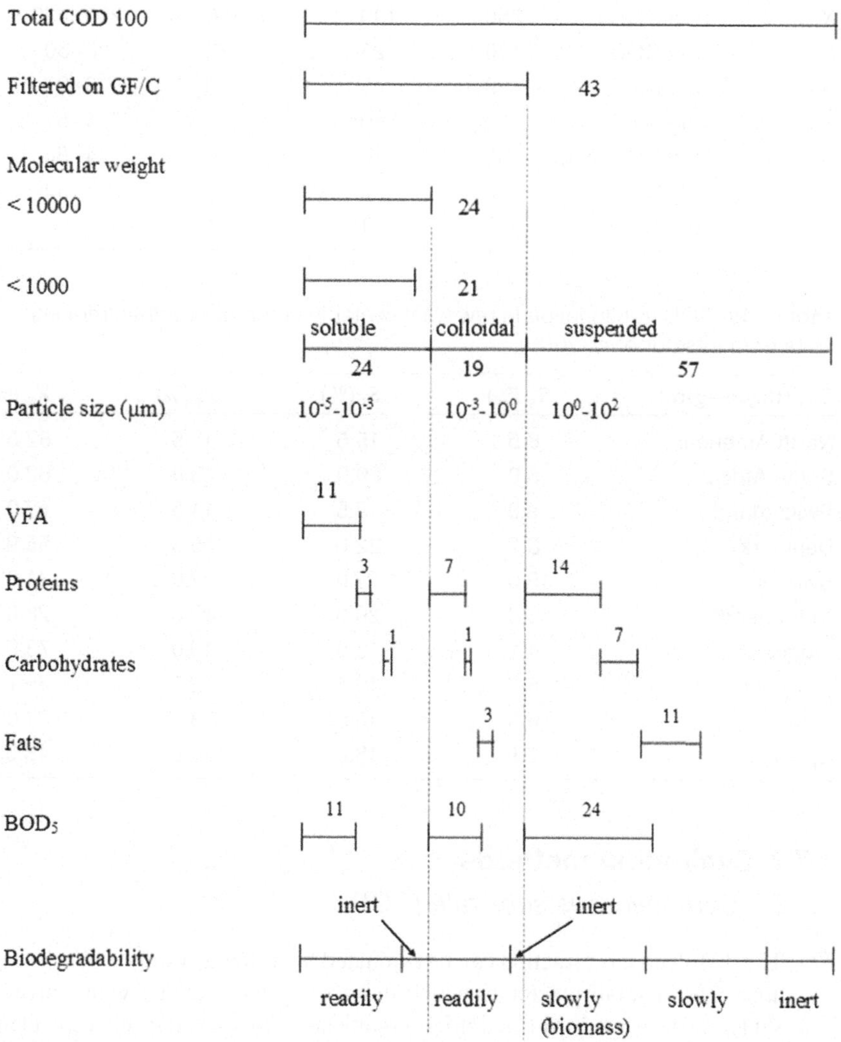

Figure 1.1 Carbonaceous substrate fractionation and biodegradability in municipal wastewater (modified from Henze *et al.* 1994).

Table 1.9 COD and nitrogen fractions in settled municipal wastewater in different countries (data from Henze *et al.* 1987; Ekama & Marais, 1984).

Parameter	Units	Country			
		Denmark	Switzerland	Hungary	South Africa
S_s	mg COD L^{-1}	125	70	100	68–135
S_I	mg COD L^{-1}	40	25	30	31–62
X_s	mg COD L^{-1}	250	100	150	210–420
X_I	mg COD L^{-1}	100	25	70	15–30
S_{ND}	mg N L^{-1}	8	5	10	3.6–9
X_{ND}	mg N L^{-1}	10	10	15	2.4–5
S_{NH}	mg $NH_3^-\cdot$ N L^{-1}	30	10	30	24–56
S_{NI}	mg N L^{-1}	2	2	3	1.5–3.5
S_{NO}	mg $NO_3\cdot$ N L^{-1}	0.5	1	1	–

Table 1.10 COD fractionation in raw wastewater in different countries/regions (data from Pasztor *et al.* 2009).

Country/Region	S_S (%)	S_I (%)	X_I (%)	X_S (%)
North America	8.5	15.5	13.5	62.5
South Africa	5.0	20.0	13.0	62.0
Switzerland	9.0	9.5	14.5	67.0
Denmark	6.2	22.6	15.3	55.9
Sweden	15.0	27.0	17.0	41.0
Netherlands	6.0	26.0	39.0	28.0
France	4.1	3.0	19.0	73.9
Germany	6.3	16.5	12.1	65.1
Italy	6.0	15.0	8.0	71.0
Spain	8.5	18.3	24.9	48.3

1.7.2 Evaluation methods

1.7.2.1 Carbonaceous substrate (COD)

S_I The dissolved inert fraction can be evaluated by utilizing a simple bench scale apparatus; a small reactor in which the wastewater is mixed with activated sludge at the same F/M (food/microorganisms) ratio as in the full scale plant. The residual COD in the effluent, measured after a contact time of 20–30 days, gives the S_I value. Control parallel tests without substrate are advisable (Henze *et al.* 2002).

S_S Two methods are actually available for readily biodegradable COD evaluation: the respiration (or biological) method (Ekama *et al.* 1986) and the flocculation method (Mamais *et al.* 1993).

The first one is accurate and reliable, but time consuming and difficult to utilize routinely. A bench scale plant consisting of an aeration basin and a secondary settler is required for the measurement. This biological plant is operated at 2.5 days MCRT (Mean Cell Residence Time) in a semi-continuous mode with intermittent feed (12 h feed on, 12 h feed off). The S_S value is determined based on OUR (Oxygen Uptake Rate) step variation (a sudden decrease) detected in the reactor after the feed period.

The S_S evaluation requires knowledge of the growth yield coefficient for the heterotrophic biomass that the authors assume equal 0.67 mg biomass (COD)/mg substrate (COD). All the details for the experimental procedure are reported in (Ekama *et al.* 1986).

The flocculation method, compared to the biological, is more rapid and easy to use in the day-to-day management control and operation of a treatment plant. The principle of this method is based on the experimental evidence that membrane filtration of a sample that has been previously flocculated (in this case by precipitating $Zn(OH)_2$ at pH 10.5) will produce a filtrate containing only truly soluble organic matter because the colloidal particles, which usually pass through the filter, are removed during the flocculation step. The measurement is carried out by the following sequence: the sample is flocculated by adding zinc sulfate, then the pH is adjusted to 10.5 by dosing sodium hydroxide and, finally, the clear supernatant (obtained after a quiescence time of a few minutes) is filtered using a 0.45 μm membrane filter.

The COD of the filtrate (COD_{sol}) is related to the S_S by the following:

$$S_S = COD_{sol} - S_I$$

Consequently, an estimation of S_I is required to evaluate S_S.

All the details of the experimental procedure are reported in Mamais *et al.* (1993).

At present, the biological determination is believed to be the safest (Henze *et al.* 1995), but due to the complexity of the measurement, a good compromise could be to employ the physical-chemical determination on the day to day practice and periodically verify the reliability of the obtained values by utilizing the respirometric test as a control.

S_A This fraction is mainly constituted by acetic acid, so a reasonable estimation can be obtained by determining the acetic acid concentration utilizing the usual chromatographic methods (gas chromatography and ion chromatography).

X_I The procedure for estimating the inert suspended organic fraction is quite complex. Experimental data obtained from continuous lab or pilot scale plants at different sludge age are required. The sludge production data are analyzed with a mathematical model of the activated sludge process and the X_I concentration is estimated by calibrating the model in order to obtain the best fitting of the sludge production data. Details on this procedure are reported in Gujer and Henze (1991).

X_S Finally, the slowly biodegradable organic fraction is determined as the difference between the total COD (COD_T) and the other fractions:

$$X_S = COD_T - S_I - S_S - X_I$$

1.7.2.2 Nitrogen and phosphorus

Nitrogen fractions can be evaluated by conventional analysis methods for Kjeldahl nitrogen (TKN including ammonia and organic nitrogen), nitrite, nitrate and ammonium (APHA, 1998).

An approximate estimation of the organic nitrogen fractions can also be obtained from the COD values considering that the nitrogen/COD ratio in the various organic fractions for different domestic wastewaters varies in quite a narrow range of 0.04–0.08 mgN $mgCOD^{-1}$ (Henze *et al.* 2002).

Phosphorus fractions determination can be easily performed with the usual well tested methods (APHA, 1998).

1.7.2.3 Micropollutants

The quantification of micropollutants in environmental samples can be an analytical challenge, because of the complexity of the matrix and their low levels of occurrence. Due to the huge and increased number of studies about the occurrence and fate of emerging contaminants in recent years (Clara *et al.* 2005; Buerge *et al.* 2009; Margot *et al.* 2013), analytical methods related to one or more compound classes of emerging contaminants have been developed in many studies (Vanderford *et al.* 2003; Gros *et al.* 2006). Furthermore, new research programs are ongoing on reliable analytical methods, which allow the rapid, sensitive and selective determination of micropollutants in environmental samples, characterized by high throughput, low limits of quantification and high precision.

Due to low environmental concentrations, a crucial step in analytical procedures is the enrichment of the sample, which is traditionally performed using off-line solid-phase extraction (SPE). With this methodology, many solid phases, eluent schemes, and final solvents with and without ion pairing reagents, buffers, and modifiers can be used for the clean-up and concentration of water samples. However, the majority of multi-group methods may include more than one extraction step, with different sorbent materials or elution solvents, fractionating target compounds in groups according to their physical-chemical properties (Gros *et al.* 2006). After extraction, a purification step is generally included in the analytical procedures, in order to minimize the matrix effects, especially in complex environmental samples.

Instrumental analysis is usually performed by gas or liquid chromatography–mass spectrometry (GC-MS or LC-MS): GC is more advisable for the analysis of non-polar and volatile compounds while LC is the favoured technique for separation of polar organic pollutants, with the advantage, if compared to GC, of shorter analysis time, required for monitoring studies. For the detection of the analytes, coupling MS–MS is progressively being used, replacing other detectors, in combination with LC (fluorescence, UV, PAD) and GC (FID, ECD) (Fatta-Kassinos *et al.* 2011). Liquid chromatography coupled to tandem mass spectrometry (LC–MS/MS) is the most applied (Huntscha *et al.* 2012) methodology due to its versatility, specificity and selectivity. These features are especially required for the analysis of pharmaceuticals, steroids, and personal care products; furthermore their detection limits are generally below 1.0 pg on column and all compounds had a reporting limit of 1.0 ng L^{-1} in water (Gros *et al.* 2006). Capillary electrophoresis (CE) has also been used for analysis of pharmaceuticals): this methodology is less complex and less expensive than GC and LC, but also less sensitive, with detection limits in the microgram per litre range (Fatta-Kassinos *et al.* 2011).

REFERENCES

Al-Salem S. S. (1987). Evaluation of the AL Samra waste stabilization pond system and its suitability for unrestricted irrigation. Paper Prepared for the Land and Water Development Division. FAO, Rome.

APHA (1998). Standard Methods for the Examination of Water and Wastewater, 20th edn.

Buerge I. J., Buser H.-R., Kahle M., Müller M. D. and Poiger T. (2009). Ubiquitous occurrence of the artificial sweetener acesulfame in the aquatic environment: an ideal chemical marker of domestic wastewater in groundwater. *Environmental Science Technology*, **43**, 4381–385.

Clara M., Strenn B., Gans O., Martinez E., Kreunzinger N. and Kroiss H. (2005). Removal of selected pharmaceuticals, fragrances and endocrine disrupting compounds in a membrane bioreactor and conventional wastewater treatment plants. *Water Research*, **39**, 4797–4807.

Clifft R. C. and Andrews J. F. (1981). Predicting the dynamics of oxygen utilization in the activated sludge process. *Journal Water Pollution Control Federation*, **53**, 1219–1232.

Dold P. L., Ekama G. A. and Marais G.v.R. (1980). A general model for the activated sludge process. *Progress in Water Technology*, **12**, 47–77.

Du B., Price A. E., Casan Scott W., Kristofco L. A., Ramirez A. J., Chambliss C. K., Yelderman J. C. and Brooks W. B. (2014). Comparison of contaminants of emerging concern removal, discharge, and water quality hazards among centralized and on-site wastewater treatment system effluents receiving common wastewater influent. *Science of the Total Environment*, **466–467**, 976–984.

Ekama G. A. and Marais G.v.R. (1984). Nature of municipal wastewaters. In: Theory Design and Operation of Nutrient Removal Activated Sludge Processes. Chap. 2, Water Research Commission, Pretoria, South Africa.

Ekama G. A., Dold P. L. and Marais G.v.R. (1986). Procedures for determining influent COD fractions and the maximum specific growth rate of heterotrophs in activated sludge systems. *Water Science & Technology*, **18**(6), 91–114.

Fatta-Kassinos D., Meric S. and Nikolaou A. (2011). Pharmaceutical residues in environmental waters and wastewater: current state of knowledge and future research. *Analytical and Bioanalytical Chemistry*, **399**, 251–275.

Gray N. F. (2004). How nature deals with waste. In: Biology of Wastewater Treatment. Chap. 1, Oxford University Press, UK.

Gros M., Petrović M. and Barceló M. (2006). Development of a multi-residue analytical methodology based on liquid chromatography–tandem mass spectrometry (LC–MS/MS) for screening and trace level determination of pharmaceuticals in surface and wastewaters. *Talanta*, **70**(4), 678–690.

Gujer W. and Henze M. (1991). Activated sludge modelling and simulation. *Water Science & Technology*, **23**(4–6), 1011–1023.

Henze M., Grady Jr. C. P. L., Gujer W., Marais G.v.R. and Matsuo T. (1987). Activated Sludge Model No 1, IAWPRC Scientific and Technical Reports No 1, IAWPRC, London, UK, 1–33.

Henze M., Kristensen G. H. and Strube R. (1994). Rate-capacity characterization of wastewater for nutrient removal processes. *Water Science & Technology*, **29**(7), 101–107.

Henze M., Gujer W., Mino T., Matsuo T., Wentzel M. C. and Marais G.v.R. (1995). Activated Sludge Model No 2. IAWQ Scientific and Technical Reports No 2, IAWQ, London, UK, 1–32.

Henze M., Harremoes P., la Cour Jansen J. and Arvin E. (2002). Wastewater treatment: Biological and chemical processes, 3rd edn. Springer Verlag Berlin, Heidelberg.

Henze M., van Loosdrecht M. C. M., Ekama G. A. and Brdjanovic D. (2008). Biological Wastewater Treatment: Principles Modelling and Design. IWA Publishing, London, UK.

Hollender J., Zimmermann S. G., Koepke S., Krauss M., McArdell C. S., Ort C., Singer H., von Gunten U. and Siegrist H. (2009). Elimination of organic micropollutants in a municipal wastewater treatment plant upgraded with a full-scale postozonation followed by sand filtration. *Environmental Science Technology*, **43**, 7862–7869.

Huntscha S., Singer H. P., McArdell C. S., Frank C. E. and Hollender J. (2012). Multiresidue analysis of 88 polar organic micropollutants in ground, surface and wastewater using online mixed-bed multilayer solid-phase extraction coupled to high performance liquid chromatography–tandem mass spectrometry. *Journal of Chromatography A*, **1268**, 74–83.

Kappeler J. and Gujer W. (1994). Development of a mathematical model for "aerobic bulking". *Water Research*, **28**, 303–310.

Mamais D., Jenkins D. and Pitt P. (1993). A rapid physical chemical method for the determination of readily biodegradable soluble COD in municipal wastewaters. *Water Research*, **27**, 195–197.

Margot J., Kienle C., Magnet A., Weil M., Rossi L., de Alencastro L. F., Abegglen C., Thonney D., Chevre N., Scharer M. and Barry D. A. (2013). Treatment of micropollutants in municipal wastewater: ozone or powdered activated carbon? *Science of the Total Environment*, **480**, 461–462.

Martin Ruel S., Choubert J. M., Budzinski H., Miège C., Esperanza M. and Coquery M. (2012). Occurrence and fate of relevant substances in wastewater treatment plants regarding Water Framework Directive and future legislations. *Water Science & Technology*, **65**, 1179–1189.

Metcalf and Eddy, Inc., Revised by Tchobananoglous G., Burton F. L. and Stensel H. D. (2003) Wastewater Engineering: Treatment, Disposal and Reuse, 4th edn., McGraw-Hill, Boston, MA.

Miége C., Choubert J. M., Ribeiro L., Eusébe M. and Coquery M. (2009). Fate of pharmaceuticals and personal care products in wastewater treatment plants: conception of a database and first results. *Environmental Pollution*, **157**, 1721–1726.

Painter H. A. (1971). Chemical physical and biological characteristics of wastes and waste effluents. In: Water and Water Pollution Handbook 1, Marcel Dekker, New York, 329–363.

Pasztor I., Thury P. and Pulai J. (2009). Chemical oxygen demand fractions of municipal wastewater for modelling of wastewater treatment. *International Journal of Environmental Science and Technology*, **6**(1), 51–56.

Raghav, M., Eden, S., Mitchell, K., Witte B. and Arroyo (2013). Water Resource Research Center, Arizona, p. 12.

Reemtsma, T., Weiss, S., Mueller, J., Petrović, M., Gonzalez, S., Barceló, D., Ventura F. and Knepper T. (2006). Polar pollutants entry into the water cycle by municipal wastewater: a European perspective. *Environmental Science Technology*, **40**(17), 5451–5458.

Regione Emilia Romagna ARPA E.R. (2002). Report on Wastewater Treatment Plants.

Schwarzenbach R. P., Escher B. I., Fenner K., Hofstetter T. B., Johnson C. A., von Gunten U. and Wehrli B. (2006). The challenge of micropollutants in aquatic systems. *Science*, **313**, 1072–1077.

Vanderford B. J., Pearson R. A., Rexing D. J. and Snyder S. A. (2003). Analysis of Endocrine Disruptors, Pharmaceuticals, and Personal Care Products in Water Using Liquid Chromatography/Tandem Mass Spectrometry. *Analytical Chemistry*, **75**, 6265–6274.

WEF (Water Environment Federation) (2008). Operation of Municipal Wastewater Treatment Plants: MoP No. 11, 6th edn, WEF Press.

Chapter 2

The microbiology of the activated sludge process

S. Rossetti, C. Levantesi and V. Tandoi

2.1 INTRODUCTION

The Activated Sludge (AS) process is an old technology, but still the most commonly used for the treatment of domestic and industrial wastewaters. New configurations have been adopted (multistage plants for nutrient removal) and new technologies are being developed (e.g. membrane bioreactors, biofiltration, sequencing batch biofilm reactors) but all derive from the traditional AS process. Knowledge of the populations operating in these systems originates from studies of AS and has helped to better operate the process, giving scientific bases to an originally empirical approach.

Despite the widespread use of activated sludge systems, this technology can suffer from operational dysfunction and instability in pollutant removal. This chapter contains updated information about the main bacterial populations operating in AS systems: heterotrophs, nitrifiers, denitrifiers, polyphosphate and glycogen accumulating organisms (PAO and GAO respectively). These microorganisms are important in terms of both their function and competition with filamentous bacteria, which often cause serious problems in the AS process. Knowledge of the identity and properties of these populations is crucial to properly address any strategy directed at modifying the composition of the biomass (i.e. promoting the growth of bacteria that play a positive role such as nitrifiers and denitrifiers, or depressing the growth of problematic organisms, such as the filamentous bacteria). The proper identification of a bacterium is a complex procedure, and for many years the identification of many microbial populations present in AS was

doi: 10.2166/9781780408637_021

performed using morphological criteria (i.e. cell shape, dimensions, Gram staining, etc.). Today, precise and rapid methods, such as Real Time PCR, fluorescence *in situ* hybridization (FISH), are available for some bacteria, but morphological identification is still necessary and useful because information about all of the populations in AS is still incomplete.

> **Focus point:** the identification of the microbial components of activated sludge is crucial for proper plant operation. Morphological microscopic identification represents an approximate, but still valid, evaluation tool (Eikelboom & van Buijsen, 1981 and updates; Jenkins *et al.* 1993 and updates; Eikelboom, 2000); however, modern and precise methods of identification are quickly developing and are already available for some important bacterial components of activated sludge.

2.2 MICROORGANISMS IN ACTIVATED SLUDGE

AS is a complex microbial community composed mainly of bacteria (the real actors of the depuration process), also sometimes containing protozoa and metazoa (i.e. rotifers, nematodes, etc.) (Table 2.1, Fig. 2.1).

2.2.1 Bacteria: cell structure

Bacteria, the main AS components, are microorganisms, between 0.2 and 3 μm in size (but in AS are often 0.5–2 μm small), with variable cell morphology. Several distinct shapes of bacteria can be recognised, including cells with a cylindrical shape (rods), with spherical or ovoid morphology (cocci), curved rods with spiral-shaped patterns (spirillum), and tetrads (four square cells linked together).

The structure of protozoa and metazoan cells, is more complex, with a defined nucleus surrounded by a membrane and several other intracellular organelles. These organisms are called eukaryotes, while bacteria and archaea are called prokaryotes.

A bacterial cell contains the following structures:

- Cell wall, a rigid structure outside the cytoplasmic membrane, provides the cell with its structural rigidity and shape and protection against osmotic lysis.
- Cytoplasmic cell membrane, the real barrier separating inside from outside of the cell.
- Cytoplasm which contains all the complex biochemical machinery for the reproduction and the survival of the bacterial cell. The major components of the cytoplasm, other than water, are macromolecules, ribosomes (containing the ribosomal RNAs), small organic molecules serving mainly as precursors of macromolecules and various inorganic ions.
- Nucleoid, bacterial DNA in an aggregated form, not surrounded by a membrane such as in eukaryotes.

Table 2.1 Main microbial components present in activated sludge.

Microorganism		Function	Reference
Floc-forming bacteria			
Heterotrophs	***Traditional aerobic bacteria*** *Zoogloea, Flavobacterium, Alcaligenes, Aeromonas, Pseudomonas, Bacillus*	Organic carbon oxidation	Seviour and Nielsen (2010)
	Denitrifiers *Azoarcus, Thauera, Pseudomonas, Bacillus, Paracoccus, Comamonas*	Nitrate reduction to nitrogen gas	Seviour and Nielsen (2010)
	Polyphosphate accumulating bacteria (PAO) '*Candidatus* Accumulibacter phosphatis' (in *Rhodocyclus* Genus)	Biological phosphorus removal	Hesselmann *et al.* (1999) Crocetti *et al.* (2000)
	G-bacteria (a morphotype tetrad forming) *Amaricoccus kaplicensis*	Carbon removal	Seviour *et al.* (2003)
	Glycogen accumulating bacteria (GAO) '*Candidatus* Competibacter phosphatis'	PAO's competitors	Crocetti *et al.* (2002)
Autotrophs	***Ammonia oxidisers*** *Nitrosomonas, Nitrosospira, Nitrosococcus, Nitrosocystis*	Ammonia oxidation to nitrite	Mobarry *et al.* (1996)
	Nitrite oxidisers *Nitrobacter, Nitrospina, Nitrococcus*	Nitrite oxidation to nitrate	Mobarry *et al.* (1996)
	Ammonia/nitrite anaerobic utilizators (Anammox) Brocadia, Kuenenia, Scalindua	Ammonia oxidation to nitrogen by respiring nitrite	Jetten *et al.* (2003)
Filamentous bacteria	M. parvicella, *Thiothrix*, *Nostocoida limicola*, Type 021N, *Sphaerotilus natans*, ecc.	Organic carbon oxidation HS^- oxidation (few) P removal (not confirmed hypothesis)	Seviour and Nielsen (2010)
Microfauna (Protozoa and Metazoa)	Flagellates and Ciliates Protozoa, *Rotifera*, *Nematoda*, etc.	Activated sludge quality Indicators (Sludge Biotic Index)	Madoni (1994)

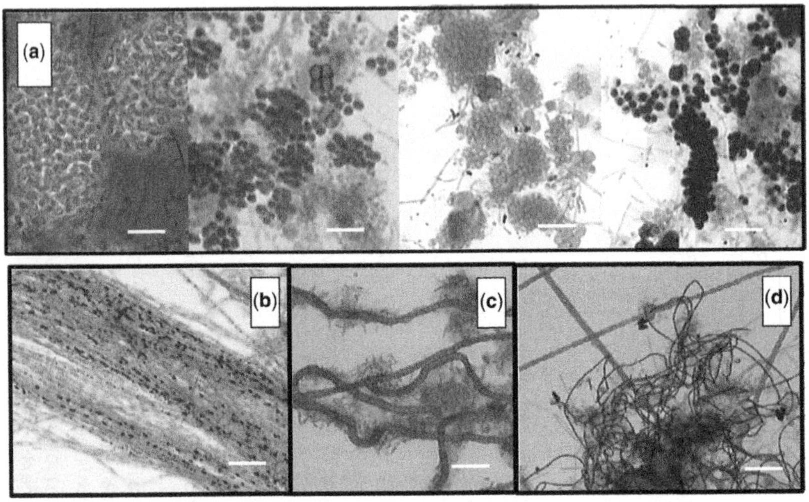

Figure 2.1 Some examples of floc former and filamentous bacteria in activated sludges: (**a**) PAO clusters (phase contrast, methylene blue staining, Gram staining and Neisser staining); (**b**) Type 1851, (Gram staining); (**c**) Alpha *Nostocoida limicola* (Neisser staining); (**d**) Microthrix parvicella (Gram staining). Bar is 5 μm.

One commonly used way to characterize bacteria (mainly utilized in the past) is to divide them into two major groups, *Gram-positive* and *Gram-negative*. The distinction between these groups is based on differences in reactions to a particular staining procedure, known as Gram staining. These differences are due to differences in cell wall structure. Gram positive bacteria are characterized by a thicker cell wall, with about 90% of the cell wall made up of a single type of molecule, peptidoglycan. The Gram-negative cell wall has a multilayered, complex structure. It contains only about 10% peptidoglycan and also contains an outer membrane made of lipopolysaccharide.

2.2.2 Inclusions

Bacteria can contain different kinds of inclusions (such as polyhydroxyalkanoates, carbohydrates, polyphosphates, or sulfur granules) and under particular growth conditions can produce biopolymeric compounds that are excreted out of the cell (esocellular polymers, ECP, known generically as *capsule* material). These inclusions play an important role in AS ecology and are involved in important phenomena, such as biological phosphorus removal or storage. Inclusions are a crucial part of selector operation as well.

2.3 THE IDENTIFICATION OF BACTERIA

Correct bacterial identification and enumeration is important for appropriate control of the AS process. The identification of bacteria requires the existence of a classification system (taxonomy) that depends on the determination of certain properties to verify the similarity or difference of a microorganism to other microorganisms that are already known and described. More than 18,000 bacterial species have been identified so far (but the number increases day by day), constituting probably around 1–10% of those in nature. A bacterial strain is assigned a formal taxonomic position, such as a new genus or species, when an accurate description of the isolate with the proposed name is published in specialized journals (such as the *International Journal of Systematic and Evolutionary Bacteriology*, IJSEM). A pure culture of the organism should also be deposited in an approved culture collection, such as the American Type Culture Collection (*ATCC*) or the Deutsche Sammlung von Mikroorganismen und Zellkulturen (*DSMZ* German Collection for Microorganisms).

The description of newly identified bacteria consists of information regarding morphology, physiology, habitat, composition of some cellular components, phenotypic and genotypic properties, and the sequence of the 16S rRNA gene (Tindall *et al.* 2010). The whole of these characteristics permits the naming of the organism, and consequently, its acceptance as a new microbiological taxon. Periodically, an approved list of bacterial names is released, formalizing newly proposed names and allowing their inclusion in *Bergey's Manual of Systematic Bacteriology*.

In addition to this traditional and rigorous classification system, a different and more practical approach has been used to describe some relevant bacteria in AS. This second approach system is based mainly on simple and easy to detect morphological properties and, conceived by Dick Eikelboom in 1975, was essential for many years (when the quick and precise molecular methods were not available) for the elucidation of the diversity of filamentous bacteria in AS systems (Eikelboom, 1975). Moreover, the development in the last decades of molecular taxonomy and of identification methods based on the detection of taxon specific nucleotide sequences, largely improved the understanding of the composition of the unculturable AS microbial population.

2.3.1 Conventional taxonomy

The identification of bacteria by conventional taxonomy was based on several characteristics, such as morphology, Gram reaction, metabolic peculiarities, cell wall and lipid chemistry, DNA guanine cytosine (G + C) total bases ratio (G + C + A + T), presence of cell inclusions and storage products, temperature and pH requirements, antibiotic sensitivity, pathogenicity, and habitat, etc. Though useful, this information does not clarify the evolutionary or phylogenetic relationships between organisms (how they developed from the previous ones during their evolution), which is a priority of new classification methods. For this reason, today a new systematic method is being used that facilitates the understanding of evolutionary relationships between all living organisms.

2.3.2 Molecular taxonomy

With the development of molecular techniques and their wide application to different research areas, a reorganization of the living world occurred. Living organisms were classified into five kingdoms (Animals, Plants, Fungi, Protists and Bacteria), which were also divided into eukaryotes (Animals, Plants, Fungi, Protists) and prokaryotes (bacteria) on the basis of the presence or absence of a nuclear membrane and certain structural characheristics. These two groups were thought to be independent and, because of the complexity of multicellular organisms, more biodiversity was initially attributed to the eukaryotes. On the contrary, phylogenetic analysis based on the molecular sequencing of particular macromolecules, ribosomes, revealed a new classification scheme for living organisms, quite different from the previous kingdom-based classification. In particular, it was determined that the sequence of the ribonucleic acid (RNA) present in ribosomes (ribosomal RNA or rRNA) was able to serve as an evaluationary chronometer. Ribosomes are the site of protein synthesis, which is a fundamental function for the survival and the function of all cells, and are made of a mixture of different sized RNA and proteins. The large molecules of prokaryotic ribosomal RNA, the 16S and 23S units (containing about 1500 and 2900 nucleotides, respectively) contain several regions of highly conserved sequences useful for obtaining proper sequence alignments, and also contain sufficient sequence variability in other regions to allow their utilization as excellent evolutionary chronometers. The 5S rRNA subunit has also been utilized in this way but its small size (about 120 nucleotides) limits the information obtainable from this molecule. The rRNA sequences have been obtained from many organisms and can be organized into a phylogenetic tree, which allows the determination of the evolutiony distance between organisms (Woese, 1987).

Molecular phylogeny revealed that the five kingdoms do not represent five major evolutionary lineages, but that life on Earth has evolved along three major lineages, two of which are prokaryotic (Bacteria & Archaea) and one is eukaryotic (*Eukarya*). With the advent of sequencing technologies which generally yield short reads, focus shifted from sequencing the full 16S rRNA gene to sequencing shorter sub-regions of the gene at high depth. Several studies critically evaluated the parameters principally determining the efficacy extent of this approach (Mizrahi-Man *et al.* 2013).

2.3.3 Molecular characterization of mixed biomass

The application of advanced molecular methods to studying the microbial communities in activated sludge plants strongly impacted the understanding of the main biological processes occurring in such systems. Among others, Fluorescence In Situ Hybridization (FISH) and quantitative PCR (qPCR) are the ones mostly applied at full scale for the biomonitoring of fastidious microorganisms (e.g. filamentous bacteria) and the main microbial functional groups (e.g. nitrifiers, denitrifiers, P-accumulating bacteria, etc.). Commercial kits based on FISH probes are already available for industrial end-users. Moreover, PCR quantitative assays are becoming increasingly employed and optimized protocols are developed for the analysis of many microbial components of activated sludge.

2.3.3.1 Polymerase Chain Reaction (PCR)

PCR is considered one of the most crucial methods in molecular microbial ecology, since it can specifically target particular taxonomic, functional markers or phylotype levels. Nevertheless, it has inherent limitations mainly caused by inefficient or preferential extraction of community DNA, variable efficiency of different extraction methods, problems in maintaining the integrity of DNA, and amplification biases during the process (Kim *et al.* 2013). Profiles obtained by PCR-based methods are a quantitative reflection of the PCR product pool and not the original community. Differences in gene copy number, primer specificity, amplification efficiency, sensitivity to template concentration and the formation of chimeric sequences may also decrease the reliability of these methods, as reviewed by Kim *et al.* (2013). Additionally, PCR-based methods present detection limits considerably higher than those obtained by cell quantification methods (Matturro *et al.* 2013).

Real Time PCR (q-PCR) has been used for the analysis of specific microbial functional groups or filamentous bacteria. Optimized qPCR protocols are available for the biomonitoring of nitrifiers, denitrifiers, PAO and some filamentous bacteria (Table 2.2).

2.3.3.2 Fluorescence In Situ Hybridization (FISH)

FISH is a widely applied technique that allows visualization of organisms in a sample, based on their phylogeny. FISH has been successfully utilized for the *in situ* monitoring of microbial population dynamics in environmental samples and it works through the use of molecular probes. These probes are synthesized in the laboratory and consist of short strands of nucleic acids (usually 15–30 nucleotides in length), complementary to 16S rRNA or 23S rRNA gene sequence regions. The oligonucleotides are usually tagged on one end with a so-called 'reporter molecule', usually a fluorochrome, which allows visualization of the probe when the specimen is observed by epifluorescence microscopy. The main oligonucleotide probes currently employed for the *in situ* detection of the main filamentous bacteria and microbial functional groups are reported in Table 2.3.

Table 2.2 Available qPCR assays for the quantification of some microorganisms/ functional genes occurring in wastewater treatment systems.

Microorganism	Reference
Filamentous bacteria:	
M. parvicella	Kaetzke *et al.* (2005), Kumari *et al.* (2009)
M. parvicella including M. calida	Vanysacker *et al.* (2014)
Thiothrix eikelboomii	Asvapathanagul *et al.* (2015)
Gordonia amarae	Asvapathanagul *et al.* (2012)
Type 021N	Vervaeren *et al.* (2005)
Nitrifiers:	
Ammonia oxidizers:	
Nitrosomonas spp.	Lim *et al.* (2008)
Nitrosospira spp.	Lim *et al.* (2008)
Nitrosococcus spp.	Lim *et al.* (2008)
Ammonia monooxygenase (amoA) gene	Rotthawe *et al.* (1997)
Nitrite oxidizers:	
Nitrospira spp.	Harms *et al.* (2003), Wang *et al.* (2012)
Nitrobacter spp.	Wang *et al.* (2012)
Nitrite oxidoreductase (nxrB) gene	Geets *et al.* (2007)
Denitrifiers:	
Thauera/Azoarcus	Liu *et al.* (2006)
Nitrite reductase (nirK, nirS, and nosZ) genes	Geets *et al.* (2007)
PAO:	
Candidatus Accumulibacter phosphatis	He *et al.* (2007)
Polyphosphate kinase 1 (ppk1) gene	He *et al.* (2007)

2.4 FILAMENTOUS BACTERIA

Filamentous organisms are characterized by their long, thread-like shape. Bacteria are the main component of filamentous organisms present in AS plants; only seldom are filamentous eukaryotes visualized (e.g. fungi). By the systematic observation of a large number of samples from different AS treatment plants, Eikelboom (1975) described (only on morphological basis) 29 different filamentous bacteria, grouped in three clusters:

(a) Bacteria ascribable to known genus and species already reported in *Bergey's Manual of Systematic Bacteriology*: i.e. *Sphaerotilus natans*, *Haliscomenobacter hydrossis*, *Beggiatoa*, *Thiothrix*, *Flexibacter.*

Table 2.3 Main molecular probes available for *in situ* identification of bacteria present in wastewater treatment systems.

Microorganism	Probe	Probe Sequence (5′–3′)	% FA	Reference
Filamentous bacteria:				
Candidatus M. parvicella	MP223	GCCGCGAGACCCTCCTAG	20	Erhart *et al.* (1997)
	MPA645	CCGGACTCTAGTCAGAGC	20	Erhart *et al.* (1997)
Candidatus M. calida	Mpa-T1-1260	TTCGCATGACCTCACGGTTT	25	Levantesi *et al.* (2006)
Thiothrix/Type021N *group*	G123T + G123T-C*	CCTTCCGATCTCTATGCA + CCTTCCGATCTCTACGCA*	40	Kanagawa *et al.* (2000)
Thiothrix nivea	TN1	CTCCTCTCCCACATTCTA	45	Wagner *et al.* (1994a)
Type 021N	21N	TCCCTCTCCCAAATTCTA	35	Wagner *et al.* (1994a)
H. hydrossis	HHY	GCCTACCTCAACCTGATT	20	Wagner *et al.* (1994a)
Type 1851	CHL1851	AATTCCACGAACCTCTGCCA	35	Beer *et al.* (2002)
S. natans	SNA	CATCCCCCTCTACCGTAC	45	Wagner *et al.* (1994a)
Acinetobacter sp.	ACA652	ATCCTCTCCCATACTCTA	35	Wagner *et al.* (1994b)
N. limicola I	LIMI91	CGCCACTATCTTCTCAGT	20	Liu and Seviour (2001)
N. limicola II	NLIMII175	GGCTCCGTCTCGTATCCG	40	Liu and Seviour (2001)
Alpha Nostocoida:				
Candidatus Alysiosphaera europaea	Noli-644	TCCGGTCTCCAGCCACA	35	Levantesi *et al.* (2004)
Candidatus Sphaeronema italicum	Nost933 + helper1010	CAGCCGAAACGGGCATGT + GAAAGCCGCCGTCTCCGG	20	Kragelund *et al.* (2006)
	Sita649 + sita-649-C*	CCW CTC CCG GAC YCC AGC + CCT CTC CCG GTC TCC	50	Levantesi *et al.* (2004)
Candidatus Alysiomicrobium bavaricum	PPx1002	AGCC*	10	Kragelund *et al.* (2006)
	PPx3-1428	GGTCTCCCCGGGCCGCGG	50	Levantesi *et al.* (2004)
Candidatus Monilibacter bavatus	MC2-649	TGG CCC ACC GGC TTC GGG CTC TCC CGG ACT CGA GCC	35	Levantesi *et al.* (2004)
Meganema perideroedes	Meg983	CGGGATGTCAAAAGGTGG	35	Thomsen *et al.* (2006)
	Meg1028	CTGTCACCGAGTCCCTTGC	35	Thomsen *et al.* (2006)
N. limicola III	NLIMIII301	CCCAGTGTGCCGGGCCAC	20	Liu *et al.* (2001)
	NLIMIII729	AGCATCCAGAACCTCGCT	20	Liu *et al.* (2001)
	NLIMIII830	CCATCGGCGAGCCCCCTA	20	Liu *et al.* (2001)

(Continued)

Table 2.3 Main molecular probes available for *in situ* identification of bacteria present in wastewater treatment systems (*Continued*).

Microorganism	Probe	Probe Sequence (5′–3′)	% FA	Reference
Nitrifiers:				
Ammonia oxidizers:				
Nitrosospira and *Nitrosomonas*	NSO190	CGATCCCCTGCTTTTCTCC	55	Schramm *et al.* (1998)
Nitrite oxidisers:				
Nitrospira	NSR1156	CCCGTTCTCCTGGGCAGT	30	Schramm *et al.* (1998)
Nitrobacter spp.	NIT3	CCTGTGCTCCATGCTCCG	40	Wagner *et al.* (1995)
Denitrifiers (some):				
Paracoccus spp.	PAR651	ACCTCTCTCGAACTCCAG	40	Neef *et al.* (1996)
Thauera spp.	Thau646	TCTGCCGTACTCTAGCCTT	45	Lajoie *et al.* (2000)
Azoarcus spp.	AZA645	GCCGTACTCTAGCCGTGC	20	Hess *et al.* (1997)
Phosphorus Accumulating Organisms (PAOs):				
Accumulibacter spp.	PAO462	CCGTCATCTACWCAGGGTATTAAC	35	Crocetti *et al.* (2000)
	PAO651	CCCTCTGCCAAACTCCAG	35	Crocetti *et al.* (2000)
	PAO846	GTTAGCTACGGCACTAAAAGG	35	Crocetti *et al.* (2000)
Glycogen Accumulating organisms (GAOs)				
Competibacter spp.	GAOQ431	TCCCCGCCTAAAGGGCTT	35	Crocetti *et al.* (2002)
	GAOQ989	TTCCCCGGATGTCAAGGC	35	Crocetti *et al.* (2002)
Defluvicoccus related organisms	TFO-DF18	GCCTCACACTTGTCTAACCG	25–35	Wong *et al.* (2004)
	DF988	GATACGACGCCCATGTCAAGGG	35	Meyer *et al.* (2006)

* Competitor probe, unlabelled.

(b) Microorganisms that, even though they have a name (i.e. M. parvicella, *Nostocoida limicola*) were still not present in *Bergey's Manual of Systematic Bacteriology* because of a paucity of the phenotypic information.
(c) Microorganisms to which it was impossible to attribute an identity and were consequently identified as a 'Type' followed by a number, such as Type 0041, etc.

The filamentous species *Sphaerotilus natans* was originally thought to be the only filamentous microorganism responsible for dysfunctions in AS systems. Because of the microscopic work that has been performed since the beginning of the 1970s, it has been possible to ascertain that filamentous bulking can be caused by several different filamentous bacteria. In particular, Eikelboom's Manual (1981), a systematic guide that enables the morphological distinction of the main filamentous bacteria present in AS, has helped with such determinations.

After more than thirty years from the publication of the first Eikelboom's Manual (1981) and an analogous manual by Jenkins and coworkers (Jenkins *et al.* 1984), it is necessary to underline the important contributions of these manuals that have led to the improvement of knowledge in this field.

These manuals have permitted morphological identification, monitoring, and quantification of filamentous bacteria, allowing comparisons of various corrective operations to control the overgrowth of filamentous bacteria. The main useful characteristics for the identification of the most common 'traditional' filamentous bacteria causing bulking and foaming problems are reported in Table 2.4.

During recent years, pure culture studies have defined the taxonomy of some of the 29 filamentous bacteria described by Eikelboom. Several organisms that were originally thought to be a single species have been identified as unique taxons, and in some cases, adequate information has been generated to properly name the organisms. This is the case for M. parvicella (Blackall *et al.* 1996; Rossetti *et al.* 2005; McIlroy *et al.* 2013), *Thiothrix* and Type 021N (Howarth *et al.* 1999; Rossetti *et al.* 2002; Aruga *et al.* 2002) as well as for the *Nostocoida limicola* diffused in urban WWTP AS (Blackall *et al.* 2000; McKenzie *et al.* 2006).

Some of the filamentous bacteria that have not been studied are of minor importance with respect to bulking (i.e. Type 0411, Type 1702, type 1852, Type 0211, etc.). In addition, it has been shown that the diversity of filamentous bacteria is much higher in industrial wastewater treatment plants (Snaidr *et al.* 2002; Eikelboom & Geurkink, 2001, 2002; Levantesi *et al.* 2004; Nielsen *et al.* 2009). Table 2.5 summarizes the main characteristics and kinetic properties of the filamentous microorganisms that have been studied, so far, in pure culture. Examples of filamentous bacteria isolated in pure culture are shown in Fig. 2.2.

However, current knowledge of the kinetic properties and metabolic versatility of filamentous bacteria cannot be considered satisfactory, mainly because only a few isolates have been studied, and often using simple substrates. This incomplete picture is primarily a result of the difficulties involved in isolation and conducting

Table 2.4 Main useful characteristics for the identification of the most common 'traditional' filamentous bacteria causing bulking and foaming problems.

	Phylogenetic Affiliation	Gram Staining	Neisser Staining	Ø µm	Sheath Presence	Sulfide Oxidation Test	Cell Shape	Key Morphological Peculiarity
Sphaerotilus natans	Betaproteobacteria	–	–	1.0–1.4	+	–	round	False branching
Type 1701		–	–	0.6–0.8	+	–	round	Attached growth
Type 0041		V*	–/+	1.4–1.6	+	–	square	Gram staining
Type 021 N (now *Thiothrix eikelboomii*)	Gammaproteobacteria	–	–	1.0–2.0	–	+	barrel	Cell shape
Thiothrix	Gammaproteobacteria	–	–	1.0–2.5	+	+	square	Sulfur granules presence/production
Type 1851	Chloroflexi	V	–	0.8	+	–	indistinguishable	Filaments bundle assembled
Type 0092	Chloroflexi	–	+	0.8–1.0	+	–	indistinguishable	Neisser staining
M. parvicella	Actinobacteria, in High G + C Gram positive bacteria	+	–	0.8	Non ascertained	–	indistinguishable	Gram staining
Nocardioforms	Several genera and species belonging to different divisions of Bacteria	+	–	1.0	–	–	indistinguishable	True branching, Gram staining
Nostocoida limicola I	Low G + C Gram positive Bacteria	+	–					Gram staining
Nostocoida limicola II (several genera and species)	High G + C Gram positive Bacteria and Alphaproteobacteria	–/+	–/+	1.2–1.4	–	–	discoid	Cell shape
Haliscomenobacter hydrossis	Cytophaga–Flavobacterium Bacteroides	–	–	0.5	+	–	indistinguishable	Filaments dimension and typical bent

* V: variable, parts of the cells are positive, parts are negative.
–/+: it can be positive or negative.

Table 2.5 Kinetic and physiological properties of filamentous bacteria isolates.

Bacterium	Aerobic Respiration	Anaerobiosis	Substrates	μ max (d^{-1})	*Ks* ($mg\ L^{-1}$)	Storage Products	Reference
Sphaerotilus natans	C	$NO_3 \rightarrow NO_2$	AL, AC, CA	6.5(1)	10.0	PHB	Richard *et al.* (1985), William and Unz (1985)
Type 1701	C	some strains: $NO_3 \rightarrow NO_2$	AL, AC, CA	2.6(1)	2.0	PHB	Richard *et al.* (1985), Williams and Unz (1985)
Haliscomenobacter hydrossis	C	–	GLU, LAT	1.2–2.2(1)	5.0	PS	Krul (1977)
Beggiatoa	C, A, M	–	AC	n.d.	n.d.	PHB, PP, S	Williams and Unz (1985)
Thiothrix spp.	M	–	AC	n.d.	n.d.	PHB, PP, S	Williams and Unz (1985)
Thiothrix, strain CT3	M	$NO_3 \rightarrow NO_2$	AC + HS^-	2.6(2+3)	n.d.	PP, S	Tandoi *et al.* (1994)
Thiothrix, strain CT3	C	$NO_3 \rightarrow NO_2$	AC	2.5(2)	n.d.	PHB	Tandoi *et al.* (1994)
Thiothrix, strain CT3	A	$NO_3 \rightarrow NO_2$	HS^-	1.8(3)	n.d.	S	Tandoi et al. (1994)
Type 021N spp.	C, M	$NO_3 \rightarrow NO_2$	CA, AC,AA,AL	3.75(2)	0.07(4)	PHB, PP, S	Williams and Unz (1985), Van Niekerk *et al.* (1987), Aruga *et al.* (2002)
Gordona amarae	C	–	CA, AC	2.5(2)	0.5(4)	PHB, PP	Blackall *et al.* (1991)
M. parvicella	C	–	AGL	1.4*(4) 0.6**(4)	n.d.	LP	Slijkhuis and Deinema (1982)

(*Continued*)

Table 2.5 Kinetic and physiological properties of filamentous bacteria isolates (*Continued*).

Bacterium	Aerobic Respiration	Anaerobiosis	Substrates	μ max (d^{-1})	*Ks* (mg L^{-1})	Storage Products	Reference
M. parvicella, strain RN1	C	$NO_3 \rightarrow NO_2$	AGL, OL	0.6(5)	n.d.	LP, PP	Tandoi *et al.* (1998)
M. parvicella, strains RN1 and 4B	C	$NO_3 \rightarrow NO_2$	AC + mixture (5)	–	n.d.	PHAs (in anaerobic and anoxic conditions)	Rossetti *et al.* (2002)
N. limicola spp.	C	$NO_3 \rightarrow NO_2$	CA, PEPT	n.d.	n.d.	PHB	Blackall *et al.* (2000)
N. limicola I	C	F					Seviour *et al.* (2002)
N. limicola II (genus *Tetrasphaera*)	C	–					Snaidr *et al.* (2002), Seviour *et al.* (2006), McKenzie *et al.* (2006)
Type 1851	C	+	CA, AC, AA	n.d.	n.d.	PHA	Kragelund *et al.* (2007a), Kohno *et al.* (2002)

*estimated in continuous reactor, **estimated in batch reactor; n.d. not determined; (1) glucose; (2) acetate; (3) HS^-; (4) oleic acid; (5) mixture of substrates constituted by glucose, yeast extract, proteose peptone, pyruvic acid, starch and casaminoacids; C, Chemoorganotroph; A, Autotrophs; M, Mixotrophs; F, Fermentative; AA, Amminoacids; AL, Alcohol; AC, organic acids; CA, carbohydrates; GLU, glucose; LAT, lactose; AGL, long chain fatty acids; OL, oleate; PEPT, peptone; PS, polysaccharides; PP, polyphosphates; LP, lipids.

experiments with these microoorganisms. It is difficult to obtain isolates, and often, they slow down and eventually stop growing in pure culture. Nevertheless, in recent decades, the limit of filamentous bacteria low cultivability was partially overcome by the use of FISH analysis in combination with different approaches for the *in situ* assessment of bacterial metabolic capacities (Nielsen *et al.* 2009). The *in situ* studies of filamentous bacteria physiology have mainly addressed the capacity to: (i) uptake different classes of carbon substrates under different redox conditions, (ii) produce extracellular enzymes for polymer hydrolysis, and (iii) store intracellularly organic and inorganic compounds (PHA, polyphosphates, sulfur). The main results of these studies are summarized in Table 2.6.

In general, the following points can be highlighted:

- All the isolated filaments are aerobic and are not able to denitrify nitrate completely to nitrogen gas (some can reduce nitrate partially to nitrite and some can uptake substrates under anoxic conditions).
- Only a few filamentous bacteria are reported to be fermentative or to grow anaerobically (*Nostocoida limicola* I, type 1851).
- Filamentous bacteria are mostly heterotrophs. Some are more versatile and can utilize different organic substrates such as carbohydrates, lipids, proteins, and alcohols (e.g. *Alpha Nostocoida*, Type 021N, *Thiothrix*), while others (mainly *Chloroflexi* and Microthrix spp.) were shown to use a limited number of carbon compounds and to have higher capacity to hydrolyze complex macromolecules.
- Some species are characterized by high metabolic versatility, such as *Beggiatoa* and *Thiothrix,* which are able to grow heterotrophically, mixotrophically (simultaneous growth on organic carbon and sulfur reduced compound), and autotrophically (growth only on sulfur reduced compounds as energy source).
- The maximum growth rates of filamentous bacteria are low (typically in the range 0.4–3.0 d^{-1}), especially if compared with growth rates characteristic of floc-forming bacteria (for instance *Acinetobacter* has a maximum growth rate at 20°C on acetate 13.5 d^{-1}, Authors data).
- All the examined bacteria store at least one intracellular polymer, such as lipid inclusions (i.e. PHA) and polyphosphates. Some organisms contain sulfur granules and others (*H. hydrossis*) store polysaccharides.
- Molecular probes have already been defined for some filamentous bacteria involved in bulking and other important AS bacteria as well; these probes should be used to complement the more traditional morphological information (Table 2.3).
- In the following paragraphs, more detailed information on relevant filamentous bacteria, namely M. parvicella, *Thiothrix* sp., and the *Nostocoida limicola* group, are presented. Additionally, taxonomic and physiological data on the bacterial population involved in nutrient removal in AS are summarized.

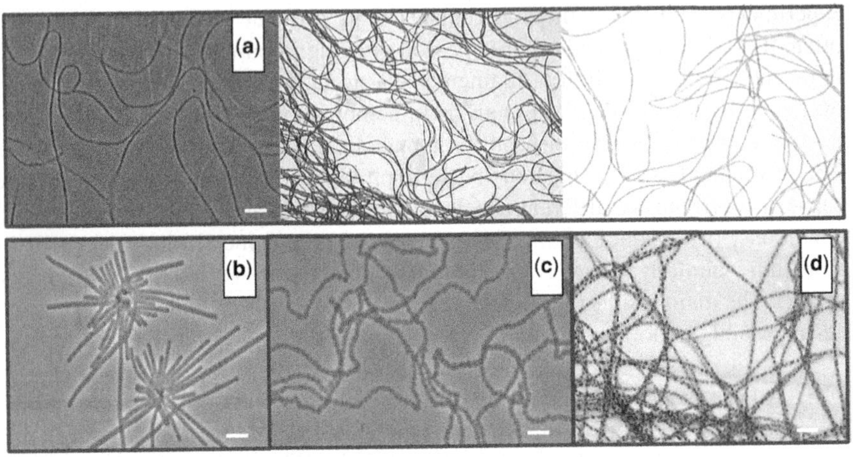

Figure 2.2 Some examples of filamentous bacteria pure culture: (**a**) *Candidatus* Microthrix parvicella, strain Bio17iso1 (phase contrast, Gram staining and Neisser staining); (**b**) *Thiothrix* strain CT3 (phase contrast); (**c**) Type 1863, *Acinetobacter johnsonii*, strain RT2 (phase contrast); (**d**) Type 1851, strain EU25 (Gram staining). Bar is 5 μm.

2.5 MICROTHRIX PARVICELLA

Microthrix parvicella is the most common cause of bulking and foaming in the AS system. Since its first detailed description (van Veen, 1973) many attempts have been made to characterize it and control its proliferation. Nevertheless, only a few pure cultures studies have been performed to date. This lack of data is mainly due to difficulties with its isolation, cultivation, and its characteristic slow growth. Information on the *in situ* uptake of substrates has been determined by MAR (microautoradiography). From this information and numerous pilot and full-scale experiments, several hypothesis have been formulated to explain its proliferation in AS systems and consequently, to define methods to control its growth. In particular, as far as the physiology of the organism is concerned, the RN1 strain, is the only isolate of the species '*Candidatus* Microthrix parvicella', for which the taxonomic position and basal physiological and kinetic characterisation have been described (Tandoi *et al.* 1998; Rossetti *et al.* 2002, 2005; McIlroy *et al.* 2013). The second described species '*Candidatus* Microthrix calida' (Levantesi *et al.* 2006) resembles the morphotype of the previous species but the trichome is thinner and it can grow at a higher temperature in pure culture studies. '*Candidatus* Microthrix calida' was found only in full-scale plants treating mainly industrial wastewater, regardless of the configuration for the biological N and P removal (Levantesi *et al.* 2006).

Table 2.6 Physiological properties of filamentous bacteria from in situ studies.

Bacterium and Probe	Uptaken Substrates	Electron Acceptor	PHA Storage	Other Storage Compounds	Surface Properties	Exoenzime Activity	References
Filamentous *Chloroflexi* (CFX1223,CFX109,CHL1851,EU25-1238)	CA, (AA,CA)	O_2,	+	n.d.	Hydrophilic	High	Kragelund *et al.* (2007a)
Type 021N spp. (21N)	AC, CA, AGL (AL, AA)	O_2	+	S	Hydrophilic	n.d.	Nielsen *et al.* (2009)
Mycolata (Gordona amarae-G.am205; *Skermania piniformis* Spin 1449*)*	(CA, AGL, AA)	O_2 ,(NO_2, NO_3)	+	PP	Hydrophobic	High	Eales *et al.* (2006), Carr *et al.* (2006), Kragelund *et al.* (2007b)
M. parvicella (MPA654)	AGL	O_2, NO_2, NO_3,	+	PP	Hydrophobic	+	Andreasen and Nielsen (1997), (2000), Nielsen *et al.* (2002)
N. limicola II *Tetrasphaera* (NLIMII175)	AGL, (AA)	O_2, NO_2, NO_3, anaerobic	–	PP, LP	n.d.	n.d.	Seviour *et al.* (2006)
Alpha Nostocoida							
Group I* (MC2-649, Nost933 + helper1010, Sita649 + *composita-649)*	AC, (AGL)	O_2 (NO_2, NO_3)	High	n.d.	Hydrophobic	(+)	Kragelund *et al.* (2005, 2006)
Group II* (PPx1002, PPx3-1428, Noli644*)*	AC, AA, CA (AGL, AL)					–	

In brackets: uptake only observed for a limited number of substrates and/or capacity observed for some filamentous subgroups; AA, amminoacids; AL, Alcohol; AC, organic acids; CA, carbohydrates; AGL, long chain fatty acids; S, sulfur; PHA, polyhydroxyalkanoates; PP, polyphosphates; LP, lipids; n.d., not determined; **Alpha Nostocoida* groups with different ecophysiology (adapted from Nielsen *et al.* 2009).

2.5.1 Identification of M. parvicella

M. parvicella is a thin (0.8–1.0 μm) long, unsheathed Gram-positive bacterium with an irregular and winding appearance. Normally, this organism does not show attached growth and it is never branched. M. parvicella often accumulates polyphosphate and can also accumulate PHAs. Due to its distinctive morphology and particularly, to its characteristic response to Gram staining, the presence of this filamentous bacterium in AS is easy to detect. Molecular probes are available and allow precise species identification of this filamentous bacterium in sludge samples.

2.5.2 Physiology of '*Candidatus* M. parvicella'

The first isolate described in detail was isolated by Slijkhuis and Deinema (1982). Unfortunately, this isolate lost its viability before the sequence of its 16S rRNA gene was determined. Two new strains of M. parvicella were isolated approximately 10 years later (Blackall *et al.* 1994; Rossetti *et al.* 1997, 2002). The limited number of strains were sufficient to describe its basic physiology and kinetic properties, and to give to the organism name the status of 'Candidatus' (Blackall *et al.* 1996), a first step in naming organisms when only few strains and information are available. Table 2.7 reports the main properties of this organism obtained in pure and mixed culture studies. Recently, several M. parvicella strains have been isolated from Chinese EBPR WWPTs (Rossetti *et al.* 2015).

Table 2.7 Summary of the growth and storage capabilities of pure and mixed cultures of M. parvicella under different metabolic conditions.

	Growth		Storage		Reference
	Mixed culture	Pure culture	Mixed culture	Pure culture	
Aerobiosis	+	+	PHA, PP (1, 2, 3)*	PHA PP (3)	(1) Tandoi *et al.* (1998)
Anaerobiosis	– (3)	– (3)	PHA (2, 3)	PHA (3)	(2) Andreasen and Nielsen (2000)
Anoxic conditions					
Nitrate	– (3)	– (3)	PHA (3)	PHA (3)	(3) Rossetti *et al.* (2002)
Nitrite	– (3)	– (3)	PHA (3)	PHA (3)	

PHA, poly-β-hydroxyalkanoates; PP, polyphosphates.

By pure culture and *in situ* physiological caracterization studies (Tables 2.5–2.6 and 2.7), M. parvicella was shown to be a metabolically versatile microorganism, capable of using several substrates and of adapting to different environmental conditions (e.g. temperatures and redox conditions). These observations have also been confirmed by a detailed description of an M. parvicella metabolic model based on whole genome sequencing (McIlroy *et al.* 2013). Moreover, more recently, by *in situ* substrate assimilation studies using nanoscale secondary-ion mass spectrometry, Sheik and collegues (2016) showed phenotypic heterogeneity between individual cells of Microthrix filaments, which might confer additional adaptive advantages in fluctuating environments. Overall, it is possible to conclude that one of the main factors promoting its growth in WWTP AS is its versatility and ability to benefit from the presence of anoxic and anaerobic zones, where it can uptake and store organic carbon preferentially as lipids. Research efforts have recently focused on this unusual storage phenomenon.

2.6 THIOTHRIX

Thiothrix is a bacterium that can grow mixotrophically, heterotrophically or lithotrophically. This filamentous bacterium is often present in AS plants treating sewage characterized by a high concentration of reduced sulfur compounds, such as sulfides and thiosulfates. These filamentous bacteria can cause bulking under such conditions (Jenkins *et al.* 2003). Bacteria belonging to the genus *Thiothrix* are characterized by distinctive morphological features, such as growth in unsheathed filaments, accumulation of sulfur granules when incubated in the presence of reduced sulfur compounds, organization in rosette-like arrays, and production of gliding gonidia from the filament apex. The filamentous bacteria of the genus *Thiothrix* were first described by Winogradsky in 1888 (cited in Odintsova *et al.* 1993) and currently, seven species of *Thiothrix* were described (Unz & Head, 2005).

It was initially hypothesized that *Thiothrix* might be autotrophic (Odintsova *et al.* 1993), but the neotype strain, *Thiothrix nivea*, was thought to be an obligate mixotroph, requiring both a reduced sulfur compound and an organic substrate for growth. All of the other strains belonging to this genus have been reported to be mixotrophic or heterotrophic (Larkin & Shinabarger, 1983). Moreover, it has been shown that an isolate from AS, strain CT3, and another strain, strain I (William & Unz, 1985) are also capable of growing autotrophically with reduced sulfur compounds as the sole energy source (Rossetti *et al.* 2003). Another chemolithotrophic strain of *Thiothrix* was obtained from a hydrogen sulfide spring and phylogenetically identified as *T. ramosa* (Polz *et al.* 1996). The type species of the genus is now *T. nivea*, the autotrophic *T. ramosa* name is still invalid, but four new species were proposed. They are *T. fructosivorans* (for the Q, I and CT3), *T. unzii* (for strain A1), *T. eikelboomii* (for the Eikelboom Type 021N strains, type strain AP3), and *T. defluvii* (for an Australian *Thiothrix* isolate called Ben57). All of the new species are from activated sludge sewage treatment plants. It is necessary to underline that the important morphotype, known as Type 021N and commonly found in activated sludge samples, has been shown to belong to the same *Thiothrix* genus (Howarth *et al.* 1999), and it is now named *Thiothrix eikelboomii*. Significative differences between *Thiothrix* strains recalling Type 021N and

Thiothrix nivea-like strains were found: the first were able to utilize several fatty acids and many carbohydrates while the second mainly fatty acids (Aruga *et al.* 2002). Type 021N morphotype was also shown to represent two additional separate species, *T. disciformis* and *T. flexilis*, within the genus *Thiothrix* (Kanagawa *et al.* 2000; Unz & Head, 2005).

2.7 NOSTOCOIDA LIMICOLA

Since the first description of *Nostocoida limicola* (Morphotypes I, II and III differing from each other for dimensions and Gram and Neisser staining) in domestic WWTPs, sequences of the 16S rRNA gene from several isolates showed that this group contained phylogenetically unrelated bacteria, including members of *Firmicutes*, *Alphaproteobacteria*, *Chloroflexi*, *Actinobacteria* and *Planctomycetes* (Blackall *et al.* 2000; Seviour *et al.* 2002; Hugenholtz & Stackebrand, 2004; Levantesi *et al.* 2004; McKenzie *et al.* 2006). On the other hand, the Gram-negative *N. limicola* seen in industrial treatment plants belong to the *Alphaproteobacteria* (Eikelboom & Geurkink, 2002; Levantesi *et al.* 2004), and these have only rarely been reported in domestic treatment plants (Seviour *et al.* 2002).

Very few physiological data are available at present on the *Nostocoida* isolates: many molecular probes are today available to identify them *in situ*, but precise correlations between these filamentous bacteria and the conditions favouring their growth, need to be investigated more deeply.

2.8 POLYPHOSPHATE ACCUMULATING ORGANISMS (PAO)

Over the past decades, design advances have taken place, resulting in plants that remove not only carbonaceous compounds and nitrogen, but also phosphorus. Microorganisms with the ability to take up and store polyphosphate are exploited for removing phosphorus from wastewater in a process known as enhanced biological phosphorus removal (EBPR).

Polyphosphate accumulating organisms (PAOs) are selectively enriched under alternating anaerobic/aerobic conditions. Wastewater containing phosphorus and soluble organic carbon compounds is mixed with sludge under anaerobic conditions. The mixture is maintained under anaerobic conditions for a period of time, then passed to an aerobic zone before going to the secondary sedimentation tank for sludge settlement and recycling. At present, neither a detailed understanding of this process nor a pure culture of PAOs is available. Nevertheless, a biochemical model has been postulated that describes EBPR (Mino *et al.* 1987; Pereira *et al.* 1996). According to this model, the PAOs have a selective advantage over other aerobic heterotrophic bacteria in the anaerobic–aerobic sequence of the EBPR system. In the anaerobic zone, PAOs utilize part of their stored polyphosphate and glycogen as sources of energy and reducing power for the uptake of soluble organic carbon (typically in the form of volatile fatty acids, VFAs) and for the synthesis and storage of poly-β-hydroxyalkanoates (PHAs) (see Chapter 5). In the aerobic zone, these same cells take up phosphorus (both that which they released and that from the influent) and store it as polyphosphate. PAOs also utilize anaerobically stored PHA as a carbon source for cell growth and for replenishing glycogen.

The most abundant identified PAO in full-scale wastewater treatment plants belong to the beta 2-subclass of *Betaproteobacteria* ('*Candidatus* Accumulibacter phosphatis') (Hesselmann *et al.* 1999; Crocetti *et al.* 2000). Phosphorus and carbon transformations indicative of EBPR have been proven to occur in these bacteria when they are in mixed EBPR sludge cultures (Crocetti *et al.* 2000). Later on, the genome of Accumulibacter by metagenomics studies of lab-scale and full scale EBPR enriched biomasses was investigated providing the basis for the analysis of key-genes involved in phosphate accumulation (Garcia Martin *et al.* 2006; Albertsen *et al.* 2011).

2.9 GLYCOGEN ACCUMULATING ORGANISMS (GAO)

Sometimes for unknown reasons, EBPR fails and the plant effluent contains high levels of phosphorus. Failures often occur despite seemingly correct conditions and typically in periods following stable, efficient EBPR. A potential reason for EBPR failure is the presence of microorganisms that use stored compounds, such as glycogen, to compete with the PAOs for carbon under anaerobic conditions. Although these microorganisms have the capacity to carry out the anaerobic and aerobic carbon transformations consistent with those of PAOs, they cannot cycle phosphorus (see Chapter 5). Because these organisms produce glycogen in the aerobic zone and apparently obtain their energy from its hydrolysis under anaerobic conditions, they are known as glycogen accumulating organisms (GAOs) (Satoh *et al.* 1994).

The GAO phenotype has been demonstrated by chemical analysis in highly enriched cultures (Fukase *et al.* 1985; Satoh *et al.* 1994; Bond *et al.* 1999) and by microscopy at a cellular level (Crocetti *et al.* 2002). A group of microorganisms belonging to Gammaproteobacteria, '*Candidatus* Competibacter phosphatis', have been shown to display *in situ* the GAO phenotype using culture-independent methodologies (Nielsen *et al.* 1999; Dabert *et al.* 2001; Crocetti *et al.* 2002). More recently, *Defluviicoccus vanus* isolates have been shown to take up glucose anaerobically with simultaneous glycogen degradation and PHA production without cycling polyphosphate (Maszenan *et al.* 2005; Wong & Liu, 2007).

Tetrad-forming organisms (TFOs), previously known as 'G bacteria' and described only on the basis of their morphology, were appointed as GAOs in EBPR by Cech and Hartman (1993). Molecular and physiological analyses of isolates with TFO morphology, however, revealed high phylogenetic diversity and provided no support for their role as GAOs (Falvo *et al.* 2001; Seviour *et al.* 2000).

2.10 NITRIFIERS

Since most of the nitrogen in wastewater influent is present either as urea (easily hydrolyzed to ammonia) or as ammonia, nitrifying bacteria play an important role in nitrogen removal in wastewater treatment plants.

Nitrifiers are extremely slow-growing microorganisms and are difficult to cultivate. Because of their sensitivity to disturbances in pH and temperature, breakdown of the nitrification process occurs frequently.

The genus *Nitrosomonas* has long been considered as the model ammonia-oxidizer. Other genera of nitrifying bacteria have been detected in nitrifying activated sludge and biofilms by molecular analysis and include *Nitrosococcus*, *Nitrosospira* and *Nitrosovibrio*. Likewise, traditionally members of the genus *Nitrobacter* were considered to be the most important nitrite-oxidizing bacteria in wastewater treatment plants. Surprisingly, this genus could not be detected in activated sludge at different nitrifying wastewater treatment plants when monitored by FISH with specific 16S rRNA-targeted probes. Using a full-cycle rRNA approach, the occurrence of novel, *Nitrospira*- like nitrite-oxidizing bacteria have been found (Juretschko *et al.* 1998; Daims *et al.* 2001). Different survival strategies have been postulated for explaining the predominance of *Nitrospira*-like bacteria over *Nitrobacter* in most wastewater treatment plants. *Nitrospira*-like nitrite-oxidizers are characterized by a low μ_{max} and are, according to results obtained by Schramm *et al.* (1999), *k*-strategists. They are thus well-adapted to low nitrite and oxygen concentrations. *Nitrobacter* was postulated to be a relatively fast-growing *r*-strategist with low affinities for nitrite and oxygen. Since nitrite concentrations are often low in most aeration tanks, *Nitrospira*-like organisms should be able to successfully outcompete *Nitrobacter* species in these systems. Recently, complete oxidation of ammonia in one organism (complete ammonia oxidation; comammox) was described for members of the genus *Nitrospira* (van Kessel *et al.* 2015). The presence of comammox-like ammonia monooxygenase genes in many published metagenomics datasets, suggested that comammox *Nitrospira* are widespread in several ecosystems including WWTPs (Daims *et al.* 2016).

2.11 DENITRIFIERS

The ability to denitrify is common among different groups of bacteria. Nevertheless, at present, no filamentous bacteria are reported to be denitrifiers. Because many floc-formers are able to denitrify, the use of specific molecular probes to detect these organisms is difficult unless probes are designed for only specific denitrifying populations (Table 2.3). Several studies have shown that members of the genera *Curvibacter* (family Comamonadaceae), *Azoarcus*, *Thauera* and *Zooglea* (family Rhodocyclaceae) are common denitrifiers in activated sludge treatment plants (Neef *et al.* 1996; Wagner & Loy, 2002; Ginige *et al.* 2004; Thomsen *et al.* 2004, 2007). Alternatively, qPCR targeting genes encoding for key-enzymes involved in the denitrification pathways, such as nitrite reductase (*nir*) nitrous oxide reductase (*nos*), was shown to be a valuable tool to monitor denitrification in WWTPS (Aoi *et al.* 2005; Geets *et al.* 2007).

REFERENCES

Albertsen M., Hansen L. B. S., Saunders A. M., Nielsen P. H. and Nielsen K. L. (2011). A metagenome of a full-scale microbial community carrying out enhanced biological phosphorus removal. *ISME Journal*, **6**, 1094–1106.

Aoi Y., Shiramasa Y., Kakimoto E., Tsuneda S., Hirata A. and Nagamune T. (2005). Single-stage autotrophic nitrogen-removal process using a composite matrix immobilizing nitrifying and sulfur-denitrifying bacteria. *Applied Microbiology and Biotechnology*, **68**, 124–130.

Andreasen K. and Nielsen P. H. (1997). Application of microautoradiography to the study of substrate uptake by filamentous microorganisms in activated sludge. *Applied Environmental Microbiology*, **63**, 3662–3668.

Andreasen K. and Nielsen P. H. (2000). Growth of Microthrix parvicella in nutrient removal activated sludge plants: studies of *in situ* physiology. *Water Research*, **34**, 1559–1569.

Aruga S., Kamagata Y., Kohno T., Hanada S., Nakamura K. and Kanagawa T. (2002). Characterization of filamentous Eikelboom Type 021N bacteria and description of *Thiothrix disciformis* sp. nov. and *Thiothrix flexilis* sp. nov. *International Journal of Systematic Evolutionary Microbiology*, **52**, 1309–1316.

Asvapathanagul P., Huang Z., Gedalanga P. B., Baylor A. and Olson B. H. (2012). Interaction of operational and physicochemical factors leading to *Gordonia amarae*-like foaming in an incompletely nitrifying activated sludge plant. *Applied Environmental Microbiology*, **78**, 8165–8175.

Asvapathanagul P., Olson B. H., Gedalanga P. B., Hashemi A., Huang Z. and La J. (2015). Identification and quantification of *Thiothrix eikelboomii* using qPCR for early detection of bulking incidents in a full-scale water reclamation plant. *Applied Microbiology and Biotechnology*, **99**, 4045–4057.

Beer M., Seviour E. M., Kong Y., Cunningham M., Blackall L. L. and Seviour R. J. (2002). Phylogeny of the filamentous bacterium Eikelboom Type 1851, and design and application of a 16S rRNA targeted oligonucleotide probe for its fluorescencein situ identification in activated sludge. *FEMS Microbiology Letters*, **207**, 179–183.

Blackall L. L., Tandoi V. and Jenkins D. (1991). Continuous culture studies with Nocardia amarae from activated sludge and their implications for *Nocardia* foaming control. *Research Journal of the Water Pollution Control Federation*, **63**, 44–50.

Blackall L. L., Seviour E. M., Cunningham M. A., Seviour R. J. and Hugenholtz P. (1994). Microthrix parvicella is a novel deep branching member of the actinomycetes subphylum. *Systematic and Applied Microbiology*, **17**, 513–518.

Blackall L. L., Stratton H., Bradford D., Sjörup C., Del Dot T., Seviour E. M. and Seviour R. J. (1996). Candidatus *Microthrix parvicella* – a filamentous bacterium from activated sludge sewage treatment plants. *International Journal of Systematic Bacteriology*, **4**, 344–346.

Blackall L. L., Seviour E. M., Bradford D., Rossetti S., Tandoi V. and Seviour R. J. (2000). Candidatus *Nostocoida limicola* II a filamentous bacterium from activated sludge. *International Journal of Systematic Evolutionary Microbiology*, **50**, 703–709.

Bond P. L., Erhart R., Wagner M., Keller J. and Blackall L. L. (1999). Identification of some of the major groups of bacteria in efficient and non efficient biological phosphorus removal activated sludge systems. *Applied Environmental Microbiology*, **65**, 4077–4084.

Carr E. L., Eales K. L. and Seviour R. J. (2006). Substrate uptake by Gordonia amarae in activated foams by FISH-MAR. *Water Science and Technology*, **54**, 39–45.

Cech J. S. and Hartman P. (1993). Competition between polyphosphate and polysaccharide accumulating bacteria in enhanced biological phosphorus removal systems. *Water Research*, **27**, 1219–1225.

Crocetti G. R., Hugenholtz P., Bond P. L., Schuler A., Keller J., Jenkins D. and Blackall L. L. (2000). Identification of polyphosphate-accumulating organisms and design of 16S rRNA-directed probes for their detection and quantification. *Applied Environmental Microbiology*, **66**, 1175–1182.

Crocetti G. R., Banfield J. F., Keller J., Bond P. L. and Blackall L. L. (2002). Glycogen-accumulating organisms in laboratory-scale and full-scale wastewater treatment processes. *Microbiology*, **148**, 3353–3364.

Dabert P., Sialve B., Delgenès J. P., Moletta R. and Godon J. J. (2001). Characterization of the microbial 16S rDNA diversity of an aerobic phosphorus-removal ecosystem and monitoring of its transition to nitrate respiration. *Applied Microbiology and Biotechnology*, **55**, 500–509.

Daims H., Nielsen J. L., Nielsen P. H., Schleifer K. H. and Wagner M. (2001). *In situ* characterization of *Nitrospira*-like nitrite-oxidising bacteria active in wastewater treatments plants. *Applied Environmental Microbiology*, **67**, 5273–5284.

Daims H., Lücker S. and Wagner M. (2016). A New Perspective on Microbes Formerly Known as Nitrite-Oxidizing Bacteria. *Trends in Microbiology*, doi: 10.1016/j.tim.2016.05.004

Eales K. L., Nielsen J. L., Seviour E. M., Nielsen P. H. and Seviour R. J. (2006). The in situ physiology of Skermania piniformis in foams in Australia activated sludge plants. *Environmental Microbiology*, **8**, 1712–1720.

Eikelboom D. H. (1975). Filamentous organisms observed in activated sludge. *Water Research*, **9**, 365–388.

Eikelboom D. H. (2000). Process Control of Activated Sludge Plants by Microscopic Investigations. IWA Publishing, London, UK.

Eikelboom D. H. and van Buijsen H. J. J. (1981). Microscopic Sludge Investigation Manual, 1st edn. TNO Research Institute for Environmental Hygiene, The Netherlands.

Eikelboom D. H. and van Buijsen H. J. J. (1983). Microscopic Sludge Investigation Manual, 2nd edn. TNO Research Institute for Environmental Hygiene, The Netherlands.

Eikelboom D. H. and Geurkink B. (2001). Catalogue of filamentous micro-organisms occurring in industrial activated sludge plants. TNO-MEP, Report R 2001/090.

Eikelboom D. H. and Geurkink B. (2002). Filamentous micro-organisms observed in industrial activated sludge plants. *Water Science and Technology*, **46**, 535–542.

Erhart R., Bradford D., Seviour R. J., Amann R. and Blackall L. L. (1997). Development and use of fluorescent in situ hybridization probes for the detection and identification of Microthrix parvicella in activated sludge. *Systematic and Applied Microbiology*, **20**, 310–318.

Falvo A., Levantesi C., Rossetti S., Seviour R. J. and Tandoi V. (2001). Synthesis of intracellular storage polymers by *Amaricoccus kaplicensis* a tetrad forming bacterium present in activated sludge. *Journal of Applied Microbiology*, **91**, 299–305.

Fukase T., Shibata M. and Miyaji Y. (1985). The role of an anaerobic stage on biological phosphorus removal. *Water Science and Technology*, **17**, 69–80.

Garcıa Martın H., Ivanova N., Kunin V., Warnecke F., Barry K. W., McHardy A. C., Yeates C., He S., Salamov A. A., Szeto E., Dalin E., Putnam N. H., Shapiro H. J., Pangilinan J. L., Rigoutsos I., Kyrpides N. C., Blackall L. L., McMahon K. D. and Hugenholtz P. (2006). Metagenomic analysis of two enhanced biological phosphorus removal (EBPR) sludge communities. *Nature Biotechnology*, **24**, 1263–1269.

Geets J., de Cooman M., Wittebolle L., Heylen K., Vanparys B., De Vos P., Verstraete W. and Boon N. (2007). Real-time PCR assay for the simultaneous quantification of nitrifying and denitrifying bacteria in activated sludge. *Applied Microbiology and Biotechnology*, **75**, 211–221.

Ginige M. P., Hugenholtz P., Daims H., Wagner M., Keller J. and Blackall L. L. (2004). Use of stable-isotope probing full-cycle rRNA analysis fluorescent in situ hybridisation-microautoradiography to study a methanol-fed denitrifying microbial community. *Applied Environmental Microbiology*, **70**, 588–596.

Harms G., Layton A. C., Dionisi H. M., Gregory I. R., Garrett V. M., Hawkins S. A., Robinson K. G. and Sayler G. S. (2003). Real-time PCR quantification of nitrifying bacteria in a municipal wastewater treatment plant. *Environmental Science and Technology*, **37**, 343–351.

He S., Gall D. L. and McMahon K. D. (2007). "Candidatus Accumulibacter" population structure in enhanced biological phosphorus removal sludges as revealed by polyphosphate kinase genes. *Applied Environmental Microbiology*, **73**, 5865–5874.

Hess A., Zarda B., Haner A., Stax D., Hohener P. and Zeyer J. (1997). In situ analysys of denitrifying toluene- and m-xylene-degrading bacteria in a diesel fuel-contaminated laboratory aquifer column. *Applied Environmental Microbiology*, **63**, 2136–2141.

Hesselmann R. P. X., Werlen C., Hahn D., van der Meer J. R. and Zehnder A. J. B. (1999). Enrichment phylogenetic analysis and detection of a bacterium that performs enhanced biological phosphate removal in activated sludge. *Systematic and Applied Microbiology*, **22**, 454–465.

Howarth R., Unz R. F., Seviour E. M., Seviour R. J., Blackall L. L., Pickup R. W., Gwyn Jones J., Yaguchi J. and Head I. M. (1999). Phylogenetic relationships of filamentous sulfur bacteria (*Thiothrix* spp. and Eikelboom Type 021N bacteria) isolated from wastewater-treatment plants and description of *Thiothrix eikelboomii* sp nov. *Thiothrix unzii* sp nov. *Thiothrix fructosivorans sp* nov and *Thiothrix defluvii sp* nov. *International Journal of Systematic Bacteriology*, **49**, 1817–1827.

Hugenholtz P. and Stackebrandt E. (2004). Reclassification of Sphaerobacter thermophilus from the subclass Sphaerobacteridae in the phylum *Actinobacteria* to the class *Thermomicrobia* (emended description) in the phylum *Chloroflexi* (emended description). *International Journal of Systematic Evolutionary Microbiology*, **54**, 2049–2051.

Jenkins D., Richard M. G. and Daigger G. T. (1984). Manual on the Causes and Control of Activated Sludge Bulking and Foaming, 1st edn. Lewis Publishers, New York, NY, pp. 1–165.

Jenkins D., Richard M. G. and Daigger G. T. (1993). Manual on the Causes and Control of Activated Sludge Bulking and Foaming, 2nd edn. Lewis Publishers, New York, NY, pp. 1–193.

Jenkins D., Richard M. G. and Daigger G. T. (2003). Manual on the Causes and Control of Activated Sludge Bulking and Foaming. 3th edn. Lewis Publishers, New York, NY, pp. 1–190.

Jetten M. S., Sliekers O., Kuypers M., Dalsgaard T., van Niftrik L., Cirpus I., van de Pas-Schoonen K., Lavik G., Thamdrup B., Le Paslier D., Op den Camp H. J., Hulth S., Nielsen L. P., Abma W., Third K., Engstrom P., Kuenen J. G., Jorgensen B. B., Canfield D. E., Sinninghe Damste J. S., Revsbech N. P., Fuerst J., Weissenbach J., Wagner M., Schmidt I., Schmid M. and Strous M. (2003). Anaerobic ammonium oxidation by marine and freshwater planctomycete-like bacteria. *Applied Microbiology and Biotechnology*, **63**, 107–114.

Juretschko S., Timmerman G., Schmid M., Schleifer K. H., Pommerening-Roser A., Koops H. P. and Wagner M. (1998). Combined molecular and conventional analyses of nitrifying bacterium diversity in activated sludge: *Nitrosococcus mobilis* and *Nitrospira*-like bacteria as dominant populations. *Applied Environmental Microbiology*, **64**, 3042–3051.

Kaetzke A., Jentzsch D. and Eschrich K. (2005). Quantification of Microthrix parvicella in activated sludge bacterial communities by real-time PCR. *Letters in Applied Microbiology*, **40**, 207–211.

Kanagawa T., Kamagata Y., Aruga S., Kohno T., Horn M. and Wagner M. (2000). Phylogenetic analysis of and oligonucleotide probe development for eikelboom type 021N filamentous bacteria isolated from bulking activated sludge. *Applied Environmental Microbiology*, **66**, 5043–5052.

Kim J., Lim J. and Lee C. (2013). Quantitative real-time PCR approaches for microbial community studies in wastewater treatment systems: applications and considerations. *Biotechnology Advances*, **31**, 1358–1373.

Kohno T., Sei K. and Mori K. (2002). Characterization of Type 1851 organism isolated from activated sludge samples. *Water Science and Technology*, **46**, 111–114.

Kragelund C., Nielsen J. L., Thomsen T. R. and Nielsen P. H. (2005). Ecophysiology of the filamentous Alphaproteobacterium *Meganema perideroedes* in activated sludge. *FEMS Microbiology Ecology*, **54**, 111–122.

Kragelund C., Kong Y., van der Waarde J., Thelen K., Eikelboom D., Tandoi V., Thomsen T. R. and Nielsen P. H. (2006). Ecophysiology of different filamentous Alphaproteobacteria species from industrial waste water treatment plants. *Microbiology*, **152**, 3003–3012.

Kragelund C., Levantesi C., Borger A., Thelen K., Eikelboom D., Tandoi V., Kong Y., van der Waarde J., Krooneman J., Rossetti S., Thomsen T. R. and Nielsen P. H. (2007a). Identity, abundance and ecophysiology of filamentous Chloroflexi species from activated sludge treatment plants. *FEMS Microbiology Ecology*, **59**, 671–682.

Kragelund C., Remesova Z., Nielsen J. L., Thomsen T. R., Eales K. L., Seviour R. J., Wanner J. and Nielsen P. H. (2007b). Ecophysiology of mycolic acid containing Actinobacteria (Mycolata) in activated sludge foams. *FEMS Microbiology Ecology*, **61**, 174–184.

Krul J. M. (1977). Experiments with *Haliscomenobacter hydrossis* in continuous culture without and with *Zooglea ramigera*. *Water Research*, **11**, 197–204.

Kumari S. K. S., Marrengane Z. and Bux F. (2009). Application of quantitative RT-PCR to determine the distribution of Microthrix parvicella in full-scale activated sludge treatment systems. *Applied Microbiology and Biotechnology*, **83**, 1135–1141.

Lajoie C. A., Layon A. C., Gregory I. R., Sayler G. S., Taylor D. E. and Meyers A. J. (2000). Zoogleal clusters and sludge dewatering potential in an industrial activated sludge wastewater treatment plant. *Water Environment Research*, **72**, 56–64.

Larkin J. M. and Shinabarger D. L. (1983). Characterization of *Thiothrix nivea*. *International Journal of Systematic Bacteriology*, **33**, 841–846.

Levantesi C., Beimfohr C., Geurkink B., Rossetti S., Thelen K., Krooneman J., Snaidr J., van der Waarde J. and Tandoi V. (2004). Filamentous Alphaproteobacteria associated with bulking in industrial wastewater treatment plants. *Systematic and Applied Microbiology*, **27**, 716–727.

Levantesi C., Rossetti S., Thelen K., Kragelund C., Krooneman J., Eikelboom D., Nielsen P. H. and Tandoi V. (2006). Phylogeny, physiology and distribution of *Candidatus* Microthrix calida, a new Microthrix species isolated from industrial activated sludge wastewater treatment plants. *Environmental Microbiology*, **8**, 1552–1563.

Lim J., Do H., Shin S. G. and Hwang S. (2008). Primer and probe sets for group-specific quantification of the genera *Nitrosomonas* and *Nitrosospira* using Real-Time PCR. *Biotechnology Bioengineering*, **99**, 1374–1383.

Liu B., Zhang F., Feng X., Liu Y., Yan X., Zhang X., Wang L. and Zhao L. (2006). *Thauera* and *Azoarcus* as functionally important genera in a denitrifying quinolone-removal bioreactors as revealed by microbial community structure comparison. *FEMS Microbiology Ecology*, **55**, 274–286.

Liu J. R. and Seviour R. J. (2001). Design and application of oligonucleotide probes for fluorescent in situ identification of the filamentous bacterial morphotype in activated sludge. *Environmenta Microbiology*, **3**, 551–560.

Liu J. R., McKenzie C. A., Seviour E. M., Webb R. I., Blackall L. L., Saint C. P. and Seviour R. J. (2001). Phylogeny of the filamentous bacterium "Nostocoida limicola" III from activated sludge. *International Journal of Systematic Evolutionary Microbiology*, **51**, 195–202.

Matturro B., Heavner G. L., Richardson R. E. and Rossetti S. (2013). Quantitative estimation of Dehalococcoides mccartyi at laboratory and field scale: comparative study between CARD-FISH and Real Time PCR. *Journal of Microbiological Methods*, **93**, 127–133.

McIlroy S. J. Kristiansen R., Albertsen M., Karst S. M., Rossetti S., Nielsen J. L., Tandoi V., Seviour R. J. and Nielsen P. H. (2013). Metabolic model for the filamentous 'Candidatus Microthrix parvicella' based on genomic and metagenomic analyses. *ISME Journal*, **7**, 1161–1172.

McKenzie C. M., Seviour E. M., Schumann P., Maszenan A. M., Liu J.-R., Webb R. I., Monis P., Saint C. P., Steiner U. and Seviour R. J. (2006). Isolates of 'Candidatus Nostocoida limicola' Blackall *et al.* 2000 should be described as three novel species of the genus *Tetrasphaera*, as *Tetrasphaera jenkinsii* sp. nov., *Tetrasphaera vanveenii* sp. nov. and *Tetrasphaera veronensis* sp. nov. *International Journal of Systematic Evolutionary Microbiology*, **56**, 2279–2290.

Madoni P. (1994). *La microfauna nell'analisi di qualità biologica dei fanghi attivi. Indice biotico dei fanghi* (Microfauna characterization for the assessment of WWTPs activated sludge quality. Bludge Biotic Index) Azienda Gas Acqua Consorziale di Reggio Emilia – Università degli Studi di Parma, Parma, Italy, pp. 1–48.

Maszenan A. M., Seviour R. J., Patel B. K., Janssen P. H. and Wanner J. (2005). *Defluvicoccus vanus* gen. nov., sp. nov., a novel Gram-negative coccus/coccobacillus in the 'Alphaproteobacteria' from activated sludge. *International Journal of Systematic Evolutionary Microbiology*, **55**, 2105–2111.

Meyer R. L., Saunders A. M. and Blackall L. L. (2006) Putative glycogen-accumulating organisms belonging to the Alphaproteobacteria identified through rRNA-based stable isotope probing. *Microbiology*, **152**, 419–429.

Mobarry B. K., Wagner M., Urbain V., Rittmann B. E. and Stahl D. A. (1996). Phylogenetic probes for analyzing abundance and spatial organization of nitrifying bacteria. *Applied Environmental Microbiology*, **62**, 2156–2162.

Mino T., Arun V., Tsuzuki Y. and Matsuo T. (1987). Effect of phosphorus accumulation on acetate metabolism in the biological phosphorus removal process. In: Biological Phosphate Removal from Wastewaters, R. Ramadori (ed.), Pergamon Press, Oxford, UK, pp. 27–38.

Mizrahi-Man O., Davenport E. R. and Gilad Y. (2013) Taxonomic classification of bacterial 16SrRNA genes using short sequencing reads: evaluation of effective study design. *PLoS ONE*, **8**(1), e53608.

Neef A., Zaglauer A., Meier H., Amann R., Lemmer H. and Schleifer K. H. (1996). Population analysis in a denitrifiying sand filter: conventional and in situ identification of *Paracoccus* spp. in methanol-fed biofilms. *Applied Environmental Microbiology*, **62**, 4329–4339.

Nielsen A. T., Liu W. T., Filipe C., Grady L., Molin S. and Stahl D. A. (1999). Identification of a novel group of bacteria in sludge from deteriorated biological phosphorus removal reactor. *Applied Environmental Microbiology*, **65**, 1251–1258.

Nielsen P. H., Roslev P., Dueholm T. and Nielsen J. L. (2002). Microthrix parvicella a specialised lipid consumer in anaerobic-aerobic activated sludge plants. *Water Science and Technology*, **46**, 73–80.

Nielsen P. H., Kragelund C., Seviour R. J. and Nielsen J. L. (2009). Identity and ecophysiology of filamentous bacteria in activated sludge. *FEMS Microbiology Reviews*, **33**, 969–998.

Odintsova E. V., Wood A. P. and Kelly D. P. (1993). Chemolithoautotrophic growth of *Thiothrix ramose*. *Archives of Microbiology*, **160**, 152–157.

Pereira H., Lemos P. C., Carrondo M. J. T., Crespo J. P. S. G., Reis M. A. M. and Santos H. (1996). Model for carbon metabolism in biological phosphorus removal processes based on in vivo ^{13}C-NMR labelling experiments. *Water Research*, **30**, 2128–2138.

Polz M. F., Odintsova E. V. and Cavanaugh C. M. (1996). Phylogenetic relationships of the filamentous sulfur bacterium *Thiothrix ramosa* based on 16S rRNA sequence analysis. *International Journal of Systematic Bacteriology*, **46**, 94–97.

Richard M. G., Shimizu G. P. and Jenkins D. (1985). Growth kinetics of *Sphaerotilus* species and their significance in activated sludge bulking. *Journal of the Water Pollution Control Federation*, **57**, 68–81.

Rossetti S., Christensson C., Blackall L. L. and Tandoi V. (1997). Phenotypic and phylogenetic description of an Italian isolate of M. parvicella. *Journal of Applied Bacteriology*, **82**, 405–410.

Rossetti S., Tomei M. C., Levantesi C., Ramadori R. and Tandoi V. (2002). M. parvicella: a new approach for kinetic and physiological characterization. *Water Science and Technology*, **46**, 65–72.

Rossetti S., Blackall L. L., Levantesi C., Uccelletti D. and Tandoi V. (2003). Phylogenetic and physiological characterization of a heterotrophic chemolithoautotrophic *Thiothrix* strain isolated from activated sludge. *International Journal of Systematic Evolutionary Microbiology*, **53**, 1271–1276.

Rossetti S., Tomei M. C., Nielsen P. H. and Tandoi V. (2005). "*M. parvicella*" a filamentous bacterium causing bulking and foaming in activated sludge systems: a review of current knowledge. *FEMS Microbiology Reviews*, **29**, 49–64.

Rossetti S., Qi R., Fan N., Levantesi C., Tandoi V. and Yang M. (2015). Causes and Bulking Control Strategies in Sewage Treatment Plants Located at High Altitude in China. MEWE Workshop, The 4th International Conference on Environmental Simulation and Pollution Control, ESPC & CSES, Beijing, 3–4 November 2015.

Rotthawe J. H., Witzel K. P. and Liesack W. (1997). The ammonia monooxygenase structural gene amoA as a functional marker: molecular fine-scale analysis of natural ammonia-oxidizing populations. *Applied Environmental Microbiology*, **63**, 4704–4712.

Satoh H., Mino T. and Matsuo T. (1994). Deterioration of enhanced biological phosphorus removal by the domination of microorganisms without polyphosphate accumulation. *Water Science and Technology*, **30**, 203–211.

Schramm A., de Beer D., Wagner M. and Amann R. (1998). Identification and activities in situ of *Nitrosospira* and *Nitrospira* spp. as dominant populations in a nitrifying fluidized bed reactor. *Applied Environmental Microbiology*, **64**, 3480–3485.

Schramm A., de Beer D., van den Heuvel J. C., Ottengraf S. and Amann R. (1999). Microscale distribution of populations and activities of *Nitrosospira* and *Nitrospira* spp. along a macroscale gradient in a nitrifying bioreactor: quantification by *in situ* hybridization and the use of microsensors. *Applied Environmental Microbiology*, **65**, 3690–3696.

Seviour E. M., Eales K. L., Izzard L., Beer M., Carr E. L. and Seviour R. J. (2006). The in situ physiology of '*Nostocoida limicola*' II, a filamentous bacterial morphotype in bulking activated sludge, using fluorescence in situ hybridization (FISH) and microautoradiography (MAR), *Water Science and Technology*, **54**, 47–53.

Seviour R. J., Maszenan A.M., Soddell J. A., Tandoi V., Patel B. K., Kong Y. and Schumann P. (2000). Microbiology of the "G-bacteria" in activated sludge. *Environmental Microbiology*, **2**, 581–93.

Seviour R. J., Liu J. R., Seviour E. M., McKenzie C. A., Blackall L. L. and Saint C. P. (2002). The "Nostocoida limicola" story: resolving the phylogeny of this morphotype responsible for bulking in activated sludge. *Water Science and Technology*, **46**, 105–110.

Seviour R. J., Mino T. and Onuki M. (2003). The microbiology of biological phosphorus removal in activated sludge systems. *FEMS Microbiology Reviews*, **27**, 99–127.

Sheik A. R., Muller E. E., Audinot J. N., Lebrun L. A., Grysan P., Guignard C. and Wilmes P. (2016). In situ phenotypic heterogeneity among single cells of the filamentous bacterium Candidatus Microthrix parvicella. *ISME Journal*, **10**, 1274–1279.

Slijkhuis H. and Deinema M. H. (1982). The physiology of M. parvicella a filamentous organism isolated from activated sludge. In: Bulking in Activated Sludge: Preventative and Remedial Methods, B. Chambers and E. J. Tomlinson (eds), Ellis Horwood Ltd, Chichester, pp. 75–83.

Snaidr J., Beimfohr C., Levantesi C., Rossetti S., van der Waarde J., Geurkink B., Eikelboom D., Lemaitre M. and Tandoi V. (2002). Phylogenetic analysis and in situ identification of Nostocoida limicola-like filamentous bacteria in activated sludge from industrial wastewater treatment plants. *Water Science and Technology*, **46**, 99–104.

Tandoi V., Caravaglio N., Di Dio Balsamo D., Majone M. and Tomei M. C. (1994). Isolation and physiological characterization of *Thiothrix* sp. *Water Science and Technology*, **29**, 261–269.

Tandoi V., Rossetti S., Blackall L. L. and Majone M. (1998). Some physiological properties of an Italian isolate of M. parvicella. *Water Science and Technology*, **37**, 1–8.

Thomsen T. R., Nielsen J. L., Ramsing N. B. and Nielsen P. H. (2004). Micromanipulation and further identification of FISH labelled microcolonies of a dominant denitrifying bacterium in activated sludge. *Environmetal Microbiology*, **6**, 470–479.

Thomsen T. R., Blackall L. L., de Muro M. A., Nielsen J. L. and Nielsen P. H. (2006). Meganema perideroedes gen. nov., sp. nov., a new filamentous alphaproteobacterium from activated sludge. *International Journal of Systematic Evolutionary Microbiology*, **56**, 1865–1868.

Thomsen T. R., Kong Y. and Nielsen P. H. (2007). Ecophysiology of abundant denitrifying bacteria in activated sludge. *FEMS Microbiology Ecology*, **60**, 370–382.

Tindall B. J., Rosselló-Móra R., Busse H. J., Ludwig W. and Kämpfer P. (2010). Notes on the characterization of prokaryote strains for taxonomic purposes. *International Journal of Systematic Evolutionary Microbiology*, **60**, 249–266.

Unz R. F. and Head I. M. (2005). Genus Thiothrix in Gammaproteobacteria. In: Bergey's Manual of Systematic Bacteriology, G. M. Garrity (ed.), Springer-Verlag, New York, NY, pp. 131–142.

van Niekerk A. M., Jenkins D. and Richard M. G. (1987). The competitive growth of *Zooglea ramigera* and Type 021N in activated sludge and in pure culture - A model for low F:M bulking. *Journal – Water Pollution Control Federation*, **59**, 262–273.

van Kessel M. A., Speth D. R., Albertsen M., Nielsen P. H., Op den Camp H. J., Kartal B.1, Jetten M. S. and Lücker S. (2015). Complete nitrification by a single microorganism. *Nature*, **24**, 555–559.

van Veen W. L. (1973). Bacteriology of activated sludge in particular the filamentous bacteria. *Antonie van Leeuwenhoek*, **39**, 189–205.

Vanysacker L., Denis C., Roels J., Verhaeghe K. and Vankelecom I. F. (2014). Development and evaluation of a TaqMan duplex real-time PCR quantification method for reliable enumeration of Candidatus Microthrix. *Journal of Microbiological Methods*, **97**, 6–14.

Vervaeren H., De Wilde K., Matthys J., Boon N., Raskin L. and Verstraete W. (2005). Quantification of an Eikelboom type 021N bulking event with fluorescence in situ hybridization and real-time PCR. *Applied Microbiology and Biotechnology*, **68**, 695–704.

Wagner M. and Loy A. (2002). Bacterial community composition and function in sewage treatment systems. *Current Opinion in Biotechnology*, **13**, 218–227.

Wagner M., Amann R. I., Kampfer B., Assmus B., Hartmann A., Hutzler P., Springer N. and Schleifer K. H. (1994a). Identification and in situ detection of Gram-negative filamentous bacteria in activated sludge. *Systematic and Applied Microbiology*, **17**, 405–417.

Wagner M., Erhart R., Manz W., Amann R., Lemmer H., Wedi D. and Schleifer K. H. (1994b). Development of an rRNA-targeted oligonucleotide probe specific for the genus *Acinetobacter* and its application for in situ monitoring in activated sludge. *Applied Environental Microbiology*, **60**, 792–800.

Wagner M., Rath G., Amann R., Koops H. P. and Schleifer K. H. (1995). In situ identification of ammonia-oxidizing bacteria. *Systematic and Applied Microbiology*, **18**, 251–264.

Wang F., Liu Y., Wang J. H., Zhang Y. L. and Yang H. Z. (2012). Influence of growth manner on nitrifying bacterial communities and nitrification kinetics in three lab-scale bioreactors. *Journal of Industrial Microbiology and Biotechnology*, **39**, 595–604.

Williams T. M. and Unz R. F. (1985). Isolation and characterization of filamentous bacteria present in bulking activated sludge. *Applied Microbiology and Biotechnology*, **22**, 273–282.

Woese C. R. (1987). Bacterial evolution. *Microbiological Reviews*, **51**, 221–271.

Wong M. T. and Liu W. T. (2007). Microbial succession of glycogen accumulating organisms in an anaerobic-aerobic membrane bioreactor with no phosphorus removal. *Water Science and Technology*, **54**, 29–37.

Wong M. T., Tan F. M., Ng W. J. and Liu W. T. (2004). Identification and occurrence of tetrad-forming Alphaproteobacteria in anaerobic-aerobic activated sludge processes. *Microbiology*, **150**, 3741–3748.

Chapter 3

Activated sludge separation problems

J. Wanner

3.1 INTRODUCTION

The activated sludge process is the most widespread method for the biological treatment of both municipal and industrial wastewaters. It is used in many process modifications and at scales that range from small package units for individual dwellings in rural areas up to large installations for big cities and industrial complexes (Wanner & Jobbágy, 2014). In all cases, the biologically treated wastewater must be separated from the activated sludge biomass to produce a high-quality final effluent. Although new separation processes like membrane filtration have been tested and used to a limited extent in pilot- and full-scale units, the gravity sedimentation is still the only economically feasible solids separation method for most municipalities and industrial enterprises. Because of this, the importance of having good activated sludge settling properties cannot be overemphasized.

Activated sludge bulking caused by filamentous microorganisms is the most common operating problem associated with poor activated sludge separation. Discussions of activated sludge bulking are found frequently in papers, books and practical manuals on wastewater treatment. Because of this, many practitioners tend to use the term 'activated sludge bulking' as a synonym for all activated sludge separation problems. This misinterpretation can lead to inappropriate control measures.

doi: 10.2166/9781780408637_053

3.2 'WELL-SETTLING' ACTIVATED SLUDGE

3.2.1 Requirements

For successful wastewater treatment using activated sludge, the biomass must be able to (i) flocculate and (ii) settle and thicken by gravity sedimentation because the treated wastewater is separated from the activated sludge in secondary clarifiers (secondary settling tanks) where the main driving force for separation is gravitation. A large and compact activated sludge floc (see Fig. 3.1) is a prerequisite for efficient activated sludge separation in secondary clarifiers. Poor separation and thickening of activated sludge in secondary clarifiers can lead to the following operating problems:

- Activated sludge biomass escapes from the secondary clarifier and deteriorates final effluent quality not only in terms of *suspended solids*, but also in BOD_5 (respiration of biomass), COD (organic nature of biomass) and in TKN and TP (the biomass content of N and P)
- A dilute return sludge stream is produced that may not allow maintenance of the desired aeration basin biomass concentration and proper control of sludge age
- A dilute waste sludge stream hydraulically overloads the sludge handling facilities

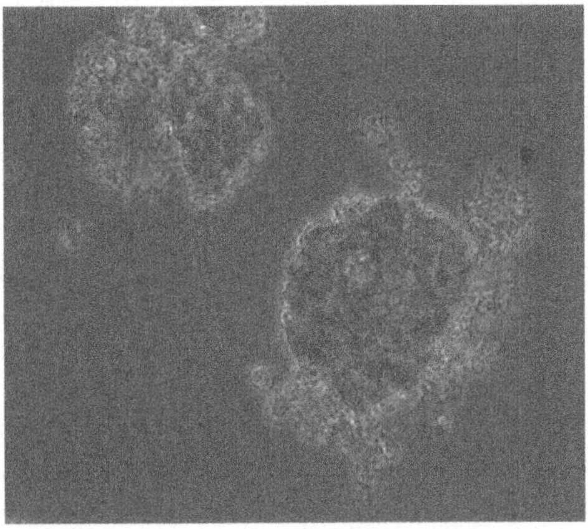

Figure 3.1 Large and compact activated sludge floc (250×, wet mount, phase contrast).

Well settling activated sludge should have the following properties (Wanner, 1994):

- It settles fast, with zone settling velocities of >3 m/h
- It does not occupy an excessive volume after settling and thickening in a secondary clarifier

- After sedimentation it leaves a clear supernatant ($X = 15$ mg L^{-1} or less)
- It does not rise after sedimentation for at least 2–3 h

3.2.1.1 Measurement of settling and thickening properties

Conditions during settling in secondary settling tanks are best simulated by zone settling velocity measurements. The height of the sludge interface is measured over time in as tall and wide a cylinder as possible. The plot of sludge layer height vs. time is called a settling curve. This curve usually has three distinct settling phases (see Fig. 3.2):

I – reflocculation
II – zone settling
III, IV – transition and compaction

The zone settling velocity is calculated from the slope of the curve in the zone settling phase (II).

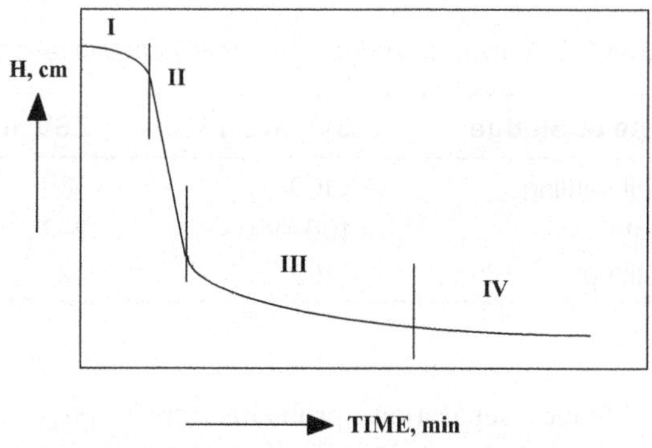

Figure 3.2 An example of an activated sludge settling curve.

In wastewater treatment practice, the commonest way to measure activated sludge settleability is by the sludge volume index SVI test. The standard sludge volume index is defined as follows:

$$\text{SVI} = \frac{V_{30}}{X} \tag{3.1}$$

where
SVI = sludge volume index [ml g^{-1}]
V_{30} = volume of settled sludge after 30 min sedimentation in a 1 L cylinder
X = concentration of activated sludge prior to settling (mixed liquor suspended solids) [g/L]

SVI values are affected by many factors Especially important are the effects of sludge concentration or the volume of settled sludge after 30 min and wall effects in the cylinder. In some countries settling test conditions are standardized. Common standardization methods include:

- Stirred sludge volume index SSVI; the measuring cylinder is equipped with a slowly rotating stirring device
- SVI at a standard suspended solids concentration; the test is performed with a defined sludge concentration (for instance 3.5 g L^{-1}, $SVI_{3.5}$)
- Diluted sludge volume index DSVI; the volume of settled sludge after 30 min sedimentation should not exceed 200 ml l^{-1}. If this occurs the test is repeated with a diluted sludge

Activated sludge can be classified on the basis of its zone settling velocity (ZSV) and SVI as shown in Table 3.1.

Table 3.1 Activated sludge settleability classification.

Type of Sludge	SVI, mL g^{-1}	ZSV, m h^{-1}
Well settling	<100	>3
'Light'	100–150	2–3
Bulking	>150	<2

When the sludge separation problems are caused by filamentous microorganisms, it is useful to quantify the amount of these microorganisms in activated sludge by a method such as that of Sezgin *et al.* (1978), where total length of filaments protruding from flocs or freely floating in the bulk liquid (TEFL) are measured. TEFL values of 10^7 μm ml^{-1} (10^4 m g^{-1}) are considered to represent the transition point between a non-bulking and a bulking activated sludge.

Because the measurement of TEFL is laborious and time consuming, for the routine examination of activated sludge, a rapid, simple method for subjective scoring of filament abundance was developed by Jenkins *et al.* (1993). The abundance of filaments is classified according to the scale given in Table 3.2.

For the foam-forming nocardioform actinomycetes, which are often irregularly shaped and frequently branched, Pitt and Jenkins (1990) recommend a counting method based on determining the number of intersections of these filaments with a hairline in the eyepiece.

Table 3.2 Subjective scoring of filament abundance according to Jenkins *et al.* (1993).

Numerical Value	Abundance	Description of Microscopic Picture
0	None	
1	Few	Filaments present, but only observed in an occasional floc
2	Some	Filaments commonly observed, but not present in all flocs
3	Common	Filaments observed in all flocs, but at low density (e.g. 1–5 filaments per floc)
4	Very common	Filaments observed in all flocs at medium density (e.g. >5–20 filaments per floc)
5	Abundant	Filaments observed in all flocs at high density (e.g. >20 filaments per floc)
6	Excessive	Filaments present in all flocs – appears more filaments than flocs and or filaments growing in high abundance in bulk solution

3.2.2 Microscopic features of well settling activated sludge

A simple microscopic examination can be very useful for the assessment of sludge separation problems. Good settling sludge is formed of roughly spherical, compact, firm flocs (Fig. 3.1) with an average size of >100 μm. Phase contrast microscopic observation of such flocs at 1000 magnifications will often reveal a 'filamentous backbone'. This 'filamentous backbone' can be better seen in Gram- or Neisser-stained samples. This suggests that large compact flocs are formed by bacterial growth on the filaments observed inside the floc. When the filaments protrude from the flocs into the bulk liquid they can cause *bulking sludge* (see below). This is referred to as poor floc *macrostructure*. In some activated sludge samples, the compact core of the flocs created by bacterial biomass is absent and the flocs exhibit a loose, open structure. When the compact core of activated sludge flocs is too small (usually 20–80 μm) the cause for poor settling is said to be due to poor floc *microstructure*. Such microstructure problems can be due either to insufficient production or overproduction of the biopolymer material (glycolax) that binds the cells together in the activated sludge flocs.

3.3 ACTIVATED SLUDGE SEPARATION PROBLEMS

Six major types of activated sludge operating problem are problems related to microbial biomass quality (Jenkins *et al.* 1993; Wanner, 1994, 2002).

3.3.1 Poor floc microstructure

The following problems depend on a floc microstructure failure.

3.3.1.1 Dispersed growth

The activated sludge microorganisms are dispersed freely in the bulk liquid as individual cells (Fig. 3.3) or small clumps with diameters of 10–20 μm. The sedimentation rate of these individual cells or bacterial clumps is too low for them to be removed efficiently by gravity sedimentation. Also no zone settling occurs in secondary settling tanks. This has two impacts on the activated sludge process:

- The separation efficiency of the secondary settling tank is very poor and the final effluent is turbid.
- Because of the poor solids separation efficiency, a significant amount of biomass escapes from the system so that only low sludge ages (Θ_X) can be maintained. A system with dispersed growth is closer to a chemostat than a continuous cultivation system with biomass recycle.

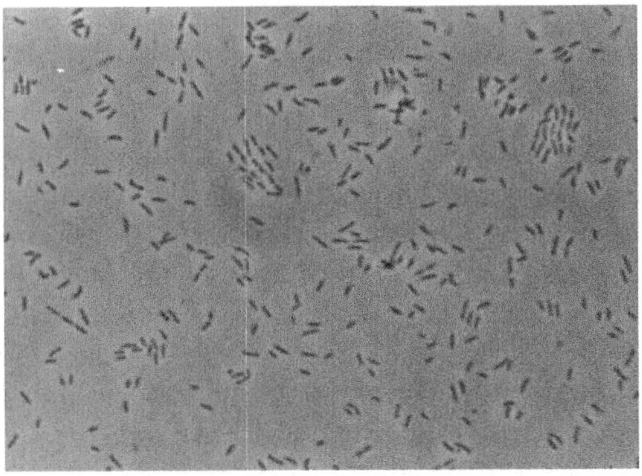

Figure 3.3 Dispersed freely growing bacterial cells (250×, wet mount, phase contrast).

The poor bioflocculation that is the root cause of dispersed growth is caused by low production of the extracellular biopolymers that create the matrix in firm activated sludge flocs. A common cause of dispersed growth is a very high organic loading (high 'food to microorganism' ratio, *F/M*) since bacteria do not need to produce a glycocalyx under these conditions. Dispersed growth can be a common problem during the start-up of activated sludge systems. Another cause is the presence of toxicity in the wastewater.

3.3.1.2 Unsettleable microflocs

The appearance that this separation problem gives is quite similar to that of dispersed growth. The final effluent from the secondary clarifier is not very clear and it contains many small particles of escaping biomass (Fig. 3.4). However, microscopic examination will reveal that the nature of the problem is different. The unsettleable particles are larger than in dispersed growth (about 50–100 μm) and are roughly spherical and compact. These microflocs result from the disintegration of firm and sound flocs.

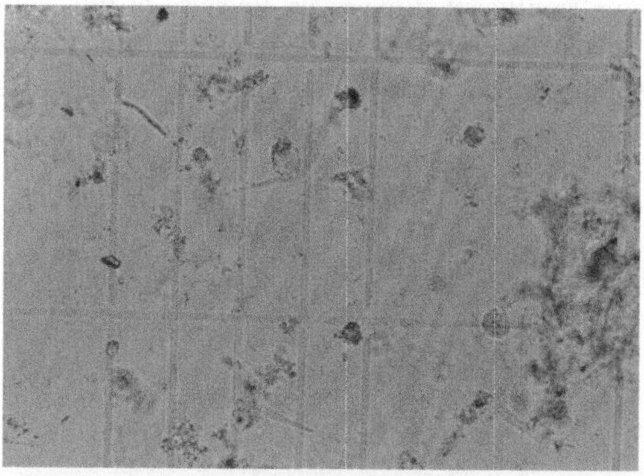

Figure 3.4 Unsettleable microflocs of activated sludge (250×, wet mount, phase contrast).

During the settling test the activated sludge rapidly separates into two fractions: large flocs and small flocs. The large flocs settle rapidly and when the SVI is calculated on the basis of the volume of these larger flocs, its value is quite low (around 50 ml g^{-1}). The supernatant in the cylinder is turbid because of the suspended small flocs and a substantial fraction of the total biomass remains in these unsettleable particles.

Reasons for floc disintegration are:

- Insufficient production of glycocalyx or its consumption by bacteria inside the flocs at low organic loadings (typical of high sludge age, extended aeration systems).
- Total absence of filamentous microorganisms which form the 'backbones' of larger flocs (Jenkins *et al.* 1993).
- Disintegration of flocs by shearing effects, for instance by high speed mechanical aerators.

3.3.1.3 Viscous bulking

The rather broad term 'viscous bulking' describes symptoms rather than causes. Viscous bulking activated sludge contains an excessive amount of extracellular biopolymers which impart a slimy, jelly-like consistency to it. Because the biopolymers are hydrophilic, the activated sludge becomes highly water retentive. Such a 'hydrous' activated sludge exhibits low settling and compaction velocities. Also, since the biopolymers are surface active, a viscous activated sludge may foam when aerated intensively (Wanner, 2002).

Biopolymer production is characteristic of most floc forming microorganisms. Under normal conditions (no toxic compounds, nutrient balanced growth), the amount of biopolymer generated is just enough for the formation of firm flocs. Zoogloeal bacteria always produce large amounts of biopolymers because their individual cells are surrounded by a significant slime layer. Because of this, some authors use the term 'zoogloeal bulking' to describe the settling problems caused by the presence of an excessive amount of zoogloeal colonies in activated sludge (Figs 3.5 and 3.6).

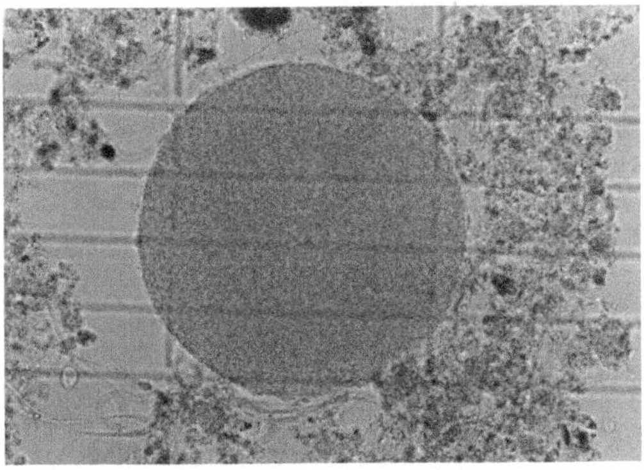

Figure 3.5 Large amorphous colony of zoogloeal bacteria (250×, wet mount, phase contrast).

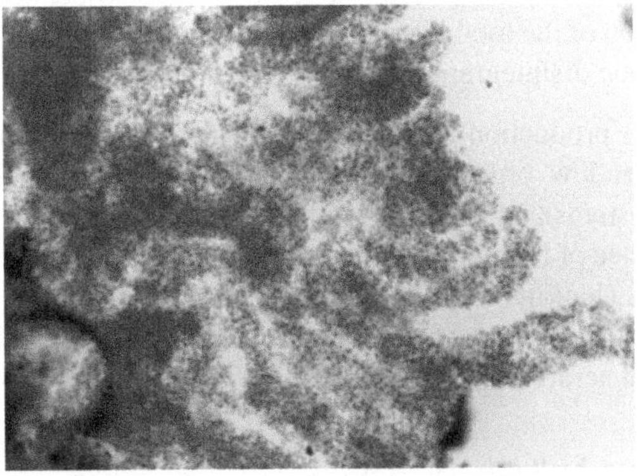

Figure 3.6 Large-fingered colony of zoogloeal bacteria (1000×, Gram stain, direct illumination).

3.3.2 Poor floc macrostructure

The following problems depend on a floc macrostructure failure.

3.3.2.1 Filamentous bulking

Filamentous bulking is a common case of poor activated sludge compaction (phase III of the settling curve – see above) which results in (Wanner, 1994):

- Low return and wasted activated sludge solids concentrations
- Difficulties in maintaining the required activated sludge concentration in aeration basins
- Poor sludge dewaterability
- Hydraulic overloading of sludge handling facilities

Filamentous microorganisms interfere with the sedimentation and compaction of activated sludge flocs in two ways (Jenkins *et al.* 1993):

(1) Some filamentous microorganisms grow inside flocs giving them a diffuse open structure. These open flocs have space to retain water in them, so that although aggregation of individual flocs is not mechanically hindered by filaments protruding from the flocs, excessive water remains 'captured' in the settled sludge.
(2) Filaments can detrimentally affect the sedimentation and compaction of activated sludge flocs which is much more common. Most filamentous microorganisms observed in activated sludges protrude from rather compact and firm flocs into the bulk liquid. These filaments, which in low numbers can form the backbones of firm flocs, in large numbers are able to mechanically prevent the compaction of individual flocs. This type of interference problem is called 'bridging' (Fig. 3.7).

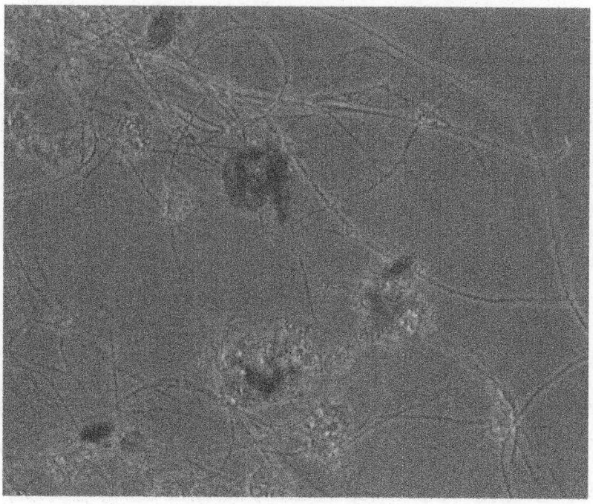

Figure 3.7 Filamentous microorganism growing in excess between activated sludge flocs (250×, wet mount, phase contrast).

3.3.2.2 Foaming caused by filamentous microorganisms

Biological foaming by 'foam-forming' filamentous microorganisms is a combination of physico-chemical and biochemical interactions leading to the stabilization of a three-phase system of air–water–microbial cells. Stabilization of these biological foams is caused by the following features of foam-forming filaments:

- Production of extracellular materials such as lipids, lipopeptides, proteins and carbohydrates which have surface active properties (biosurfactants)
- The cell walls of foam-forming microorganisms being strongly hydrophobic

The formation of stable foams in the aeration basins of activated sludge plants can create several operating problems:

(1) Aesthetic problems, such as slippery pathways covered by escaping foam.
(2) Floating biomass in the secondary clarifier that can deteriorate final effluent quality.
(3) Accumulation of a significant amount of biomass into the foam that is not active in the treatment processes; loss of ability to control activated sludge age.

Biological foam, especially on the surface of a secondary clarifier, may look like floating activated sludge. To distinguish between biological foam and floating sludge, microscopic examination can be very useful. The abundance of filamentous microorganisms in biological foams is usually much higher than in the corresponding mixed liquor. On the contrary, the microscopic appearance of floating sludge is almost the same as the mixed liquor (unless bubbles of nitrogen gas are present). Especially when caused by M. parvicella, the surface foam on the aeration tanks, and secondary clarifiers can look like a monoculture of the filament (see Fig. 3.8).

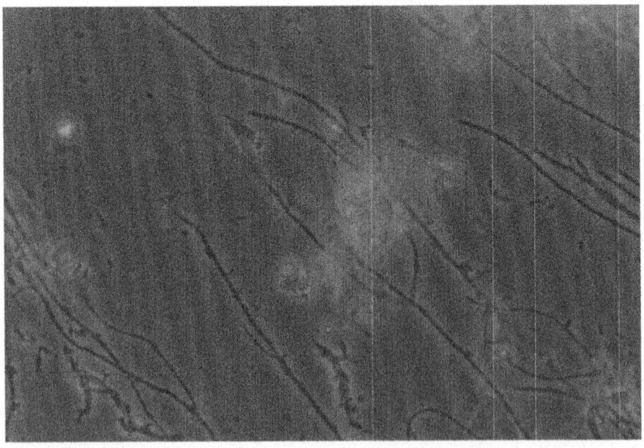

Figure 3.8 Microscopic picture of biological foam caused by Microthrix parvicella (1000×, phase contrast).

There are no unified and generally accepted methods for the quantification of foaming problems. Probably the most exact method for predicting the amount of foam formed on aeration basins and the extent of the resulting problems is the Scum Index (SI) suggested by Pretorius and Läubscher (1987). SI [%] is calculated as follows:

$$SI = \frac{\text{mass of biomass in foam}}{\text{total mass of biomass}} \times 100 \tag{3.2}$$

The portion of the biomass in the foam is estimated by flotation using a standard aeration rate of 10 m^3/m^3 h in in a batch flotation cell (80 mm diameter; 500 mm high). The floatation is repeated several times until all foam-forming microorganisms have been transferred into the scum. SI can be used for predicting the operating problems that might occur using the information presented in Table 3.3.

Table 3.3 Scum index and expected operational problems according to Pretorius and Läubscher (1987).

SI%	Extent of Problem
0–0.5	Negligible
0.5–6	Small
>6–10	Medium
>10–15	Serious
>15	Catastrophic

Plant operators can monitor the tendency of activated sludge to foam using the method suggested by Kocianova *et al.* (1992). The percentage of aeration basin and secondary clarifier surfaces covered by foam is regularly recorded. The observations should be made at the time of the day when loading conditions and operating procedures are similar. The authors found a good correlation between percentage coverage and other parameters reflecting the presence of foam-formers such as foam accumulation rate and the hydrophobicity of the activated sludge.

Recently, the measurement of hydrophobicity has become more common because this parameter is directly connected with the phenomenon of biological foaming. The hydrophobicity of mixed liquor and foam can be measured by using the method of Rosenberg *et al.* (1980) as modified by Kocianova *et al.* (1992) – the so-called MATH test (Microbial Adherence To Hydrocarbons). The assay is based on adhesion of hydrophobic microorganisms to a hydrophobic surface of a solvent like hexadecane. The results of the MATH assay as modified by

Kocianova *et al.* (1992) are affected by the electrical surface charge properties of the water/hexadecane system because the droplets of hexadecane formed during the 'extraction' of hydrophobic microorganisms from the aqueous phase are highly negative (Medrzycka, 1991), presumably due to the adsorption of anions.

Foaming intensity depends strongly on the concentration of surface active compounds. Nowadays products containing non-ionic detergents are used both in households and in industry. The concentrations of non-ionic detergents are not usually measured by wastewater laboratories because of the complexity of the analytical procedures. Because of this the concentration of non-ionic detergents in the influent wastewater should be checked when some unexpected foaming occurs.

Extracellular polymers (ECPs) produced under certain conditions by activated sludge microorganisms can be also responsible for the formation of biological foam, especially in aeration basins. The presence of ECPs is not measured commonly in activated sludge. If frequent foaming periods occur, it is recommended that the measurement of ECPs should be performed regularly. If these show a tendency to increase in concentration some operational countermeasures should be taken, for example, balancing the influent BOD with nitrogen and/or phosphorus, or checking to determine whether toxic compounds are present in the raw wastewater. The analytical procedures for the determination of carbohydrates, proteins and DNA (the main constituents of ECPs) can be found in specialized literature.

3.3.3 Other reasons

3.3.3.1 Rising sludge

When this is the problem, the water surface in the secondary clarifier is covered by patches (or in severe cases there is complete coverage) of floating activated sludge. If the phenomenon is observed in a glass cylinder, two phases can be distinguished:

- The activated sludge settles rapidly and a rather compact bottom layer of settled sludge and a clear supernatant are formed.
- After a certain period of time part or all of the settled and thickened sludge starts to float and move up to the water surface. (If this occurs within 30 min it can interfere with the SVI measurement.)

The floating material is full of gas bubbles derived from denitrification which takes place in the settled and thickened layer of activated sludge. Because the settled sludge has a high biomass concentration, the dissolved oxygen is quickly depleted and if the activated sludge is from a nitrifying plant, anoxic conditions are established. The bubbles of nitrogen liberated during denitrification act as sludge 'carriers'. Thus, the primary reason for this kind of separation problem does not lie in the activated sludge itself but in:

- Insufficient denitrification in the activated sludge system.
- Poor performance of settled sludge removing equipment in secondary clarifiers, which allows the sludge to remain too long in the clarifier.

3.4 SUMMARY

The primary conditions for successful activated sludge system performance are good separation and thickening properties of activated sludge. The quality of the final effluent from an activated sludge system is determined by the efficiency of activated sludge separation in the secondary clarifiers. Six types of activated sludge separation problem can cause the deterioration of secondary clarifier final effluent. Some practitioners used to call all these separation problems 'activated sludge bulking'. This is incorrect and does not allow us to determine the causes of a given separation problem or to choose a proper control method. This section has described the characteristic features of individual separation problems. The importance of microscopic examination of activated sludge for the identification of separation problems is stressed and has been illustrated by microphotographs. Quantification of separation problems together with careful microscopic examination of activated sludge should always be the first step in solving and rectifying activated sludge separation properties. For foaming problems the conventional methods of quantification of separation problems, like SVI and ZSV measurement, should be accompanied by the measurement of biomass hydrophobicity and by the estimation of the non-ionic detergent and extracellular polymer concentrations.

ACKNOWLEDGMENT

Microphotographs were taken by Olga Krhutkova.

REFERENCES

Jenkins D., Richard M. G. and Daigger G. T. (1993). Manual on the Causes and Control of Activated Sludge Bulking and Foaming, 2nd edn., CRC Press, Lewis Publishers, Boca Raton, FL, USA.

Kocianova E., Foot R. J. and Forster C. F. (1992). Physicochemical aspects of activated sludge in relation to stable foam formation. *Journal of the Institution of Water and Environmental Management*, **6**, 342–350.

Medrzycka K. B. (1991). The effect of particle concentration on the zeta potential in extremely dilute solutions. *Colloid & Polymer Science*, **269**, 85–90.

Pitt P. and Jenkins D. (1990). Causes and control of *Nocardia* in activated sludge. *Research Journal of the Water Pollution Control Federation*, **62**, 143–150.

Pretorius W. A. and Läubscher C. J. P. (1987). Control of biological scum in activated sludge plant by means of selective flotation. *Water Science & Technology*, **19**(5–6), 1003–1011.

Rosenberg M. Gutnick D. and Rosenberg E. (1980). Adherence of bacteria to hydrocarbons: A simple method for measuring cell-surface hydrophobicity. *FEMS Microbiology Letters*, **9**, 29–33.

Sezgin M., Jenkins D. and Parker D. S. (1978). A unified theory of filamentous activated sludge bulking. *Journal Water Pollution Control Federation*, **50**, 362–381.

Wanner J. (1994). Activated Sludge Bulking and Foaming Control. Technomic Publishing Co. Inc., Lancaster, PA, USA.

Wanner J. (2002). Control of filamentous bulking in activated sludge. In: Encyclopedia of Environmental Microbiology, G. Bitton (ed.), John Wiley & Sons Inc., New York, USA, pp. 1306–1315.

Wanner J. and Jobbágy A. (2014). Activated sludge solids separation. Chapter 10 In: Activated Sludge – 100 Years and Counting, D. Jenkins and J. Wanner (eds), IWA Publishing, London, pp. 171–194, ISBN 9781780404936.

Chapter 4

Aeration tank and secondary clarifier as one system

J. Wanner and M. Torregrossa

4.1 INTRODUCTION

Traditionally the activated sludge process is treated as a two stage process. The first stage (aeration basin) is described in terms of biological processes while the description of the second stage concentrates on processes of biomass separation and thickening. The modern approach to the activated sludge process looks at both the biological and the separation stage as one system. The aim of this chapter is to explain the interactions between the two parts of the activated sludge system. For information on the the principles of design and construction of secondary clarifiers, the reader should consult the basic literature on secondary clarifiers from recent years (Albertson, 1992; ATV-DVWK Standard A 131E, 2000; Ekama *et al.* 1997; Parker *et al.* 2014). The activated sludge system comprising a biological reactor and a separation/thickening stage can be schematically described by Fig. 4.1.

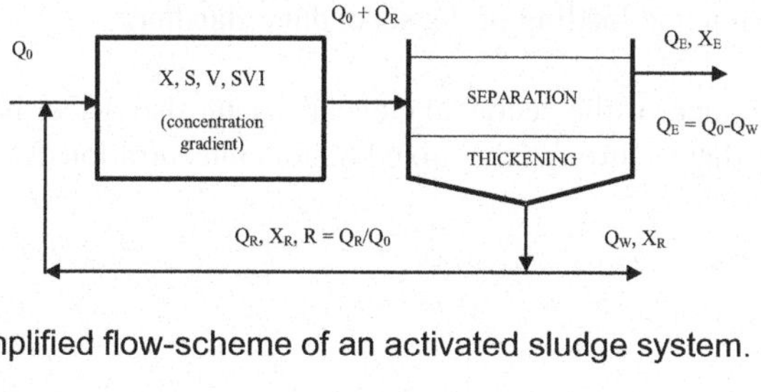

Figure 4.1 Simplified flow-scheme of an activated sludge system.

doi: 10.2166/9781780408637_067

4.2 AERATION TANK AND SECONDARY CLARIFIER INTERACTIONS

4.2.1 Activated sludge process

The major technological parameter affecting the performance of activated process in the reactor (aeration basin) is the biomass retention time, or sludge age defined as follows:

$$\Theta_X = \frac{V \cdot X + M_{FC}}{Q_W \cdot X_R + Q_E \cdot X_E} \qquad (4.1)$$

Where:

Θ_X biomass retention time, sludge age (d)
M_{FC} amount of activated sludge biomass stored in secondary clarifier (kgSS)
V reactor volume (m^3)
X biomass concentration in the reactor ($kgSS \cdot m^{-3}$)
Q_W waste sludge flow rate ($m^3 \cdot d^{-1}$)
X_R concentration of thickened sludge ($kgSS \cdot m^{-3}$)
Q_E effluent flow rate ($m^3 \cdot d^{-1}$)
X_E effluent biomass concentration ($kgSS \cdot m^{-3}$)

Equation (4.1) shows that the biomass retention time in the entire activated sludge system is directly affected by:

- the storage capacity of secondary clarifier M_{FC} (depends on the construction of the clarifier and its operating mode)
- the concentration of thickened sludge X_R, which is a function of both activated sludge quality and the design and construction of the secondary clarifier
- the concentration of biomass in the effluent X_E, which is partially influenced by the activated sludge separation characteristics (see Chapter 3) and by the construction and loading of the secondary clarifier.

The performance of the secondary clarifier can also affect biomass retention time indirectly, that is, through the mixed liquor concentration (X) that it is possible to maintain

$$X = \frac{R}{1 + R} \cdot X_R \qquad (4.2)$$

where R is the return sludge ratio.

The concentration of thickened sludge, X_R, depends on both the activated sludge settling properties and on the performance of the secondary clarifier.

In the past, the German design procedure A131 used the simple relationship of Kalbskopf:

$$X_R = \frac{1200}{SVI} \tag{4.3}$$

(X_R in $g \cdot L^{-1}$ and SVI in $mL \cdot g^{-1}$)

At present, more sophisticated relationships for the prediction of recycle flow concentration X_R can be found in the literature (Ekama *et al.* 1997).

In general, the poor thickening in the secondary clarifier leads to problems with:

- control of biomass retention time (see above)
- hydraulic overloading of sludge handling facilities (because of low X_R the waste sludge flow rate Q has to be increased)

The loss of control over the sludge age Θ_X can cause various separation problems. Because many activated sludge systems are operated as BNR plants, problems with sludge age control may cause instabilities in nutrient removal processes, especially in nitrification.

4.2.2 Secondary clarifier

4.2.2.1 General performance, thickening function

The performance of a secondary clarifier is affected by three major parameters:

- q_A surface overflow rate ($m^3 \cdot m^{-2} \cdot h^{-1}$)

$$q_A = \frac{Q_0}{A_{SC}} \tag{4.4}$$

  where Q_0 is the influent flow rate ($m^3 \cdot h^{-1}$) and A_{SC} is the surface area of secondary clarifier (m^2)
- Θ hydraulic retention time (h), obtained by the ratio:

$$\Theta = \frac{V_{sc}}{Q_0} \tag{4.5}$$

  where V_{sc} is the volume of secondary clarifier (m^3)
- N_A 'solids loading rate' or 'solid flux', defined as the mass of suspended sludge solids entering the clarifier per unit clarifier area and per time unit, ($kgMLSS \cdot m^{-2} \cdot h^{-1}$):

$$N_A = \frac{X \cdot Q_0 \cdot (1 + R)}{A_{SC}} \tag{4.6}$$

Because these parameters are affected by the settleability of the activated sludge the numerical values of clarifier design criteria, in various design manuals, are usually limited to a range of activated sludge quality. For instance, the Czech Design Recommendation CSN 75 6401 states that the design criteria are valid only for SVI less than 150 $mL \cdot g^{-1}$ and for recycle ratios less than 1.5.

4.2.2.1.1 Solids flow theory

With particular reference to the thickening function that a secondary clarifier must guarantee, some important pieces of information are summarized below, useful both in the design stage and during operation of this treatment unit.

The incoming sludge suspension in the clarifier flows in a downward direction owing to two parallel circumstances: the return sludge extraction from the bottom of the clarifier and the gravitational sedimentation. So, the downward velocity of the liquid phase is given by the sum of the downward velocity of the liquid phase, V_{zs}, called 'zone settling velocity' and the velocity of sludge abstraction, u.

The zone settling velocity is estimated in good approximation by the follow exponential equation valid for concentrations of mixed liquor suspended solids more than 1 $g \cdot L^{-1}$:

$$V_{zs} = V_0 \cdot e^{(-nX)} \tag{4.7}$$

where V_0 ($m \cdot h^{-1}$) and n ($m^3 \cdot kgMLSS^{-1}$) are constants readily obtainable by liner least-squares regression of $\log V_{zs}$ against X over a range of concentration from 1 to 12 $g \cdot L^{-1}$.

Pointing with R the sludge recirculation ratio, the velocity u is given by the ratio between the return sludge flow rate, $R \cdot Q_0$, and the cross-sectional area of the clarifier, A_{SC}.

Given the above, the solids loading rate, N_A, can be expressed by the equation:

$$N_A = N_{A1} + N_{A2} = X \cdot V_{zs} + X \cdot u = X \cdot V_0 \cdot \exp^{(-n \cdot X)} + R \cdot Q_0 \cdot A_{SC}^{-1} \tag{4.8}$$

Graphically, the overall equation is represented in Fig. 4.2 obtained after the construction of the curves representing individual summand equations in summary equation (4.8).

In correspondence with the minimum point of the curve N_A, the value of the solids flow limiting ($N_{A,lim}$) can be determined on the ordinate axis. This value represents the maximum of solid flow value applicable to a secondary clarifier having a specified size.

While the solids stream N_{A2} is adjustable by varying the flow rate of the sludge recirculation pumps, the solids flux G_1 is consequent to the characteristics of the sludge settleability. In Fig. 4.3 are represented solids flow curves as a function of the SVI parameter.

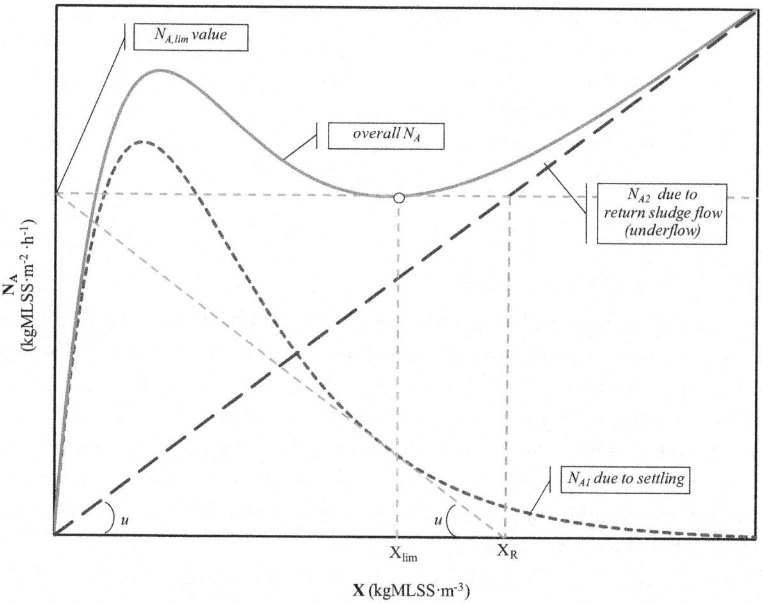

Figure 4.2 Solids flux curves.

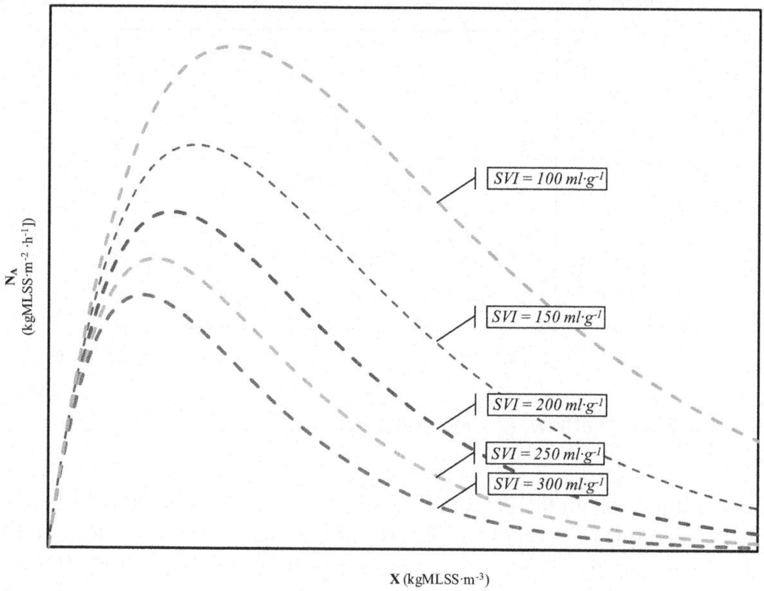

Figure 4.3 Solid flux curves as a function of SVI.

This highlights that at the same X concentration, when the sludge shows values of optimum SVI (i.e. 80 ÷ 100 $ml \cdot g^{-1}$), the N_{A1} solids flow curve allows us to operate within limit values of solids flows that are higher than those resulting from high SVI values. Such a circumstance implies a satisfactory level of safety for the process of solid–liquid separation. On the contrary, for high SVI values, the curve G_1 subtends a smaller area, thus reducing considerably the safety factor.

4.2.2.1.2 State point analysis

In this regard the approach referred to as the '*State point analysis*' is very useful. The state point is also called the operating point of the clarifier. This is a tool that, starting from the solids flow theory, allows us to have both an immediate evaluation of the operating conditions of the secondary clarifier and of values of the X concentration in the reactor. The representative point of the '*state*' is determined in the intersection of the straight line of solids flow with overflow with the underflow straight line, which is normally represented in a specular way with respect to that shown in Fig. 4.2.

In the graph X vs. N_A (Fig. 4.4), to draw the straight line overflow it is necessary to identify the point which has the following coordinates: $x = X$ and $y = (Q_R \cdot A_{SC}^{-1}) \cdot X^{-1}$ being $Q_R = R \cdot Q_0$

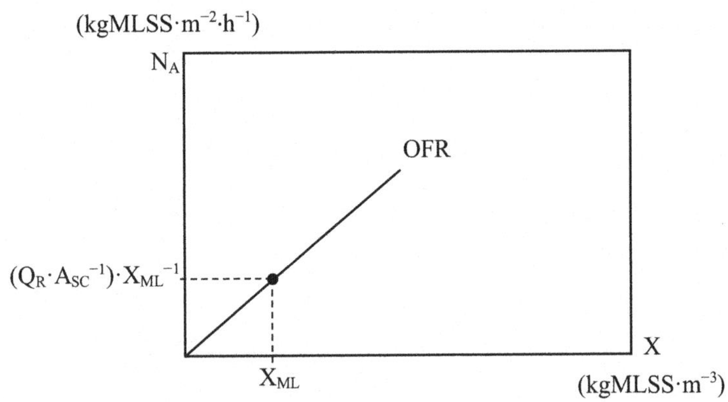

Figure 4.4 Solid overflow rate line construction.

Then a line through the origin of the axes and the point previously stated must be drawn. The line of overflow, so determined, is named Overflow Rate (OFR).

The underflow line (Fig. 4.5), called Underflow Rate Operating (UFR), will pass through the previously determined point, and will have a slope equal to $-(Q_R \cdot A_{SC}^{-1})$.

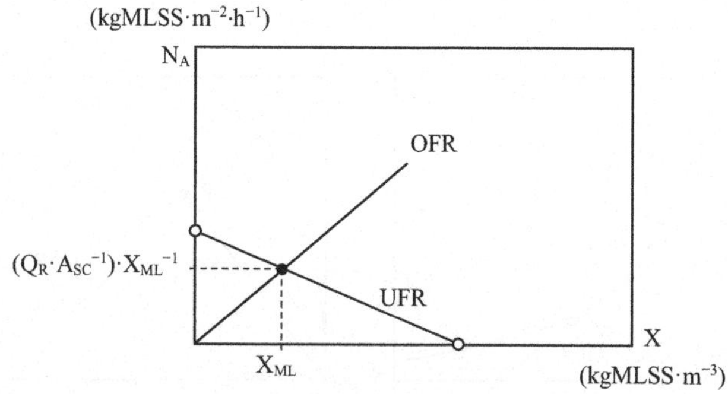

Figure 4.5 Underflow rate operating line construction.

To carry out the analysis of the state of the point it is still necessary to plot the curve of the gravity solid flow N_{A1} for established sludge sedimentation characteristics (Fig. 4.6).

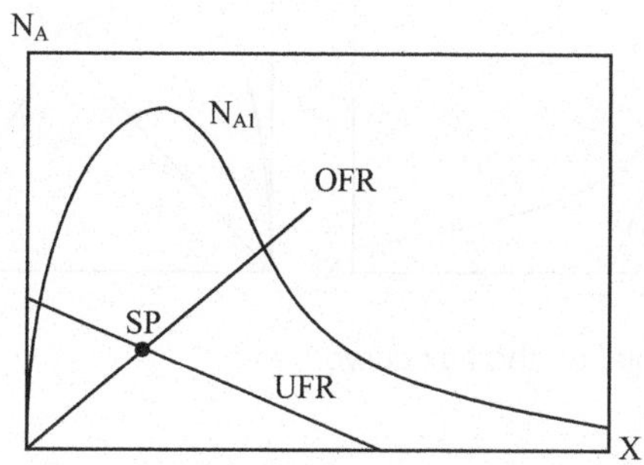

Figure 4.6 Diagram for analyzing the state of point.

Using the diagram of Fig. 4.6 it is possible to identify six situations in which the point can be:

(1) below the N_{A1} curve. In this case the UFR line can take three possible locations:
   (1.1) below the descending limb of the flux curve (Fig. 4.6);
   (1.2) tangential to the descending limb of the flux curve (Fig. 4.7a);
   (1.3) above the descending limb of the flux curve (Fig. 4.7b);
(2) on the N_{A1} curve. In this case the UFR line can take two possible locations:
   (2.1) below the descending limb of the flux curve (Fig. 4.8a);
   (2.2) above the descending limb of the flux curve (Fig. 4.8b);
(3) outside the N_{A1} curve (Fig. 4.9).

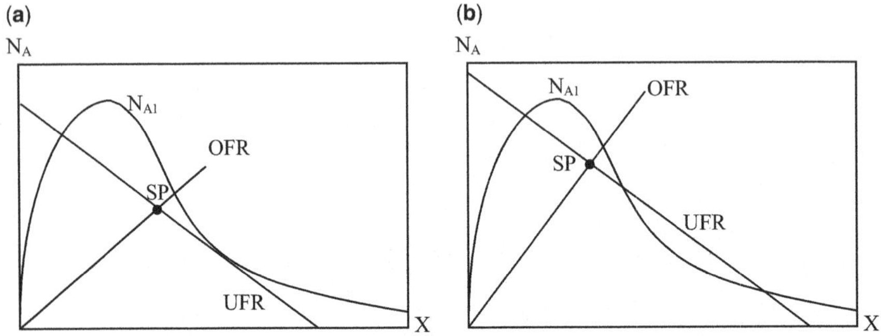

Figure 4.7 State point within the flux curve.

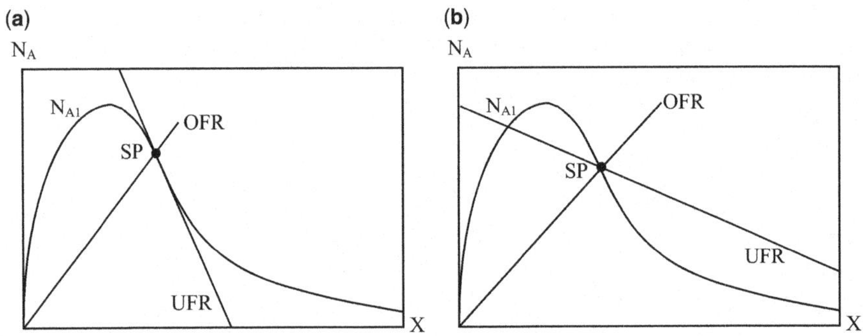

Figure 4.8 State point on the flux curve.

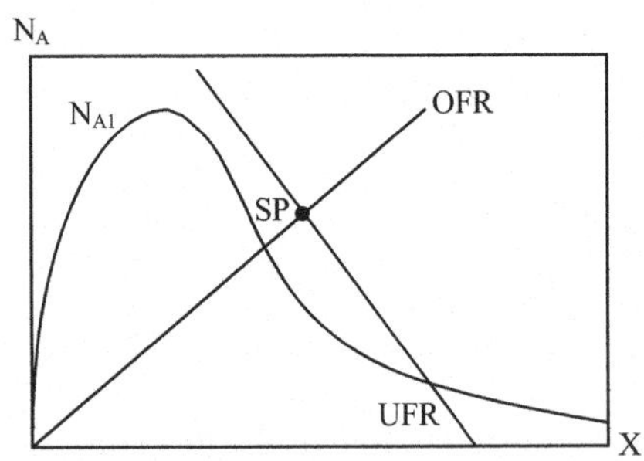

Figure 4.9 State point outside the flux curve.

The state of the point represented in Fig. 4.6 is that of a normal operating condition and sufficiently precautionary for the proper functioning of the secondary clarifier.

Instead, the graph of Fig. 4.7a is representative of a critical condition. In such circumstances, the state of the point is below the solid flux curve but the urderflow line is tangential to the solid flux curve. This means that the load of solids which reaches to the final clarifier is at the limit of the ability of this to retain solids. In these conditions, only an increase of the recirculation ratio of the sludge can reduce the critical condition.

Even more critical is the condition represented by the Fig. 4.7b graph, in which, although the state of the point is inside the zone bounded by the solid flux curve, the underflow line is located, in part, above the descending limb of that curve. This is one of the possible overload conditions of solids on the clarifier and, also in this case, it is necessary to increase the sludge recirculation ratio to reduce the overload of solids.

When the state of the point is located right on the solid flux curve, if the underflow line is below the descending limb of the latter (Fig. 4.8a) it represents a critical load condition, therefore it is necessary to reduce the intake of solids with mixed liquor or reduce the concentration of solids in the biological reactor or transform the configuration of the latter according to a step-feed scheme.

Instead, in the event that the underflow line was above the descending limb of the solid flux curve (Fig. 4.8b) there is a solids overload.

In the last case, when the state of the point is outside the area bounded by the curve of the solid flow (Fig. 4.9), this is also an overload condition.

With regard to the above, in most cases it is necessary to increase the recirculation ratio of the sludge, but this involves, inevitably, a reduction of the return sludge concentration (Fig. 4.10).

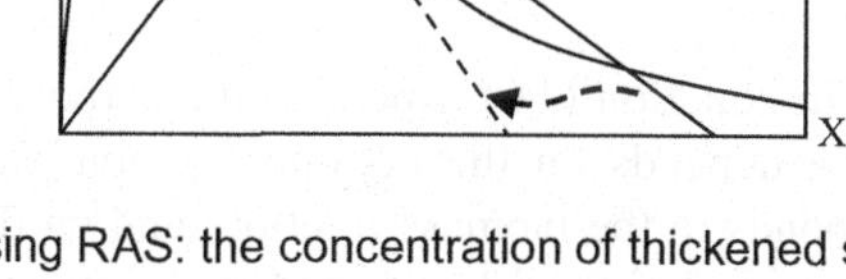

Figure 4.10 Increasing RAS: the concentration of thickened sludge decreases.

In all cases it is necessary that the operating point is below the solid flow curve, $N_{A1,}$ and therefore the UFR curve also. In addition, to ensure good settling conditions, it is recommended to maintain the state of the point below the so-called 'safe flux curve' obtained by reducing by 80% the value of each point of the solid gravity flow curve (Fig. 4.11).

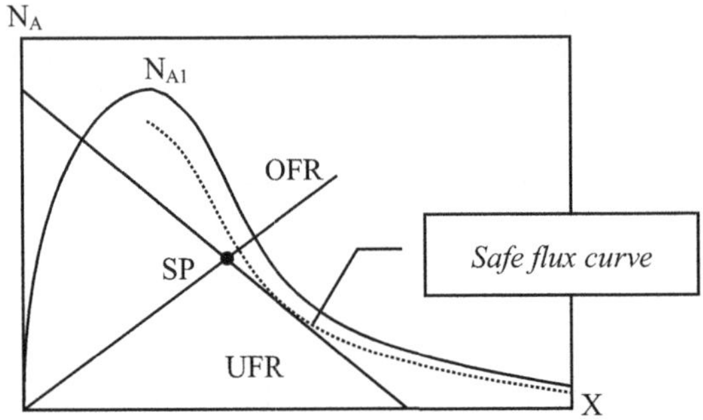

Figure 4.11 Safe flux curve.

4.2.3 Separation function

Some activated sludge separation problems can also affect secondary clarifier effluent quality (X_E). Elevated effluent biomass concentration can result from:

- dispersed growth of activated sludge or the formation of unsettleable microflocs
- rising sludge (denitrification) or floating biological foam when the effluent weir is not protected by baffles

Increases of biomass concentration X_E in the final effluent leads to deterioration of other effluent parameters.

4.2.4 BOD$_5$

The BOD$_5$ of the final effluent consists of two fractions: soluble and particulate. The particulate BOD$_5$ is caused by the respiration (mostly endogenous) of microorganisms which form the suspended solids in the final effluent.

$$(\mathrm{BOD}_5)_{\mathrm{TOT}} = (\mathrm{BOD}_5)_{\mathrm{SOL}} + k \cdot X \tag{4.9}$$

The coefficient k in equation (4.6) expresses the activity of biomass (x). The amount of respiration depends on the organic fraction X_O of total suspended solids (which corresponds to the biomass fraction) and on the biomass retention time (Θ_X). Based on a correlation of data from Czech activated sludge plants, the following empirical equation can be used for the determination of k (Chudoba *et al.* 1991):

$$k = \frac{0.7 \cdot (X_0/X)}{1 + 0.044 \cdot \Theta_X} \tag{4.10}$$

4.2.5 COD

The residual COD consists of two fractions – soluble and particulate. Soluble COD consists of biologically non-degradable compounds which:

- were either present in raw wastewater
- or, resulted from microbial activity of activated sludge microorganisms

Most of the particulate COD results from the oxidation of biomass in the suspended solids (X_0) during the COD test. For a rough calculation of residual COD the following empirical relation can be used:

$$1 \text{ mg } X_0 \approx 1.4 \text{ mgCOD}$$

4.2.6 Nitrogen and phosphorus

The biomass fraction of activated sludge effluent suspended solids X contains 6–8% organic nitrogen (measured as Kjeldahl nitrogen), 1.5–2% phosphorus for conventional systems, 4–6% phosphorus for BNR plants incorporating enhanced biological phosphorus removal.

4.3 THE IMPACT OF AERATION BASIN EQUIPMENT AND OPERATION ON THE PERFORMANCE OF SECONDARY CLARIFIERS

4.3.1 Mechanical vs. diffused-air aeration

Both mechanical and diffused-air aeration systems are used for mixing and aerating activated sludge mixed liquor. The effect of various types of aerators on the quality of activated sludge had been a topic of discussion for many years. Konicek and Burdych (1988) showed that diffused air systems were superior to mechanical aeration by:

- microscopic pictures of activated sludge flocs in aeration basins
- a statistical evaluation of the relationship between total effluent suspended solids concentration X and type of aeration
- an evaluation of the size distribution of microparticles in the effluent from activated sludge systems aerated by both types of aeration devices

In the 1990s, diffused-air aeration using fine bubble diffusers became common in new and upgraded activated sludge plants mostly because of economic reasons. These full-scale plants confirmed the results of Konicek and Burdych (1988). In most modern activated sludge systems the presence of activated sludge microflocs in secondary clarifier effluents does not result from the action of aeration system.

Figures 4.12 and 4.13 clearly demonstrate the difference between the turbulent environment created by mechanical aerators (4.11) and the mild conditions created by fine-bubble diffused-air aerators (4.13).

Figure 4.12 Aeration by vertical-shaft mechanical aerator (aeration turbine).

Figure 4.13 Aeration by fine-bubble diffused-air system is friendly to flocs.

4.3.2 Mixed liquor mixing

BNR activated sludge plants require both the aerated (aerobic) and the non-aerated conditions (anaerobic and/or anoxic). To keep the activated sludge in suspension and in contact with the wastewater, the aerated zones (oxidation–nitrification zones) are equipped with aeration devices (air-bubble diffusing plant or surface mechanical aerators), while those non-aerated reaction zones must be mechanically stirred.

The energy input required for mixing by fine bubble aeration (e.g. nitrification zone with diffused air aeration) can be practically controlled by controlling the so-called volumetric intensity of aeration I_V ($m^3 \cdot m^{-3} \cdot h^{-1}$):

$$I_V = \frac{Q_{AIR}}{V_{AB}} \tag{4.11}$$

Where Q_{AIR} is the volumetric rate ($m^3 \cdot h^{-1}$) of air blown into the aeration basin with volume V_{AB} (m^3).

For keeping the activated sludge flocs in proper suspension, the value of I_V should range between 0.5 and 1.0 $m^3 \cdot m^{-3} \cdot h^{-1}$. If the value is below the range, the activated sludge can start settling in the aeration basin, when the value is above the range, the system is overaerated. This means that concentration of dissolved oxygen is higher than necessary and the aeration is accompanied by turbulence in the aeration basin. The overaeration thus results in destruction of activated sudge flocs and reduction of oxygen transfer efficiency (decrease in alpha factor).

When the activated sludge zone is mechanically mixed (e.g. anaerobic or anoxic zone in nutrient removal activated sludge systems), then the specific energy input should be about 15 $W \cdot m^{-3}$ of installed power of the submerged propeller.

Nowadays it is common to use submerged mixers with both horizontal and vertical shafts. The primary consideration in the design of these mixers is the avoidance of excessive shear stresses during contact of the activated sludge flocs with the body of the propeller. Typically low-speed, large-diameter propellers are used. The shape of the propeller blades is such that maximum energy efficiency is provided without side-effects such as cavitation, which would lead to floc disintegration.

4.3.3 Degasification and reflocculation in aeration tanks

Aeration tanks with fine-bubble diffused -air aerators can be rather deep. Depths around 4 m are most common but aeration tanks that are up to 12 m deep are becoming more and more common. While deep tanks can improve oxygen transfer efficiency they can also cause separation problems in the secondary clarifier because the mixed liquor entering the secondary clarifiers is supersaturated with air. On entering the shallower secondary clarifier, small air bubbles are released that attach to the activated sludge flocs and cause them to float. The result of this problem is similar to 'rising sludge' (see Chapter 3.3.3.1).

The remedy for this problem is quite simple and consists of a degasification and reflocculation zone that can be incorporated directly into the aeration tank system. There are two approaches for the construction of degasification chambers. In the first case the activated sludge mixed liquor flows in a very thin layer from the intensively aerated environment of the aeration basin over a weir. The abrupt change of the hydrostatic pressure between the aeration basin and this thin layer causes a rapid release of air bubbles, which are then stripped-off by the coarse-bubble aeration in the degasification chamber. The gentle mixing by these large bubbles also contributes to the reflocculation of activated sludge flocs into aggregates larger than those in the aeration basin. The second approach, (Fig. 4.14) exploits the hydraulic mixing that results from the flow of mixed liquor.

Figure 4.14 Hydraulically mixed degasification chamber.

The degasification chamber is separated from the aeration basin by a baffle that reaches almost the bottom of the aeration basin. The mixed liquor is forced to flow from the bottom of this chamber to the surface, so that the hydrostatic pressure drop causes the air bubbles' release. The gentle hydraulic mixing in the chamber also contributes to the formation of large activated sludge flocs.

In both cases, the channel connecting the aeration basin with the secondary clarifier(s) should avoid highly turbulent flow conditions.

4.4 FEATURES OF SECONDARY CLARIFIER CONSTRUCTION

Construction details of secondary clarifiers are described in the specialized literature on secondary clarifiers. Besides the IAWQ/IWA Scientific and Technical Report No. 6 (Ekama *et al.* 1997) and IWA publication on 100 years of activated sludge process (chapter 11 by Parker *et al.* 2014), much valuable information can be found on the web pages of design, engineering, manufacturing and construction companies involved in this business by using general Internet search engines with keywords such as secondary clarifiers and activated sludge separation.

The web pages of national and international professional organizations in the field of wastewater treatment such as EWA (European Water Association, www.ewa-online.eu), IWA (International Water Association, www.iwahq.org.uk), WEF (Water Environment Federation, www.wef.org) and specialized web pages on activated sludge (www.scitrav.com/wwater/) can also be useful.

This chapter only describes those construction features of secondary clarifiers that are important in handling poorly settling activated sludge.

4.4.1 Inlet structure with a flocculation zone

4.4.1.1 Principles of a flocculation zone

In the past, the inlet structure of secondary clarifiers was designed similarly to those in primary clarifiers. The main purpose of this structure (called *the feed well*) was to dissipate the energy of the incoming mixed liquor and distribute evenly the flow throughout the cross-section of the clarifier. Usually the retention time in the feed well was not long enough for reflocculation of small activated sludge flocs. Current construction of secondary clarifiers typically provides a longer time for flocculation. This can be achieved in rectangular clarifiers by an additional compartment between the feed well and the sedimentation zone. In circular clarifiers the center feed well can be increased in diameter or an additional concentric flocculation zone provided (see Fig. 4.15).

Figure 4.15 Flocculation zone with tangential flow into the feed well.

Flocculation process principles and design parameters are described together with construction recommendations for both circular and rectangular clarifiers in the specialized literature (e.g. Albertson, 1992 and IWA Scientific and Technical Report No. 6, Ekama *et al.* 1997).

4.4.1.2 Conventional flocculation zones with tangential flow regime

Flocculation zones in circular clarifiers are usually rather shallow so that the retention time required for flocculation is achieved by increasing the diameter of the zone. Mixing is provided by the tangential horizontal flow regime (Fig. 4.16).

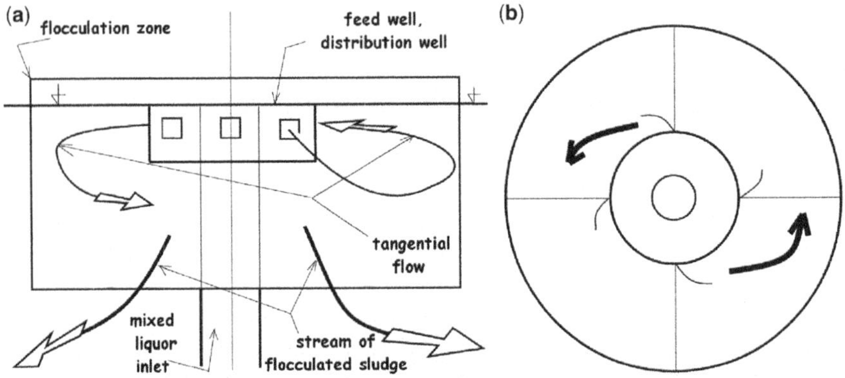

Figure 4.16 Schematic view of flocculation zone with a predominantly tangential flow regime (**a**) horizontal cross-section, (**b**) vertical cross-section.

This arrangement has a tendency to allow a thick layer of scum (caused by hydrophobic filamentous microorganisms, fats and other floating materials) to accumulate on the flocculation zone (Fig. 4.17), If the scum is not removed regularly it may putrefy and produce unpleasant odors.

Figure 4.17 Scum layer trapped in the ring of the flocculation zone.

4.4.1.3 *Novel flocculation zone with deflectors and a vertical flow regime*

This major drawback of the conventional flocculation zone can be eliminated by changing the flow direction of the stream entering it (Fig. 4.18). The incoming mixed liquor is directed by the upper deflector towards the water surface so that it prevents the formation of a scum layer. This type of flocculation zone is much deeper and has a smaller diameter than a flocculation zone with a horizontal inlet flow.

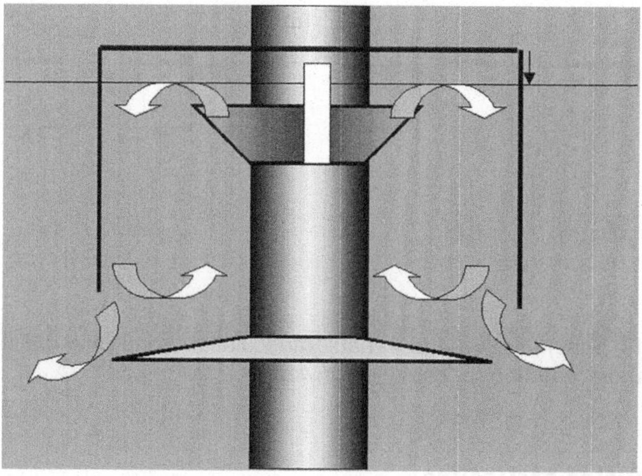

Figure 4.18 Flocculation zone with a vertical inlet flow (courtesy by KUNST, spol. s r.o., Hranice, Czech Republic).

4.4.1.4 *Flocculation zone with deflectors with variable profile*

Recently a new concept of flocculation zone has been developed in a way to adapt continuosly to the variation of the flow value fed in the tank.

The device that implements this new possibility has been patented to installation in secondary clarifiers with horizontal flow with both radial and longitudinal with the aim to increase the solids separation efficiency. It allows variation of the section area and also the share of input of the mixed liquor at the exit of the flocculation zone.

This is implemented by means of a deflector with variable surface and this variation, adjusted according to the flow fed to the tank, allows a continuous adaptation to the incoming hydraulic load and guarantees the horizontal flow of mixed liquor that is always put inside the sludge blanket without causing formation of vortices.

Therefore, the input flow is always subjected to the action of the filtering sludge blanket, so, in this way, the concentration of the effluent sludge is reduced with the consequent increase of the separation efficiency of the secondary clarifier.

In Fig. 4.19, the scheme (a) shows the case of the position of the deflector when the sludge blanket is low (e.g. during the dry weather), while the scheme (b) represents the case in which the sludge blanket rises greatly (e.g. during a heavy storm) and goes up towards the free surface. In both cases, and of course in all the intermediate situations, resuspension of the sludge bed is reduced to a minimum.

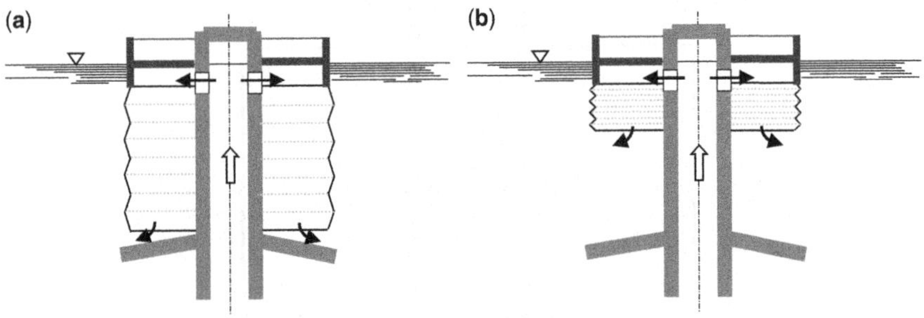

Figure 4.19 Schematic view of variable flocculation zone of a radial secondary clarifier: (**a**) dry weather, (**b**) heavy stormwater.

4.4.2 Outlet structure

4.4.2.1 Peripheral vs. internal effluent launders

The main function of the outlet structure is to produce high quality final effluent containing low concentrations of suspended solids. One source of effluent suspended solids is from density currents. Usually the density currents are formed by differences in the densities of the mixed liquor and the sludge blanket at the bottom of the clarifier. These density currents can extend to the peripheral wall of the clarifier and turn the current towards the water surface. There are two scenarios in which the density current can reach the water surface.

4.4.2.1.1 Circular clarifier, peripheral effluent launder (with inboard weir)

Peripheral effluent launders are preferred by some clarifier designers because they allow the entire surface of the clarifier to participate in sedimentation. However this type of effluent launder is very susceptible to the effects of density currents. Albertson (1992) described how to deflect density currents back to the separation zone of the clarifier using a peripheral deflector (Stamford baffle), as depicted in Fig. 4.20.

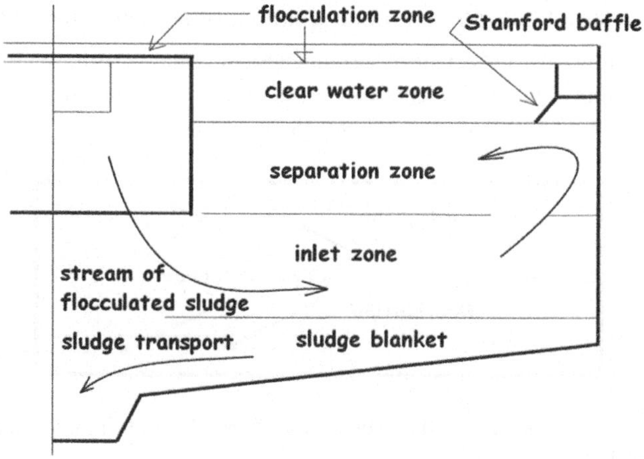

Figure 4.20 Schematic description of the effect of a Stamford baffle in a circular clarifier with a peripheral effluent launder.

4.4.2.1.2 Circular clarifier, internal effluent launder (with inset double-sided weirs)

Another approach to the design of the clarifier consists of installing the effluent launders on a structure placed about 1/3 of the clarifier diameter from the outside wall. With this weir placement a deflector (Stamford baffle) is not necessary because the density current turns back towards the separation zone away from weirs. In some designs two sets of internal launders have been used to increase effluent weir length. An example of double-sided weir is shown in Fig. 4.21.

Figure 4.21 Example of secondary clarifier with internal effluent launder with double-sided weir.

4.4.2.1.3 Effluent launders in rectangular clarifiers with longitudinal flow

Figure 4.22 shows the flow pattern through a rectangular clarifier with *longitudinal* flow. To prevent density currents affecting effluent suspended solids levels, the effluent launders should be oriented parallel to the flow.

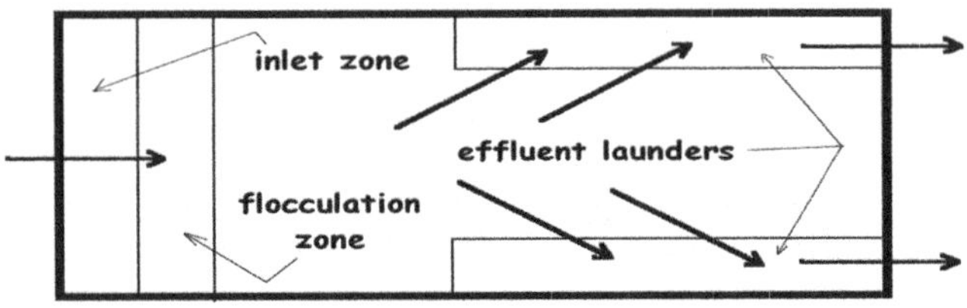

Figure 4.22 Scheme of a hydraulic flow through rectangular clarifier with a longitudinal flow.

4.4.3 Scum baffles

The role of scum baffles is to protect the final effluent from the escape of floating biomass. As explained in Chapter 3, the floating biomass in secondary clarifier results from:

- denitrification in the sludge blanket at the bottom of the clarifier
- flotation of foam formed by hydrophobic filamentous microorganisms and other trapped biomass

When the effluent weirs are not protected, the floating biomass escapes to some degree into the final effluent and deteriorates its quality. This can be prevented by inserting a scum baffle in front of the effluent weir (Figs 4.23 and 4.24). The installation of scum baffles also requires the installation of a skimming device to prevent the captured floating material from accumulating on the surface of the secondary clarifier.

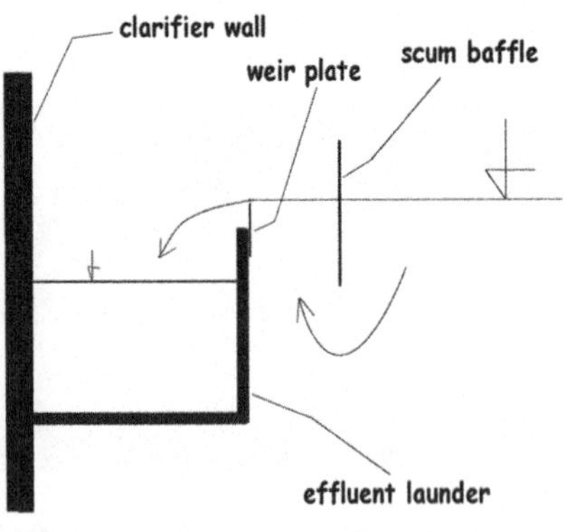

Figure 4.23 Effluent launder protected by a scum baffle.

Figure 4.24 The importance of a scum baffle.

4.5 EFFICIENT SCUM REMOVAL FROM THE SURFACE OF SECONDARY CLARIFIERS

4.5.1 Conventional scum boxes

The removal of floating biomass from the surface of circular secondary clarifiers is one of the most difficult tasks for the clarifier designer. Conventional scum removal devices such as those used in primary circular clarifiers (Figs 4.25 and 4.26) are not suitable, because:

- the skimmer usually pushes the scum around the surface of the clarifier
- the surface area of the scum removal box is usually much smaller than the area of floating scum so that during one rotation of the travelling bridge only a small amount of the floating material is removed from the clarifier surface
- biological foam caused by filamentous microorganisms is much denser and more viscous than the 'common scum' on primary clarifiers so that the scum box often becomes clogged.

(a) (b)

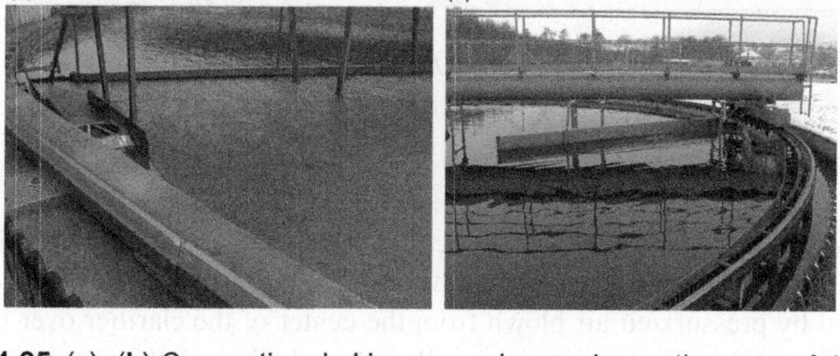

Figure 4.25 (**a**), (**b**) Conventional skimmer and scum box – the area of the scum box is too small to capture all of the floating material.

(a) (b)

Figure 4.26 (**a**), (**b**) Scum boxes clogged by viscous biological foam.

4.5.2 'Travelling' scum boxes

Substantial improvement in scum removal efficiency can be achieved by installing the scum box on the travelling bridge together with the skimmer. In this way, the scum is skimmed to the box as the bridge rotates rather than only once each rotation. The captured scum is continuously pumped out of the box (Fig. 4.27).

(a) (b)

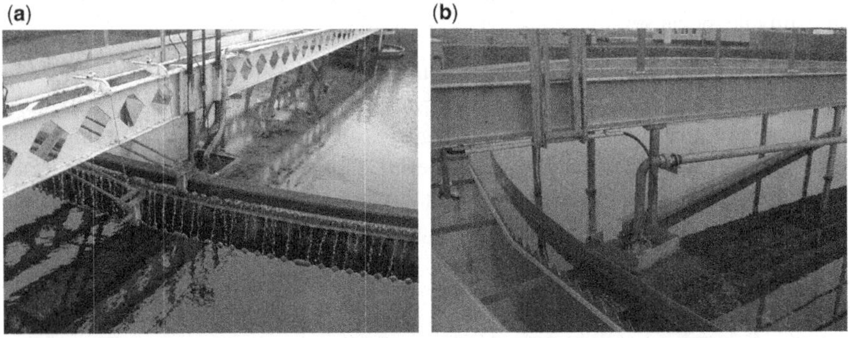

Figure 4.27 (**a**), (**b**) Examples of a 'travelling' scum box.

4.5.3 Pneumatic systems

These systems take advantage of the fact that floating material is much lighter than water and can be moved by small movements of the surface. This movement is generated by pressurized air blown from the center of the clarifier over the water surface (Fig. 4.28). The air creates small waves which push the scum towards the peripheral scum launder.

Figure 4.28 Pneumatic collection of scum in a circular secondary clarifier.

The same effect can be achieved by installing a simple ventilator on the traveling bridge, which pushes the floating biomass along the skimmer to the scum box (Fig. 4.29).

Figure 4.29 Additional installation of a ventilator in front of the skimmer.

4.6 REMOVAL OF SETTLED AND THICKENED SLUDGE

4.6.1 Effect on the final effluent quality

There are two traditional lines of thought concerning the amount of thickened activated sludge that should be allowed to accumulate in the secondary clarifier:

(i) the amount of sludge in the bottom layer should be minimized because:
  - a thick sludge layer allows denitrification leading to sludge rising
  - in EBPR plants, phosphate release may occur in the thick sludge layer

(ii) the sludge blanket may play a positive role because:
  - it exhibits a filtration effect, especially when the flocculated stream is introduced into the sludge blanket (see Fig. 4.19). This allows unflocculated microflocs to be 'captured' in the settled sludge
  - it represents an additional inventory of sludge which can be used to control F/M
  - denitrification can contribute significantly to overall nitrogen removal efficiency

Modern secondary clarifiers (both circular and rectangular) are constructed with the side wall depths of approximately 4 m. This allows operation with a sludge blanket. A procedure for the estimation of optimum sludge blanket thickness is presented in the ATV-DVWK Standard ATV A131 E (2000).

Secondary clarifier operation with an optimized sludge blanket depth is important with respect to new legislation in many (European) countries which requires the treatment of some wet weather flows. Storage of sludge in the clarifier can be used to prevent sludge wash-out during the storm flows. The sludge removal system must have adequate capacity to prevent an excessive accumulation of the sludge in the secondary clarifier because this can lead to problems such as sludge rising and phosphate release.

4.6.2 Mechanical scrapers

Mechanical scrapers are used in both circular and rectangular clarifiers. The pricipal role of this device is not to 'scrape' the sludge (i.e. to push it mechanically towards the hopper) but to keep it in a liquid state. Secondary clarifiers with mechanical scrapers have sloped floors so that the main sludge transport mechanism is gravity flow. However, activated sludge is a biological material with lot of extracellular biopolymers. Because it contains biopolymers, sludge is highly thixotropic so that the settled sludge quickly loses its ability to flow unless it is mechanically mixed at regular intervals. Because the major role of the scraper is to provide the energy necessary to overcome the thixotropic behavior of the sludge the following design features are important:

- *circular clarifiers*: scrapers are mounted on a traveling bridge over the whole diameter of the clarifier; in this way the sludge blanket is mechanically disturbed twice in each rotation of the bridge.
- *rectangular clarifiers*: scrapers are mounted on an endless chain which provides almost continuous movement of the sludge blanket.

4.6.3 Vacuum sludge removal

Vacuum sludge removal is used widely in the USA but not in Europe where if used it is usually in rectangular clarifiers. The advantage of vacuum removal is that the thickened sludge is removed from the place where it settled rather than having to flow to a central sludge collection well or hopper. For this reason the bottom of the clarifier can be flat. The advantage of simpler construction is offset by the more complicated operation of the vacuum suction system. Sludge removal should be as frequent as possible because the thixotropic nature of the sludge will make its vacuum removal difficult. Initially the vacuum is created by a vacuum pump but subsequently the vacuum is created by the difference in water surface levels between the clarifier and the mixed liquor channel (Fig. 4.30).

Figure 4.30 Rectangular secondary clarifier with the suction pipe system for vacuum removal of settled and thickened sludge.

4.7 OPERATION OF AERATION TANK – SECONDARY CLARIFIER SYSTEM FOR BULKING AND FOAMING CONTROL

The experience of the last decade stressed the importance of proper understanding of the relationships between the activated sludge process in anaerobic, anoxic, and oxic zones on one side and secondary clarifiers for sludge separation and thickening on the other side. The modern construction of secondary clarifiers and their proper operation can correct some of the activated sludge separation problems to a great extent. The secondary clarifier is no more a 'black box' in the control systems of the activated sludge process thanks to the continuous measurement of the concentration of individual important fluxes (suspended solids in mixed liquor – effluent – the return and waste sludge stream) and thanks to the continuous observation of the position of the sludge blanket in the clarifier. The development has brought also new possibilities to plant engineers for how to correct bad activated sludge properties which developed because of errors in plant design, unfavorable changes in wastewater composition or for other reasons.

4.7.1 Use of chemicals in activated sludge process

The main role of the activated sludge process during his long history was to remove organic pollution and nitrogen (via nitrification and denitrification). The removal of organic pollution and nitrogen is solely based on biochemical processes by activated sludge bacteria. The removal of phosphorus is more complex. A small fraction of phosphorus is removed via synthesis of new biomass; in EBPR activated sludge systems (EBPR – enhanced biological phosphorus removal) part of the phosphorus is removed by accumulation to PAOs. However, by these 'biological' phosphorus removal processes it is difficult to reach the final effluent concentration of total phosphorus of about 1 mg L^{-1}. In many countries in the world, the effluent limit 1 $mg \cdot L^{-1}$ of total P is no more considered to be sufficient enough for controlling the process of eutrophication of natural surface waters. The only reliable way to achieve total effluent concentration in the order of magnitude of 0.X $mg \cdot L^{-1}$ is simultaneous phosphate precipitation. When the legislation requirements are even more stringent (effluent total P concentration in the order of magnitude of 0.0X $mg \cdot L^{-1}$) the tertiary phosphate precipitation is necessary. The difference between simultaneous and tertiary phosphate precipitation can be seen in Fig. 4.31.

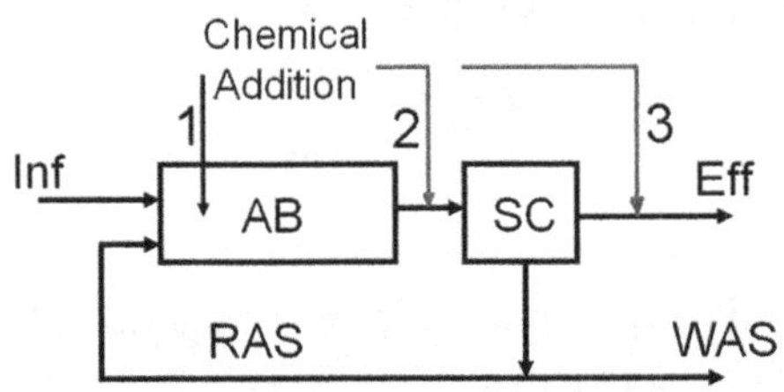

Figure 4.31 Schematic description of simultaneous (1 & 2) and tertiary (3) phosphate precipitation.

The advantage of simultaneous precipitation is that the existing separation stage (secondary clarifier) can be used for separation of precipitated phosphates together with activated sludge. However, the dosage of chemicals and therefore the phosphorus removal efficiency is limited by possible negative impact of chemicals on activated sludge bacteria; especially harmful are high dosages of ferric salts for nitrification bacteria. Therefore, extremely low total phosphorus concentrations can be achieved only by tertiary precipitation. Of course, tertiary treatment requires extra investment for new volumes for rapid and slow mixing of the precipitating agent with secondary effluent and for separation of phosphate precipitation (sedimentation, flotation, and filtration).

4.7.1.1 Use of iron salts

Both ferrous (Fe^{2+}) and ferric (Fe^{3+}) mineral salts can be used for simultaneous phosphate precipitation. The choice depends mostly on the price of the original chemical. In some countries, cheap ferrous sulfate is available and the difference in the price in comparison with ferric salts can cover the increased consumption of air for the oxidation of ferrous to ferric ions in the aeration basin. Phosphate can be efficiently precipitated only by ferric ions because the solubility of ferrous phosphate is much higher than solubility of ferric phosphate. This is also the main disadvantage of simultaneous phosphate precipitation in wastewater treatment plants with anaerobic digestion of waste activated sludge. The waste activated sludge (see Fig. 4.30) brings the ferric phosphate to the anaerobic digester where ferric ion is reduced to ferrous and released phosphate is returned with reject water back to the main water line.

Long-term simultaneous precipitation with ferric salts performed at numerous wastewater treatment plants world-wide also proved a positive effect of continuous ferric salts dosage on activated sludge settling properties. It is apparent that long-term dosage of ferric salts in rather small dosages (based on molar Fe^{3+}/P ratio around 1.5–2.0) contributes to the development of more compact flocs and can to a certain extent compensate the adverse effect of the presence of filaments. On the other hand high dosages of ferric salts can result in overproduction of slimeous polymers and in inhibition of nitrification bacteria.

4.7.1.2 Use of aluminium salts

Simultaneous precipitation of phosphate as depicted in Fig. 4.31 can be performed also with aluminium (Al^{3+}). In this case the insoluble product is aluminium phosphate which is stable even under anaerobic conditions. Thus the precipitation by Al^{3+} does not create any return flux of phosphorus in reject water.

In addition, in the experiments performed in a pilot-scale model of the full-scale plant Liberec (Czech Republic) it was observed that the M. parvicella foam repeatedly disappeared when ferric sulfate was replaced with aluminium salts (Al^{3+}) for phosphorus precipitation (Váňa, 1996; Váňa & Wanner, 1996, 1998). This phenomenon was recently studied and confirmed by several laboratories, e.g. by Paris *et al.* (2005) among the first.

Excessive occurrence of M. parvicella foam is a typical seasonal problem which appears in the winter months. Therefore some plants combine the dosage of aluminium salts during the winter months with cheaper ferric salts dosed during the rest of the year. For instance, the Central Wastewater Treatment Plant of the City of Prague with a capacity of about 1.6–1.7 million PE has been using this mode of operation for many years (Fig. 4.32). The dosage effective against the M. parvicella growth ranges between 2 and 15 g Al per 1 kg dry matter per day for about the first 3 weeks of dosing; later the dosage can be reduced by about 30%. Some suppliers of chemicals produce also a mixed inorganic coagulant containing both ferric and aluminium salts in a ratio suitable for phosphate precipitation and M. parvicella control, which is dosed also in the Prague Central WWTP (Soukup *et al.* 2015).

Figure 4.32 Photograph of corridor-type aeration basins of Prague Central WWTP. Left –with M. parvicella *foam*; right – without foam.

4.7.1.3 Oxidizing agents

Section 5.3.1 describes the use of oxidizing agents for controlling activated sludge settling properties, especially by controlling the growth of filamentous microorganisms. Ozone is one of the strongest oxidizing agents but its application in wastewater teratment was limited by price. However, ozone is becoming more commonly used today for tertiary treatment (ozonation of effluent for removal of emerging organic pollutants) and thus the use of ozone for bulking and foaming control is becoming also more feasible. In their report, Soukup *et al.* (2015) summarized positive experiences with ozone for bulking and foaming at wastewater treatment plants in Brussels (Belgium), Himmerfjärden (Sweden) and Bottrop (Germany). Currently, the application of ozone into the stream of return activated sludge is also being considered by Veolia for the Central WWTP Prague. The main advantage of ozone in comparison with aluminium salts is in its broader impact because ozone can also control filaments other than just M. parvicella.

4.7.1.4 Use of organic flocculants

A very strong toll of modern secondary clarifiers is the presence of flocculation zone. As described earlier, the main role of the zone is to provide optimum conditions for the formation of stable and well settling flocs. The process of flocculation is quite a complex and delicate process in which apart from physical parameters (retention time, mixing regime, temperature, density, etc.) physico-chemical parameters such as viscosity can play a significant role. To achieve good compactness of activated sludge it is necessary to have a certain amount of polymers in the water environment. These polymers can be either naturally produced by bacteria (*glycocalyx*) or can be added artificially. Operational experience from the Prague Central WWTP has proved the positive role of synthetic kationic high-molecular weight polymers which are added directly to the flocculation zone of a secondary clarifier (Soukup *et al.* 2015). The dosing pumps of kationic flocculants can be coupled with flowmeters and thus polymers are dosed only when the increased inflow will cause increased loading of the secondary clarifiers. The addition of polymers increases the compactness of settled sludge and creates temporary spare capacity in the thickening part of the clarifier.

4.7.2 Operation of secondary clarifiers

Modern secondary clarifiers are constructed with a minimum side-wall-depth of 3.5 m and more; this provides volume enough for efficient settling and thickening of activated sludge even with higher SVIs. The secondary clarifiers are designed with very conservative values of hydraulic surface loading and solids flux – this prevents clarifier overloading even in the case of more diluted (bulking) sludge. If necessary, the clarifiers are equipped with very efficient scum removal systems which prevent the escape of foam from the system. All these features are elements of the passive safety of the activated sludge process–secondary clarifier system. However, thanks to the modern instrumentation, the secondary clarifiers can be also involved in the active safety of modern treatment systems.

4.7.2.1 Automatic sludge blanket level detector

Secondary clarifiers can be included in the control and automation system of an overall activated sludge system with the signal from sludge blanket level detector. These detectors can continuously and fully automatically check the current position of the level of the sludge blanket in the secondar clarifier. This kind of real time information can be helpful for:

- Optimizing energy usage
- Automation of thickened sludge pumping
- Optimizing thickened sludge concentration, which reduces the costs of sludge dewatering, enables better adjustmemt of flocculant dosages, etc.
- Maintaining preferred sludge depth in the clarifier and thus minimizing the risk of overflows and similar problems.

4.7.2.2 Suspended Solids Sensors

These sensors can be used for continuous online measurement of suspended solids in liquids. The sensors are available with different ranges of suspended solids concentrations varying from units of milligrams per litre up to about 20,000 $mg \cdot L^{-1}$. Therefore by means of these sensors we can measure suspended solids concentrations in all streams entering or leaving a secondary clarifier, as depicted in Fig. 4.1. In combination with sludge blanket level detection and measurememt of sludge blanket concentration, the plant engineer can obtain all the necessary inputs for exact calculations of sludge age according to Eq. (4.1).

REFERENCES

Albertson O. E. (1992). Clarifier Design. In: Design and Retrofit of Wastewater Treatment Plants for Biological Nutrient Removal, C. W. Randall, J. L. Barnard and H. Stensel (eds), TECHNOMIC Publ. Co. Inc., Lancaster, PA, USA, pp. 185–254.

ATV-DVWK Standard A 131E (2000). Dimensioning of Single-Stage Activated Sludge Plants. GFA Publisher Hennef, Germany.

Chudoba J., Dohanyos M. and Wanner J. (1991). Biological Wastewater Treatment (In Czech). SNTL (State Publishing Company of Technical Literature), Prague.

Ekama G. A., Barnard J. L., Gunthert F. W., Krebs P., McCorquodale J. A., Parker D. S. and Wahlberg E. J. (1997). Secondary Settling Tanks: Theory Modelling Design and Operation Scientific and Technical Report No. 6. IAWQ/IWA, London, UK, pp. 1–216.

Konicek Z. and Burdych J. (1988). Effect of activated sludge process on secondary sedimentation tank efficiencies. *Water Science & Technology*, **20**(4–5), 153–163.

Paris S., Lind G., Lemmer H. and Wilderer P. A. (2005) Dosing Aluminum Chloride to Control Microthrix parvicella. *Clean Soil Air Water 33*, **3**, 247–254.

Parker D. S., Günthert W. and Wilén B.-M. (2014). Secondary Clarifiers. Chapter 11 In: Activated Sludge – 100 Years and Counting, D. Jenkins and J. Wanner (eds), IWA Publishing, London, ISBN 9781780404936.

Soukup B., Rosenbergová R., Todt V. and Chudoba P. (2015). Methods of Microthrix parvicella growth control in practice. Proc. of the 20th CzWA Seminar on New Methods and Procedures in the Operation of Wastewater Treatment Plants, Moravská Třebová, 14–15 April 2015, Czech Republic.

Váňa M. (1996). Evaluation of the appearance of filamentous microorganisms at the WWTP Liberec and verification of the control of Microthrix parvicella filaments by dosing aluminium coagulants in a pilot-plant model. Report for the Czech Grant Agency, project No. 206/94/1183, The Control of Activated Sludge Separation Properties in Nutrient Removal Activated Sludge Systems. Inst. Of Chemical Technology, Prague.

Váňa M. and Wanner J. (1996). Report on The Liberec and Jablonec Wastewater Treatment Plant, Newsletters of IAWQ specialist group on Design and Operation of large Wastewater Treatment Plant, March 1996.

Váňa M. and Wanner J. (1998). The effect of aluminium sulphate dosing on the population of activated sludge at the WWTP Sedlec. Proc. conf. Progressive Methods of Wastewater Treatment, Slovak Association of Wastewater Experts, Tatranská Štrba, 2–5 June 1998, Slovak Republic.

Chapter 5

Bulking and foaming control methods

V. Tandoi, M. Majone and S. Rossetti

5.1 INTRODUCTION

Bulking and foaming, the first more so, are serious dysfunctions of the activated sludge process, usually due to the proliferation of specific bacteria (e.g. filamentous bacteria), which cause poor separation of activated sludge in the settler, loss of suspended solids in the final effluent, and eventually lower biomass concentration in the biological tank. When these detrimental bacteria find the proper conditions for their growth, they can constitute more than 50% of the biomass (abundance class: 6, in a rank between 1 and 6). If the plant arrives, at this point the situation is very hard to recover, because tons and tons of detrimental bacteria are present, and can be eliminated only by wasting a large amount of sludge and its substitution with non-filamentous bacterial biomass. Moreover, this will be effective only if the conditions for filamentous bacteria proliferation have been found and eliminated.

Before starting an investigation to solve a bulking case (foaming can be prevented by selective collection and elimination of the scum), it is fundamental to have clearly established that substantial filamentous bacteria proliferation is the cause of plant dysfunction. The other common reasons are, rising sludge, overareation and consequent flotation of the activated sludge, viscous bulking due to nutrient deficiency, activated sludge flocs breakage (due to unusually high shearing forces), pin point floc, etc. (see Chapter 3). A mistake at this stage, in individuating the correct route to follow, can have dramatic consequences, wasting further time to

 Activated Sludge Separation Problems:
Theory, Control Measures, Practical Experiences
Simona Rossetti, Valter Tandoi and Jiri Wanner
doi: 10.2166/9781780408637_099

find the proper solution. Some steps are crucial, to find out the proper pathway to identify the dysfunction causes:

(1) Start with a detailed flowsheet of the plant, including unit volume, dimensions, mass flows, etc.
(2) Calculate the actual operating conditions (organic load, TKN load, hydraulic retention time, clarifier overflow rate, sludge age, excess sludge waste, excess sludge production, pollutants removal efficiencies, etc.)
(3) Compare the actual operating conditions with those forming the basis of the design criteria (plant correct, too small, too big)
(4) Define the type of dysfunction (poor settling characteristics, rising sludge, sludge flotation, biological foaming, bulking, insufficient pollutants removal, etc.)
(5) Start a systematic microscopic characterization of the activated sludge
(6) From the frameworks depicted, list some hypotheses on the dysfunction causes and on remedial action
(7) Follow over time the effect of the remedial action chosen with the aid of microscopic characterization
(8) Remember that if your goal is to suppress filamentous bacteria proliferation, you have to wait for a complete replacement of the activated sludge, and this will depend on the sludge age and fluid dynamics (under complete mixing, three sludge ages are necessary to substitute the activated sludge).

5.1.1 Microscopic characterization of the activated sludge

As seen in Chapter 2, many relevant microbial populations are present in the activated sludge, including filamentous bacteria, poly-P and GAO, denitrifiers, nitrifiers, generic heterotrophic floc forming bacteria, etc. A complete characterization of microbial components and their quantification in terms of mass and activity is not possible. For practical purposes, simple traditional microscopic characterizations, which plant operators without specific microbiological expertise can also do, provide fundamental information for the operation of the plant, giving a very high benefit/effort ratio. Figure 5.1 reports the common series of information that is recorded periodically at the plant according to specific manuals (Eikelboom & van Buijsen, 1983; Jenkins *et al.* 2004). Figure 5.2 visualizes the filament abundance difference in normal activated sludge (class 2–3) and a bulking sludge (class 5–6). More complex biomolecular tools permit us to have a detailed picture of the main microbial populations present, and the hope is that in the future these techniques will be applied in WWTPs, using external specialized laboratories and facilities. Such activated sludge characterizations, even periodically done, provide a deep view of the fundamental actors of the depuration process (the activated sludge microbial components), which can be very helpful in a knowledge-based plant operation.

Sample number:
Sampling date:

Sampled at:
Observation date:

Filament Abundance:

0	1	2	3	4	5	6
None	**Few**	**Some**	**Common**	**Very Common**	**Abundant**	**Excessive**
no filaments in the flocs	rare filaments in the flocs	filaments in some flocs	1-5 filaments/floc	5-20 filaments/floc	> 20 filaments/floc	Filaments >50% biomass

Filament effect on floc structure: ☐ Little or None ☐ Bridging ☐ Open Floc Structure

Morphology of floc ☐ Firm ☐ Weak ☐ Round ☐ Irregular ☐ Compact ☐ Diffuse

	<150 μm	150-500 μm	> 500 μm
Floc distribution: (%)			

Presence of:

Free cells in suspension	______	Neisser positive Clumps	______
Zoogloeas	______	India Ink test	______
Spirochetes	______	Chlorine damge to filaments	______
Inorganic/Organic Particles	______	Other:	______

Filamentous Micro-organism Summary:

	Rank	Abundance		Rank	Abundance
Nocardioforms			*Microthrix parvicella*		
Type 1701			Type 0581		
Sphaerotilus natans			Type 0092		
Type 021N			Type 0803		
Thiothrix spp.			Type 1851		
Type 0041			Type 0961		
Haliscomenobacter hydrossis			Type 0675		
Nostocoida limicola			Other:		
Fungi			Other:		

X: Dominant; O: Secondary

Remarks:

Figure 5.1 Microscopic characterization sheet (Jenkins *et al.* 2004).

5.1.2 Biological foam

Biological foaming is due mainly to the proliferation and trapping of hydrophobic bacteria in the plant, such as M. parvicella and a heterogeneous group generally called nocardioforms, but more properly named Mycolata. The mechanism of foam formation is due to the combination of three factors: (1) air bubbles, (2) hydrophobic bacteria (Mycolata and often M. parvicella), (3) surfactants (either residual content in the wastewater and/or compounds produced by the same activated sludge microbial components). The solution for facing the problem consists in simple rules:

(1) Collect the scum, by a proper mechanical device and avoid its recirculation in the plant ahead
(2) Use oxidizing agent (such as hypochlorite) spraying it in the areas where the foam accumulates

(3) Monitor regularly by microscopic observation the amount of Mycolata in the activated sludge and in the foam. Monitoring will drive the proper action to take, quantifying the Nocardioforms presence, by microscopically counting the number of intersections with a line on the slide (Jenkins *et al.* 2004). When a threshold value is overcome, apply the containment strategies above or, for worse cases, change operational parameter, such as decreasing the sludge age (when possible), installing a selector, etc.

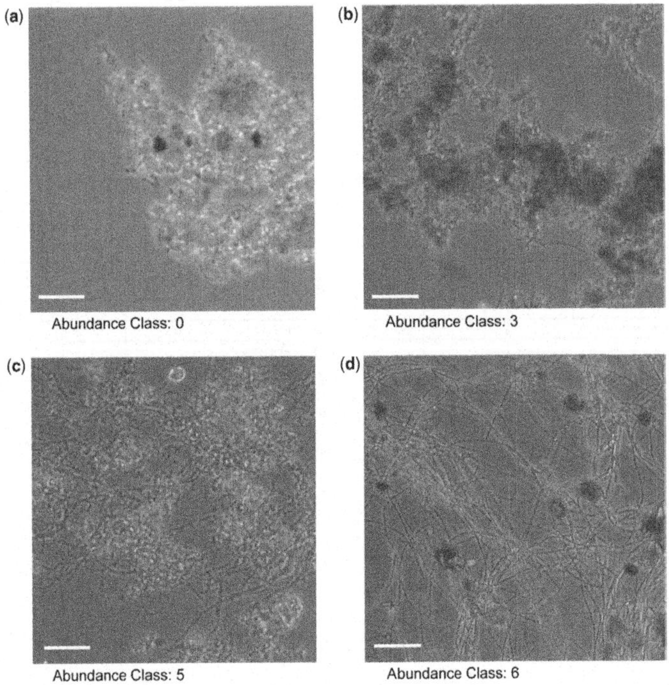

Figure 5.2 Activated sludge with different level of filamentous bacteria abundance: none (**a**), common (**b**), abundant (**c**) and excessive (**d**). Observation in phase contrast at 100× magnification. Bar is 50 μm.

It is important to stress as the group of Mycolata is constituted of a large number of bacteria belonging to different genera and species, such as *Gordonia, Nocardia, Tsukumarella*, etc. (Seviour & Nielsen, 2010) that tentatives to find out specific remedies to contrast the growth of specific foam forming bacteria have not been successful, so far, at full-scale level.

Specific equipment for selective foam forming microorganism removal has also been realized (classifying selctors), developing the principles of flotation, traditionally applied for separation of ores from minerals (Jenkins *et al.* 2004).

5.1.3 Bulking

Traditionally, bulking studies have been conducted more systematically, associating identified filaments to specific conditions, in either pure and mixed cultures, bench, pilot or full scale. Filamentous bulking control methods are usually divided into specific and non-specific methods.

Specific methods are aimed at removing the cause for filament proliferation, and are therefore targeted to a specific microorganism or group of microorganisms, whereas non-specific methods are aimed at reducing filament levels without any effect on the causes of their growth. Therefore, they are almost independent from filament type.

The use of specific control methods is usually the more desirable method, because only when the cause of bulking is addressed can the plant be operated with a limited filament proliferation. Non-specific methods have a temporary effect; they are useful when the cause of the filamentous bulking cannot be determined immediately when a rapid resolution of the bulking problem is needed.

Examining bulking causes and corresponding specific control methods, other than non-specific control methods, it is necessary to take into account two crucial points: from one side the problem of excess sludge disposal at WWTP (described in the next paragraph) and from another side the storage phenomena in bacterial biomass (described in Section 5.2.1.4). On this basis, several engineering solutions have been designed to modify the microbial composition of the biomass, and to disfavour filamentous bacteria growth.

5.1.4 The problem of excess sludge production and its disposal

The activated sludge process produces excess sludge that must be treated and disposed of. A proper treatment line comprehends several steps: thickening, anaerobic or aerobic digestion, dewatering (mechanical and thermal). This line is costly: sludge treatment and disposal often account for 50% of the plant's operating cost (Bertanza *et al.* 2015). The consequence is that plant operators have a tendency to work at high sludge age, which increases both suspended solids concentration and the oxygen demand in the biological tank (either for active or endogenous respiration of the biomass), and consequently the sludge blanket height in the secondary clarifiers. In these conditions, the plant operates in a situation of instability, with possible frequent suspended solids loss in the final effluent. Hence, a proper evaluation of the operation of an activated sludge system must include also the complete data set of the sludge disposal line.

Recently, for proper comparisons of alternative technologies to decrease the sludge production, a detailed approach has been established to monitor the plant, based on energy and mass (on COD basis) balances (Bertanza *et al.* 2015). The sludge production (Figure 5.3) in the reference plant examined (19.110 $m^3\ d^{-1}$ influent)

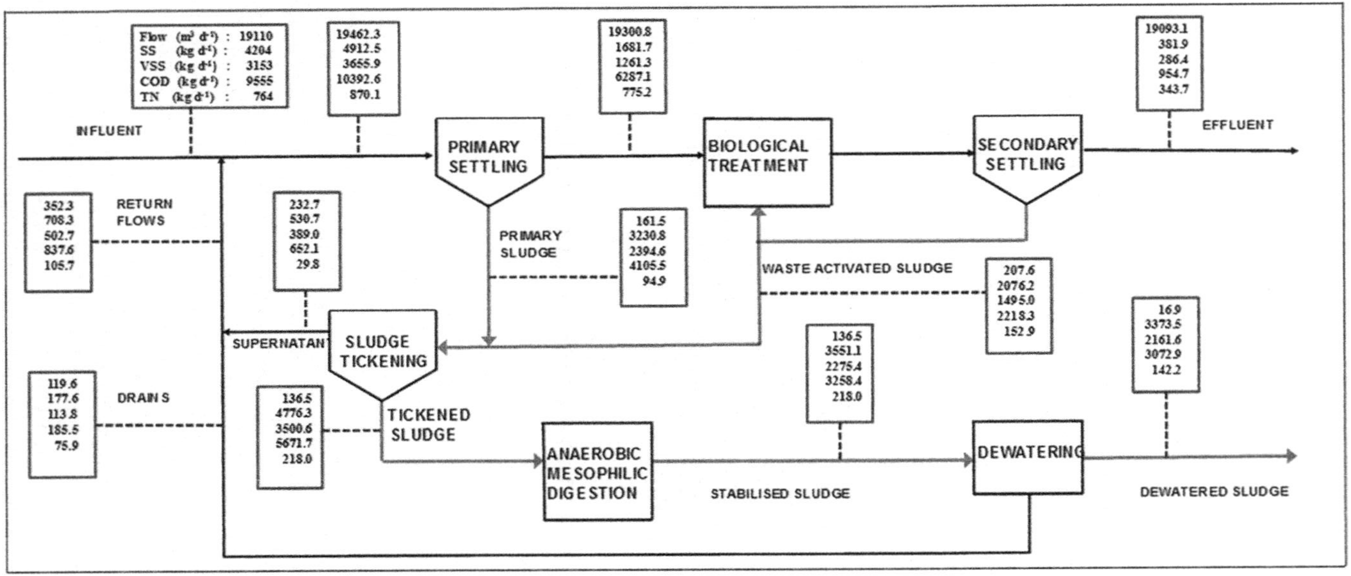

Figure 5.3 Mass balance on COD basis in a municipal WWTP, as reference for alternative sludge reduction technologies comparison (Bertanza *et al.* 2015), with permission of the Editor).

was 0.251 kg VSS kg^{-1} COD removed, corresponding to 0.357 kg COD kg^{-1} COD removed: in the upgraded plant (including sludge dynamic thickening and sludge aerobic post-retament) the sludge production was more than halved. On this basis, it is possible to properly compare sludge reduction technologies (Mininni *et al.* 2015).

Such balances are crucial to fix the optimal plant operating conditions: if the digestion step is under-designed, return flows might contain large amount of M. parvicella or Nocardioforms, contributing to increasing their number in the aeration basin, worsening bulking and foaming problems.

5.2 SPECIFIC CONTROL METHODS

Although the causes of filamentous bacterial proliferation in activated sludge plants have been studied for many years, precise conclusions about specific factors, that favour their growth, cannot be drawn at present. Table 5.1 summarizes the present status of knowledge, of the various factors, which have been reported as major causes of bulking. For each of these causes, the usually adopted specific method for bulking control is also reported.

Table 5.1 Summary of causes of filamentous bulking, related filaments and suggested cure methods (Jenkins *et al.* 2004).

Cause of Filamentous Bulking	Most Common Filaments	Suggested Cure Method
Low F/M	*H. hydrossys, Nocardia* sp., *Thiothrix* spp., Types 0041, 0675, 0092, 0803, 0914, 1851	Use of plants with substrate gradient (SBR, plug-flow systems or selectors)
Low dissolved oxygen	Type 1701, *S. natans, H. hydrossys*	Increase dissolved oxygen level in the plant
Nutrient deficiency	*Thiothrix* spp., Type 021N, *N. limicola* II	Addition of the lacking nutrient
Fatty acids in wastewater, low F/M, low temperatures	M. parvicella	Use of anoxic or anaerobic selectors

5.2.1 Bulking due to low (F/M) ratio

In activated sludge processes characterized by low F/M ratio (0.05–0.10 kg COD kg^{-1} VSS d^{-1}), the concentration of COD in the aerobic tank is very low, especially in the case of a completely mixed tank reactor (Figure 5.4). According to the kinetic selection theory developed long ago by Chudoba and co-workers (Chudoba *et al.* 1973), at these low substrate concentrations (Figure 5.5), the growth of filaments is favoured with respect to floc-forming bacteria, because they have high and low affinity for the substrate, respectively (i.e. low and high substrate semi-saturation coefficient (K_s), respectively). On the other hand, due to their higher specific maximum growth rate, floc-forming bacteria are favoured with respect to filamentous bacteria when substrate concentration is high.

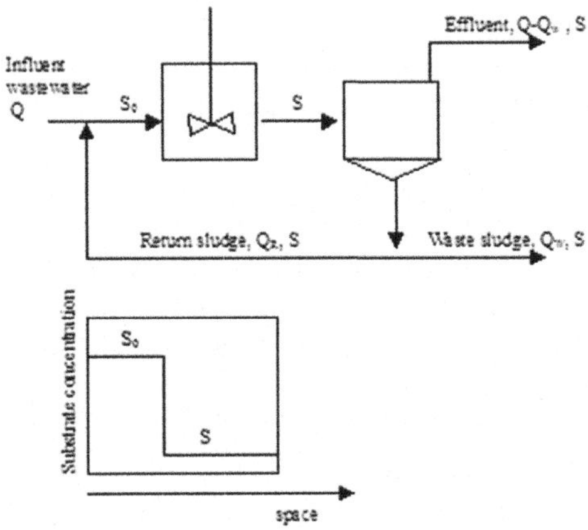

Figure 5.4 Conventional activated sludge process with continuously mixed aeration tank and substrate profile inside the aeration tank.

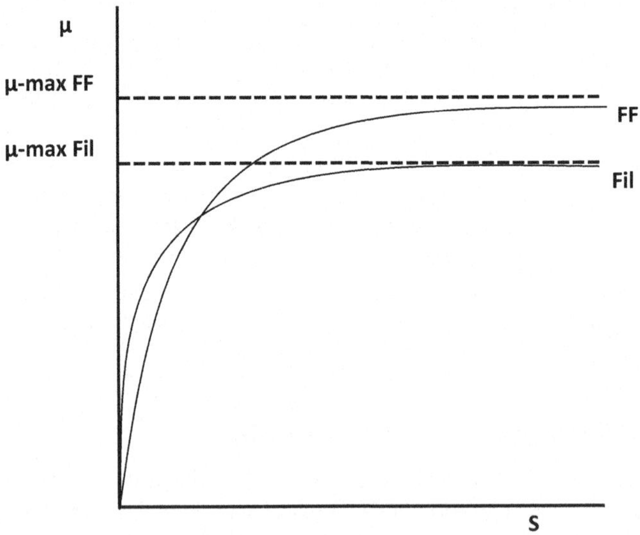

Figure 5.5 μ-max and K_s for filamentous and floc formers bacteria (Chudoba *et al.* 1973): in the activated sludge microbial competition filamentous bacteria are *K*s-strategists, while floc former bacteria are μ-max strategists.

A different theory for explaining the selection for filaments at low substrate concentration is the 'diffusion-based' selection (Martins *et al.* 2003). According to this theory, due to substrate diffusion limitations inside the flocs, floc-forming microorganisms actually face lower substrate concentration than filaments, which extrude from the flocs. Because substrate diffusion is driven by the substrate gradient from the bulk and through the floc, at low bulk substrate concentration this phenomenon is more important and would lead to a lower growth rate of floc-formers than of filaments.

In this sense, because lower F/M values are associated with higher COD removals, the bulking associated with low substrate concentration is also linked to a good activated sludge performance. In order to solve (or avoid) the problem without affecting substrate removal in the plant, the most useful strategy is to create a substrate concentration gradient inside the aeration tank reactor (or at least the presence of zones with different substrate concentrations). This goal can be achieved by modifying the activated sludge process configuration, from the completely mixed condition to one of three methods (Figure 5.6):

- Continuous plug flow reactors (CPFR)
- Completely mixed reactors with a contact zone placed ahead of the main reactor (CMR)
- Sequencing batch reactors (SBR).

In all three methods, the aim of the process configuration is to achieve the removal of most of the substrate when it is present at high concentrations, where floc-formers are kinetically favoured, and to remove only a small fraction of the substrate at the low concentrations where filament growth is favoured. In SBR systems, the desired substrate gradient is achieved in time, while in plug-flow and in selector systems the substrate gradient is achieved in space.

5.2.1.1 Continuous plug-flow reactors

In this type of reactor, a substrate profile is created along the reactor. The substrate gradient causes selection of floc-formers over filaments: floc formers are favoured in the first zones of the reactor, near to the inlet, where substrate concentration is high, whilst filaments are favoured at the reactor outlet, where substrate is present at low concentration. Because most of the substrate is removed when its concentration is high and more floc-formers are produced, the overall result of this reactor is a kinetic selection that favours floc-formers. However, large-scale 'corridor-type' reactors, which are often used in activated sludge plants and often considered as most similar to plug-flow reactors, usually have some longitudinal mixing due to the presence of aeration in aerobic zones, and mechanical mixing in anoxic or anaerobic zones (Wanner, 1994a). Therefore, they are often not effective for bulking control: best results are obtained with a compartmentalization of the corridor type basin. In this case, the system can be considered and designed as a selector.

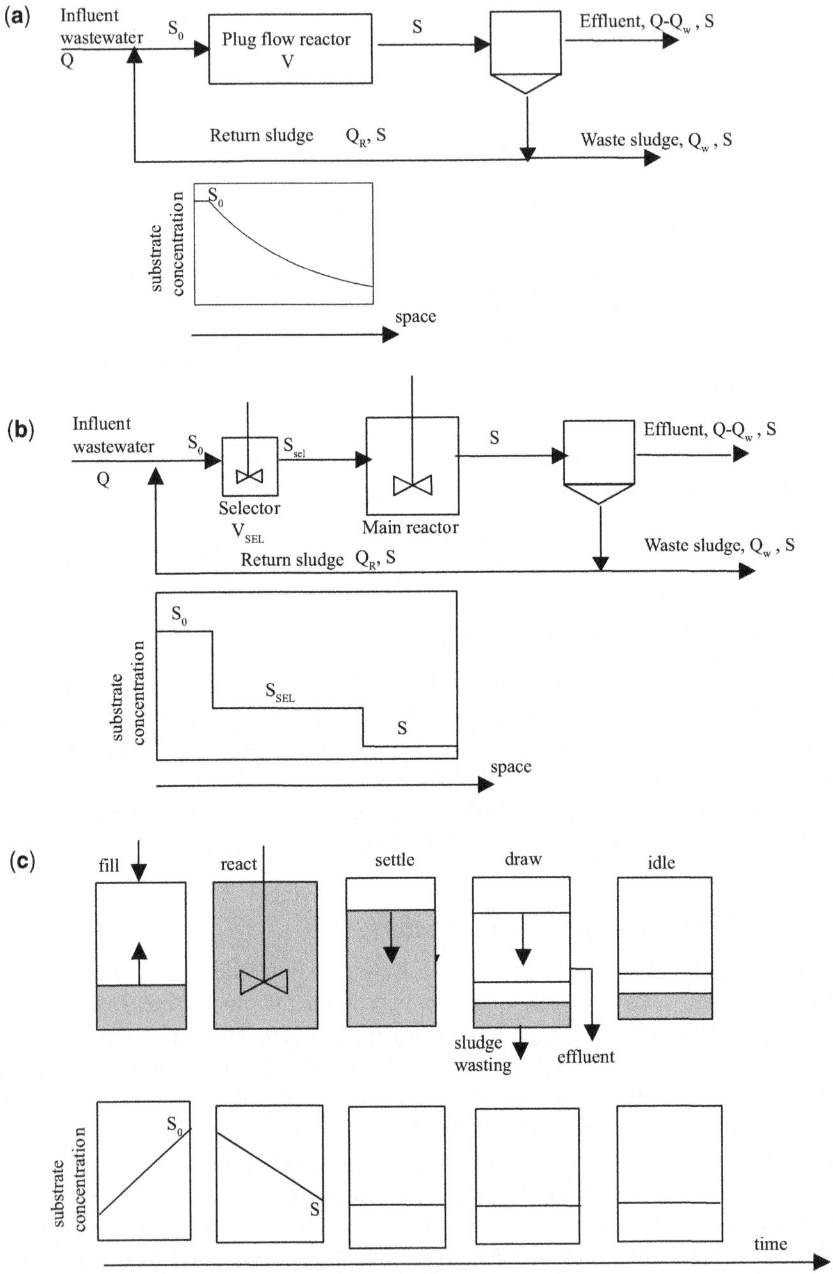

Figure 5.6 Activated sludge process flowsheets with a substrate gradient (**a**) plug-flow reactor, (**b**) mixed reactor plus selector, (**c**) sequencing batch reactor.

5.2.1.2 Selectors

This is the most common scheme to face filamentous bulking and the one that most closely resembles the original selector idea of Chudoba and co-workers (1973). The term selector refers simply to a smaller reactor placed ahead of the main biological reactor, where the above-described kinetic pressure is established. In the selector, biomass removes COD, but due to the small selector volume and thus the low substrate-biomass contact time, COD concentration does not reach the low levels that favour filament growth. In the main biological reactor a low COD concentration is present but the lower amount of residual COD to be removed does not allow the filamentous organisms to grow enough to cause bulking.

Proper selector design is critical. Too large a volume can cause too high a COD removal in the selector and thus lead to a COD concentration below the critical value that favours filament growth. On the contrary, too small a volume could cause too much substrate removal in the main reactor, where filamentous organism proliferation might occur.

Selectors can be designed using either biokinetic models or empirical methods. In general, empirical criteria are preferred, because of the uncertainty in kinetic parameters for the biokinetic models. Examples of both types of design criteria will be shown.

(a) Design based on a biokinetic models

The design criteria, proposed by van Niekerk *et al.* (1988), were based on a mathematical description of the selector–aeration basin system in which filament and floc-former growth rates were described by the Monod relationship. A higher floc-former growth rate at high substrate concentrations and higher filament growth rate at low substrate concentrations was used. The problem of determining the optimum range of selector residence time which allows a higher growth rate for floc-formers than for the filaments was solved by setting up materials balances for the selector, the reactor and the settling tank and solving the resulting equations.

Another mathematical model of floc-former/filament competition was developed by Kappeler and Gujer (1994a, b). This model takes also into account the different fractions of the influent COD and the inoculation of floc-formers and filaments from the sewer system. As in the previous described model, the Kappeler and Gujer model allows for a selector design, which minimizes the amount of filaments in the main reactor. Knowing the kinetic and stoichiometric parameters, simulations can be made for various selector volumes with a range of several design parameters such as sludge age, filament inoculation, etc. Figure 5.7 shows the results of model calculation for optimum selector size. It is evident that in the typical ranges of activated sludge operating parameters the optimum volume of the selector is 5–10% of the overall volume of the system (oxidation tank + selector). As already pointed out, the main drawback of this design methodology is the large amount of kinetic and stoichiometric parameters that need to be known.

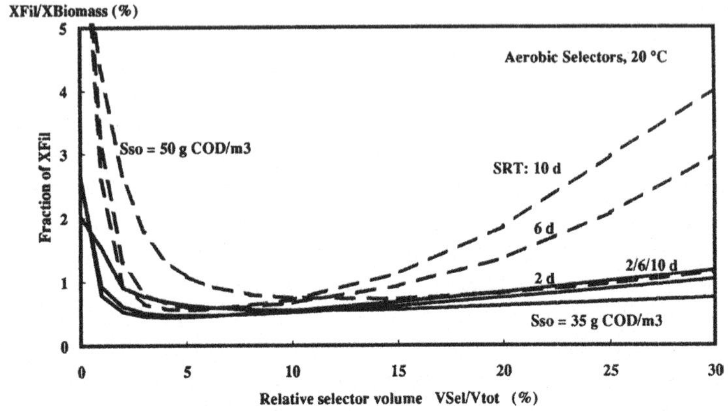

Figure 5.7 Calculations of the optimum selector size with the model by Kappeler and Gujer (1994b).

(b) Design based on empirical criteria

Several workers (see Table 5.2) have proposed empirical selector design criteria. Most useful are parameters that specify the selector contact time (*t*) and the selector loading (B_X or activated sludge loading in the selector), defined as follows (see Figure 5.6).

Table 5.2 Proposed empirical design criteria for aerobic selectors.

(a)	***t* (min)**	**References**
	15	Daigger *et al.* (1985)
	10–12	ATV (1989)
	10–15	Eikelboom (1991)
	15	Pujol (1992)
	>10	Rensink and Donker (1991)
	10–30	Prendl and Kroiss (1998)
	11	Guida *et al.* (2001)
	10–13	Martins *et al.* (2003)
(b)	**B_X (kg(BOD_5 or COD) $kg^{-1}SS\ d^{-1}$)**	**References**
	>20 (COD)	Lee *et al.* (1982)
	3 (BOD), 20 (COD)	ATV (1989)
	>3	Chudoba and Wanner (1989)
	5–6 (BOD)	Linne *et al.* (1989)
	3.2–4.9 (BOD)	Daigger and Nicholson (1990)
	3 (BOD)	Albertson (1991)
	2–5 (BOD)	Eikelboom (1991)
	4.2 (COD)	Still *et al.* (1996)
	1.8–2.7 (BOD)	Nakhla and Lugowski (2003)

Selector contact time

$$t = \frac{Q + Q_R}{V_{SEL}}$$

Selector loading

$$B_X = \frac{Q \cdot S_0}{V_{SEL} X}$$

Once the value of t or B_X has been chosen, the selector volume can be calculated.

A selector effect can be enhanced if the selector volume is compartmentalized: in this way, the substrate gradient along the system is enhanced. As an example of design criteria for multiple compartment selectors, Jenkins *et al.* (1993) propose the following loading values for the three-compartment selector:

- First compartment: 12 kg COD kg^{-1} MLSS d^{-1}
- Second compartment: 6 kg COD kg^{-1} MLSS d^{-1}
- Third compartment: 3 kg COD kg^{-1} MLSS d^{-1}

Duine and Kunst (2001) found that growth of Type 021N and Type 0961 in an industrial WWTP could only be avoided with a three compartment aerobic selector with an HRT of 5–8 minutes per compartment. A corresponding completely mixed selector was not successful in contrasting these filaments.

(c) Enhancing the selector effect: anoxic and anaerobic selectors

The selector effect can be enhanced if the selector is operated under anoxic or anaerobic conditions. The improvement in selector performance is expected because, with few exceptions, the filaments responsible for bulking (and listed in Table 5.1) cannot use electron acceptors other than oxygen. The exceptions are a few filaments, which can grow under anoxic and anaerobic conditions the so-called 'all zone growers' (Wanner & Grau, 1989). Examples of these are M. parvicella and Type 0092.

The implementation of anoxic or anaerobic selectors has some limitations: in order to use an anoxic selector, nitrate must be made available through nitrification whereas, to achieve rapid COD removal in anaerobic selectors, EBPR must be established (including Bio-P competitors metabolism; see paragraph 5.2.1.4). The suggested design criteria for anoxic selectors are reported in Table 5.3.

Table 5.3 Reported criteria for anoxic selector design.

Parameter	Suggested Value	References
Contact time	10–30 min	Hsu and Wilson (1992)
Activated sludge loading in the selector	2–6 kg COD kg^{-1} MLSS d^{-1}	Wanner (1993)
Activated sludge loading in the selector	0.8–1.2 kg BOD_5 kg^{-1} MLSS d^{-1}	Marten and Daigger (1997)
Contact time	17–22 min	Andreasen *et al.* (1999)

The same HRT that is used in an EBPR system can be used for the design of anaerobic selectors. Bortone *et al.* (1995) obtained good results with an anaerobic selector designed for an HRT of 45 min and Chevakidagarn *et al.* (2001) showed good results with an anaerobic selector HRT of 55–65 min. A similar anaerobic HRT (60 min) was shown to be effective in a study by Lee and Oleszkiewicz (2004).

5.2.1.3 Sequencing batch reactors

Sequencing batch reactor (SBR) technology is becoming more and more widespread in the field of wastewater treatment (Wilderer *et al.* 2001). The operation of a typical SBR is shown in Figure 5.6c. The typical sequence of phases is feed, reaction (oxic or anoxic or combination thereof), settle, effluent and sludge withdrawal, and idle. SBRs offer many advantages over continuous flow systems, especially the possibility to perform the whole treatment in a single tank.

The number of cycles per day and the length of the feed period during each cycle are specific parameters for SBR design and operation, which are all related to filamentous bulking control.

Indeed, at a fixed organic load rate, the number of cycles per day determines the organic load rate per cycle, i.e. the extent of the temporal substrate gradients, which are useful for bulking control. By increasing the number of cycles per day, the organic load rate per cycle decreases and so the substrate gradients imposed on the biomass become weaker (although more frequent). In general, to obtain a proper selection in favour of floc-forming microorganisms, the number of cycles per day should not be too high; positive effects of SBRs on bulking control have been reported with 4 or 6 cycles per day (Dionisi *et al.* 2001; Martins *et al.* 2003).

As far as the feed period is concerned, SBRs can select floc formers over filaments if the feed period is short enough and/or if feeding occurs under non-aerated conditions. With too long an aerated feed period the substrate will be removed simultaneously with feeding and the temporal substrate gradient necessary to avoid filament proliferation will not be achieved. A study by Martins *et al.* (2003), showed that, in a lab-scale acetate-fed reactor, a feed period length of less than 15 min (6.2% of the total HRT) was needed to select for floc-formers and to obtain a low SVI: with a feed period length 15 min or more (8.2% or more of the HRT) *Sphaerotilus natans* proliferated and a high SVI was obtained.

5.2.1.4 Role of storage phenomena in microbial competition in substrate gradient processes

Substrate gradient processes are now quite diffused in all their possible configurations, but their fundamental action mechanism cannot be considered sufficiently understood, and they still work on a trial and error basis. The substrate gradient processes had been investigated during the eighties–nineties, when the enormous diversity of microbial components of the activated sludge was just appearing, through the new biomolecular tools developed. The conclusions of periodic or substrate gradient studies in that time were associated with the Eikelboom' morphotypes and with the few known floc former bacteria.

As described previously, the positive effect of selectors, plug-flow configurations and SBRs on bulking control was first explained by the different relationship between growth rates and substrate concentration for the two populations (Chudoba *et al.* 1973; Figure 5.5), or by the role of substrate diffusion in the flocs (Martins *et al.* 2003). A different model for explaining the effect of systems with substrate gradients on population dynamics is based on storage phenomena. Storage polymers are intracellular reserve substances, which are synthesized when the external substrate is in excess of that needed for cellular synthesis and are consumed when the external substrate is exhausted. The most common storage polymers are the polyhydroxyalkanoates (PHA), which are usually synthesized from volatile fatty acids, and glycogen, from carbohydrates. Storage usually occurs when normal cellular synthesis (i.e. synthesis of RNA, proteins and DNA) is prevented or slowed by some growth limitation: these growth limitations can be due to the lack of some macronutrient, such as nitrogen or phosphorus or even sulfur (in this case the limitation is called external) or to the lack of the full metabolic apparatus (such as enzyme and RNA levels) needed to grow at maximum rate (in this case the limitation is called internal). In activated sludge plants treating domestic wastewaters, external limitations are unlikely, but might be possible for industrial waste. Internal limitation can occur because of the very low growth rates and the frequently changing environment in which biomass is grown (Majone *et al.* 1999). The more dynamic the growth conditions are, the more likely storage phenomena will occur, depending on feeding regime and sludge age (Çığgın *et al.* 2011a, b). Thus, in plants with selectors or plug-flow regimes (which can be in aerobic or anoxic or anaerobic conditions), the preliminary mechanism of substrate uptake usually is through storage; then growth will occur when conditions change.

The relevance of storage as the first substrate uptake mechanism in activated sludge plants has been recognized from a modelling point of view (Gujer *et al.* 1999; Bucci *et al.* 2012). The particular behaviour observed in biological reactors, in alternating substrate concentration and/or redox conditions has generated the recognition during the nineties of new categories of bacteria showing special metabolisms (Figure 5.8): (1) heterotrophic accumulating bacteria (the traditional aerobic storing bacteria), (2) polyphosphate accumulating organisms (PAOs), (3) glycogen accumulating organisms (GAOs). The metabolism of these groups of bacteria, operating in alternate conditions, can be summarized as:

(1) HAB (Figure 5.8a). Heterotrophic bacteria, operating in a dynamic substrate regime under fully aerobic conditions, store organic carbon in the form of PHA, which is then utilized in the main aeration basin. The aerobic storage

is quite diffused, and common either to floc formers or to filamentous bacteria: it occurs for example in a WWTP with a small aerobic selector up stream of the main aeration basin;

(2) PAOs (Figure 5.8b). These particular bacteria, in anaerobic conditions, are able to utilize the hydrolysis energy of polyphosphate to uptake and synthetize PHA, using the fermentation of internal carbohydrates (ICH-RPS), such as glycogen, to derive the necessary reducing power for PHA synthesis. The Poly-P is restored, and the glycogen reserve as well, in the further aerobic reactor at the expense of PHA, allowing also the growth. The identity of PAOs has for a long time been unclear. *Accumulibacter phosphatis* has been shown to be the typical PAO: their metabolism is evident for example in an EBPR WWTP with alternate anaerobic/aerobic zones or in small anaerobic zones in front of the main aeration basin;

(3) GAOs (Figure 5.8c). This group of bacteria lacks the potential energetic resource of Poly-P, but they are able to store the organic carbon as PHA, by using internal carbohydrates (ICH-RPS) fermentation energy (deriving also the necessary reducing power for PHA synthesis). Under aerobic conditions, glycogen inclusions are restored together with the organism growth. The metabolism of this group of bacteria is clearly detrimental in EBPR plants, reducing the proliferation of poly-P bacteria, but it has a positive role if floc-former GAOs compete with filamentous bacteria.

More recently, it has been shown that similar metabolisms can appear for PAOs and GAOs also when anoxic conditions are present (Shoji *et al.* 2003; Zeng *et al.* 2003; Seviour & Nielsen, 2010): then the list of 'particular' groups of bacteria in activated sludge systems elongates to include DPAOs and DGAOs (D is for denitrifying). It is worthwhile remembering too that in the past also intracellular inorganic phosphates precipitates were supposed to contribute to anaerobic release of phosphate (Arvin, 1985; Appeldoorn *et al.* 1992); this kind of information unfortunately has been forgotten, and chemical–physical contributions in EBPR are now completely overlooked.

The complexity of the bacterial interconnected metabolisms is even more highlighted considering that filamentous bacteria might behave as GAOs and PAOs. A GAO metabolism has been attributed to the filamentous bacteria '*Candidatus* Monilibacter batavus' (Levantesi *et al.* 2004) and cluster III Defluviicoccus members (McIlroy *et al.* 2010). A possible PAO metabolism has been recently attributed even to M. parvicella (Wang *et al.* 2014), which has a large part of the Poly-P bacteria enzymes (McIlroy *et al.* 2013).

The role of storage in microbial competition relies on the fact that various populations have different storage capabilities and rates (Wanner, 1994b). Storage ability seems to be widespread among floc-formers; consequently, the substrate gradients (with space or time) which encourage storage phenomena can also favour floc-formers over filaments.

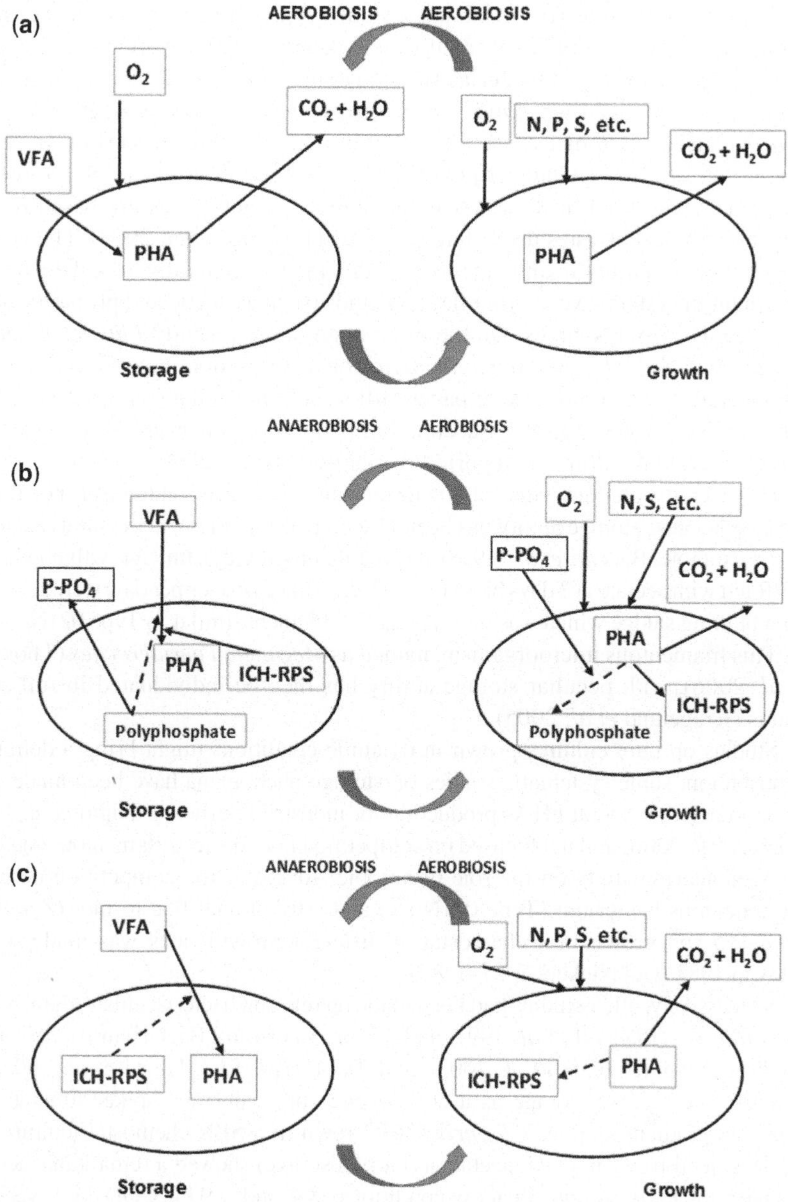

Figure 5.8 Organic carbon storage mechanisms by (**a**) aerobic heterotrophic bacteria, (**b**) poly-P bacteria, (**c**) glycogen accumulating organisms (GAO). (VFA: Volatile Fatty Acids; PHA: polyhydroxyalkanoates; ICH-RPS: Internal Carbohydrates – Reducing Power Source).

As an example, referring to lab-scale studies with aerobic SBRs by using either synthetic or true substrates, several different genera were most often individuated. They include a variety of bacteria affiliated to the genera *Amaricoccus*, *Thauera*, *Azoarcus*, *Paracoccus*, *Defluvicoccus*, *Alcaligenes*, *Comamonas*, *Zooglea* and *Achromobacter* (Dionisi *et al.* 2006; Serafim *et al.* 2006; Lemos *et al.* 2008; Queiros *et al.* 2014; Albuquerque *et al.* 2013). According to the literature, the differences observed in the relative abundance of each genus are based on the type of carbon substrates used (Jiang *et al.* 2011; Lemos *et al.* 2008). The genera *Amaricoccus*, *Thauera* and *Azoarcus* were indeed generally found to be the dominant in systems where volatile fatty acids (such as acetic and propionic acid) are used as carbon sources. On the other hand, more recently *Plasticicumulans acidivorans* and *Thauera selenatis* were found to predominate in systems fed with lactate as carbon source or paper mill wastewaters (Jiang *et al.* 2011, 2012). An overview on the main PHA-accumulating microorganisms so far described in mixed microbial cultures is reported in Queiròs *et al.* (2015).

However, the predominance of floc formers under storage conditions is not always the case because storage ability has been observed also in many filamentous bacteria, As an example, Beccari *et al.* (1998) showed that highly dynamic growth conditions (SBR fed with acetate at 3 days biomass residence time, 4 cycles per day) were selecting for a bulking sludge which was dominated by a filament similar to Type 021N.

This filamentous microorganism, named as *Meganema perideroedes* (Thomsen *et al.* 2006), with peculiar storage ability has then be individuated in full scale plants (Kragelund *et al.* 2005).

Studies on pure cultures grown in dynamic conditions might bring a definitive contribution: some systematic studies on storage phenomena have been made with an approach directed at PHAs production for industrial purposes (Pagni *et al.* 1992; Frigon *et al.* 2006) and not focused on competition floc formers/filamentous bacteria.

A pioneering study on the role of aerobic storage in the competition between a filamentous bacterium (Type 021N sp.) and a traditional floc former (*Zoogloea ramigera* sp.) grown in a chemostat at different growth rates was made in the USA in 1987 (van Niekerk *et al.* 1987).

Successively, other studies had been made on relevant bacteria, either filamentous, *Nocardia* sp. (Blackall *et al.* 1991) and a Tetrad Forming Bacterium *Amaricoccus kaplicensis* sp. (Aulenta *et al.* 2003), and *Thiothrix* CT3 (Majone *et al.* 2007), to quantify the aerobic storage as a consequence of a substrate spike, simulating a substrate gradient regime. *Thiothrix* CT3, grown in a SBR chemostat (culture age 12 days, feed frequency 4/d, acetate as carbon source) showed a remarkable acetate uptake and PHB storage in feast conditions: 8.4 and 2.9 kg COD kg^{-1} VSS d^{-1} (PHB yield: 0.35). This filamentous bacterium was able to increase 15.5 times the acetate uptake rate in comparison with the pseudo-steady state value (0.54 kg^{-1} COD kg^{-1} VSS d^{-1}), showing a very high capacity of carbon substrate uptake: the value 8.4 kg^{-1} COD kg^{-1} VSS d^{-1} represents an organic uptake rate more than one order of magnitude higher than even high loaded plants. In other words it would not be surprising if this filamentous bacterium was able in an aerobic selector to sequester part of the available carbon source (acetate or a short chain fatty acid that it is able to metabolize), winning the competition with the floc formers. Floc formers would also suffer from the substrate diffusional resistances, explaining why aerobic selectors are not suggested to eliminate *Thiothrix*, when the readily biodegradable substrate fraction is high.

Studies of this kind would provide precise quantitative information on the filamentous/floc-former competition, but unfortunately there are few available, due mainly to the difficulties in growing pure cultures of filamentous bacteria, especially in conditions of a substrate profile regime.

In particular, studies on filamentous bacteria pure cultures, growing in alternate anaerobic/aerobic conditions are still missing: the only experience with a pure culture growing in an anaerobic/aerobic chemostat was made on *Acinetobacter calcoaceticus* strain, showing its incapacity to uptake acetate during the anaerobic phase (Tandoi *et al.* 1998a).

Nowadays, the diversity of a large part of the activated sludge microbial components has been widely clarified, but still the knowledge cannot be considered complete and sufficient for proper WWTP operation. We know for example that much of the Eikelboom filamentous bacteria morphotype identity has been revealed, but we do not know their basic metabolism and role in activated sludge. We know also that the typical Poly-P bacterium is *Accumulibater phosphatis* but also that other Poly-P bacteria might exist and that several bacterial species belong to the group 'Competibacter' with a different metabolic diversity (McIlroy *et al.* 2014): it is hard to study a competition between Poly-P and GAOs whose identity is still unknown.

The research needs suggest investigating more deeply storage phenomena in properly identified floc formers and filamentous bacteria for proper operation of the WWTP with a gradient substrate regime. In the meantime, available analytical methods for the main storage compounds in activated sludge may provide useful information for operating the activated sludge process. Quantitative measurements of the main storage polymers, PHAs, Total carbohydrates (how glycogen is estimated) and phosphorus forms (organic, polyphosphate and inorganic phosphate), combined with specific staining and FISH techniques would permit us to test directly in the full-scale plants, the efficacy of modifications introduced (several selector configurations). This is the way to obtain valuable *in situ* information, explaining microbial components' capacity to colonize the activated sludge and in turn the possibility to favour or disfavour their proliferation.

5.2.2 Bulking due to low dissolved oxygen concentrations

One of the most frequent causes of filamentous bulking is low dissolved oxygen (DO) concentration in the aeration tank.

Similar to the low F/M case (as reported in Section 5.2.1 above), a model for the effect of DO on microbial competition can be set-up by assuming that DO is the limiting substrate and that biomass growth rate is dependent on DO concentration according to a Monod-like kinetic equation. Experimental data on pure cultures of Type 1701 and *Sphaerotilus natans* show values of the Monod half-saturation constant for DO in the range 0.01–0.03 mg L^{-1}, whereas typical values for pure cultures of floc-formers are about 0.07 mg L^{-1} (Jenkins *et al.* 2004). Given that the usual DO concentration in activated sludge plants is approx. 2 mg L^{-1}, it is clear that both populations can grow at their maximum rate and therefore DO concentration should not play a role in population dynamics. However, the role of DO can be relevant because the above given half-saturation constants are intrinsic values that do not take into account the diffusional resistances to DO diffusion within the flocs. Taking diffusional resistance into account, the DO concentration inside the floc is much lower than in its external part and thus growth of floc-formers inside of the floc can be DO-limited, whereas growth of filaments which protrude from the floc in the bulk liquid is not. This phenomenon introduces 'apparent' half-saturation constants, whose values are higher than the intrinsic values for both populations, but much higher for the floc-formers than for the filaments.

The relevance of diffusional resistance in determining the DO profile inside the floc is also dependent on the rate of substrate removal by the floc, which is in turn dependent on the organic loading. A correlation (Table 5.4) between organic loading and minimum DO concentration needed to avoid filamentous bulking was developed by Palm *et al.* 1980. Values of DO concentrations required to avoid filament proliferation can be calculated and can be provided for proper aerator design.

Table 5.4 Minimum dissolved oxygen values necessary to avoid filamentous bulking as a function of organic load rate (Palm *et al.* 1980).

Organic Load Rate (kg COD kg^{-1} MLSS d^{-1})	Dissolved Oxygen (mg L^{-1})
0.3	1
0.5	2
0.75	3
0.9	4

5.2.3 Bulking due to low nutrient concentration

The term 'nutrient' in activated sludge systems usually refers to chemical elements, other than carbon, oxygen, sulfur and nitrogen that are essential for biomass synthesis. Nutrients needed in major amounts are nitrogen (N) and (P) phosphorus: the balanced BOD:N:P ratio in the influent stream is usually assumed to be 100:5:1. Table 5.5 shows the usual nutrient requirements of activated sludge. Domestic wastewaters are normally rich in nutrients (hence, N and P often have to be removed by additional processes), whereas industrial wastewaters can be lacking in one or both of these nutrients or even micronutrients, such as Cu, Co, Ni, etc. (elements needed in traces).

Table 5.5 Nutrient requirements of activated sludge process (Grau, 1991).

Nutrient	Requirement ($g\ kg^{-1}$ BOD Removed)
N	50
P	10
Fe	12
Ca	6.2
K	4.5
Mg	2.0
Mo	0.43
Zn	0.16
Cu	0.15
Co	0.13
Na	0.05

Data reported by Wagner *et al.* (1982) and Richard *et al.* (1985) show that filament growth is enhanced with respect to floc-former growth when N and P levels are low. The study by Richard *et al.* (1985), shows that the filamentous bacterium Type 021N, has a high N uptake rate when it grows under N-deficient conditions. A study performed by Simpson *et al.* (1991) relates filaments proliferation to the lack of metals: SVI in the studied plants quickly decreased after solutions of metal salts were added to the plant. In one case, the filaments that were successfully controlled by metal addition were identified as M. parvicella and Type 021N. When bulking can be definitely related to nutrient deficiency, the control method is to add the deficient nutrient and it is easily implemented. A successful case study is reported by Switzenbaum *et al.* (1992) where the proliferation of *Nostocoida limicola* II due to P deficiency in the plant was controlled by increasing the sludge age so that the rate of nutrient utilization by the biomass decreased and P was no longer limiting.

5.2.4 Bulking due to fatty acids in the influent stream: control methods for Microthrix parvicella

Microthrix parvicella, reported in this book as M. parvicella (but its correct name is '*Candidatus* Microthrix parvicella') is a filamentous bacterium that causes either bulking or severe foaming problems in domestic WWTPs worldwide and whose growth is particularly difficult to prevent. It has attracted research interest for long time, since it was first described during seventies: it is a fastidious bacterium, difficult to cultivate (it presents in pure culture the phenomenon of growth paucity (scarcity) (see the Review by Rossetti *et al.* 2005 and Chapter 2). Microthrix parvicella is easy to identify in the activated sludge by traditional microscopy for its distinctive properties: dark coiled filament, 0.8 μm diameter, strongly Gram positive, not branched. Nevertheless, it has to be taken into account that another species of Microthrix has been described: '*Candidatus* Microthrix calida'. '*Candidatus* Microthrix calida' has similar morphology to M. parvicella, but it is distinguisible by its minor dimensions 0.3–0.7 μm (Levantesi *et al.* 2006): moreover, '*Candidatus* Microthrix calida' grows preferentially at higher temperatures (as its names tells: *calida*, warm). In case of necessary insights, FISH has to be applied. At present, only a few strains of M. parvicella isolated from a domestic plant in Italy are available (Rossetti *et al.* 2005): only recently, some Chinese strains have been isolated from activated sludge of EBPR plants in China (Rossetti *et al.* 2015).

Most of the work performed with pure cultures of this bacterium shows that its growth is enhanced by low temperatures and by the presence of long chain fatty acids in the influent. Pure culture studies have shown that it is only able to grow under aerobic or microaerophilic conditions, whereas it cannot grow under anoxic or anaerobic conditions (Tandoi *et al.* 1998b). It has been reported (Nielsen *et al.* 2002) that this micro-organism can take up and store long chain fatty acids under anaerobic conditions and this ability should explain its proliferation in P-removal plants. Genome sequencing of M. parvicella strain RN1 has confirmed the presence of the biochemical mechanisms to store anaerobically PHAs and the presence of a large part of the enzymes typical of Poly-P bacteria (McIlroy *et al.* 2013).

The most frequently used technique for controlling M. parvicella has been the use of selectors. Eikelboom reported that aerobic selectors are not effective (Eikelboom, 1994), but good results were obtained with a plug-flow aerobic or an anoxic/aerobic reactor (Mamais *et al.* 1998). The best results were obtained using anoxic and anaerobic reactors in series (Kruit *et al.* 2002) or anoxic selectors alone (Davoli *et al.* 2001). In the first case, the growth of M. parvicella was avoided with a process including an anaerobic reactor (contact time 80 min) followed by an anoxic selector (contact time 30 min) or with a process including a four-compartment anaerobic selector (total contact time 35 min) followed by a two-compartment anaerobic reactor (contact time 55 min). Both of the processes described above perform full nutrient removal and are based on a sequence of anoxic and aerobic reactors. The experiences described by Davoli *et al.* (2001) showed that M. parvicella's growth was controlled by anoxic selectors with contact times of 25–69 min.

Twenty years ago, it had been proposed that the addition of polyaluminium chloride interferes strongly with the growth of the filament M. parvicella by causing either a flocculating or a specific toxic effect (Eikelboom, 1997). This approach was later successfully applied in several full-scale plants at a dosage of from 1.5 to 4.5 g Al^{3+} kg^{-1} MLSS d^{-1}, depending on SRT (Roels *et al.* 2002; Badoer *et al.* 2003, 2015; see also Chapter 6, Belgium, Czech Republic, France, Italy and Spain). This strategy is now largely adopted worldwide.

5.2.5 Microbial and enzymatic preparations

The addition of microbial and enzymatic preparations to control filamentous bulking has been much more limited than the use of biocides and flocculating agents. Inamori *et al.* (1991) reported that two ciliated protozoa were able to feed selectively on filaments like Type 021N or *Sphaerotilus natans*. Yaguchi *et al.* (1991), showed that a lytic enzyme specifically active against Type 021N could be produced by soil microorganisms. Recently, another experience was reported (Fialkowska & Pajdak-Stos, 2012) showing that a culture of *Lecane inermis* (*Rotifera*) could feed selectively on M. parvicella. This kind of 'Biological Control' has to be proved at full-scale level.

A new emerging and promising field is related to the use of Phages (viruses) specific for target filamentous bacteria. Such an approach appears quite promising, as 'Biological Control' without any chemical addition, but at present there are no full-scale applications. Recently a phage active against the foam forming Actinomycete *Skermania piniformis*, was isolated (Dyson *et al.* 2016): phage SPI 1 targets seven of the nine strains of *S. piniformis* held in the authors collection, but none of the other 73 Mycolata strains of different genera.

Many commercial microbial and enzymatic preparations exist but their effectiveness has not been studied in a controlled and systematic fashion; this does not mean that in individual specific cases the use of enzymes cannot be successful. Recently a wide and comprehensive biomolecular approach has been developed to answer the fundamental question on the bioaugmentation: do the added bacteria remain and proliferate in the activated sludge system? Dueholm *et al.* (2015) tested the capacity of a *Pseudomonas* strain able to degrade toluene to colonize an activated sludge, determining its number by a q-PCR and FISH, its activity by RT-PCR and its metabolic capacity by stable isotope technique. The result was that the strain, after a quick positive response, was washed out or grazed by protozoa. While the experience might suggest that such action could be effective in confined reactors, such as in an MBR system, the biomolecular approach provides a valuable tool for customers to ascertain the effective benefits of using these commercial products.

5.3 NON-SPECIFIC CONTROL METHODS

Often the factors leading to sludge bulking are unidentified, so that the use of non-specific control methods is widespread. The use of non-specific control methods can be the first method of controlling bulking, before a cause–effect relationship is found, and a specific control method implemented. Nonspecific control methods usually consist of the addition of chemicals, such as:

- Oxidizing agents (chlorine, ozone, hydrogen peroxide)
- Weighting and flocculating agents (salts of iron and aluminium, lime, polymers and talc)
- Specific biocide

5.3.1 Oxidizing agents

The aim of oxidizing agents (biocides) is to kill filamentous organisms without affecting floc-formers. This is possible because filaments protrude from the floc and are therefore more subject to the toxic agent. Because of its potentially deteriorative effects on floc-formers, the correct choice of addition point and of the amount of toxicant added should be carefully chosen. According to the scheme of Figure 5.9, three main locations for biocide addition are possible:

- Directly into the aeration tank;
- Into a side stream recycle from the aeration tank;
- Into the return sludge

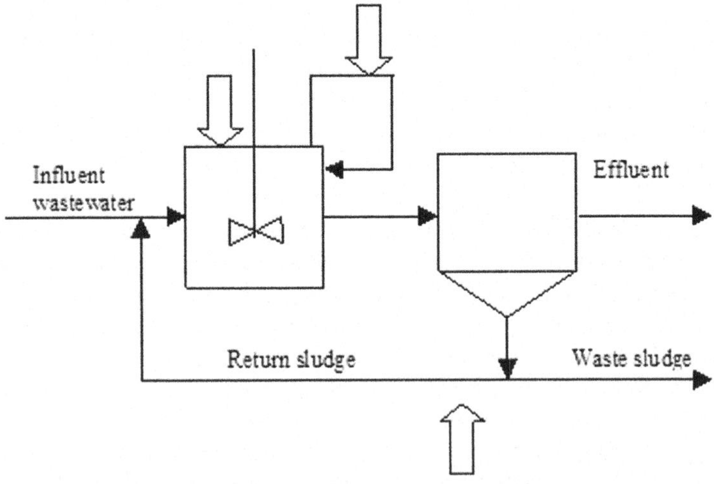

Figure 5.9 Possible locations (indicated by the arrows) for addition of oxidizing agents.

The best performance is usually obtained by adding the toxicant to the return sludge stream, because of the higher amount of solids, which are exposed to it. Almost all of the data reported in Table 5.6 refer to addition of the toxicant to the return sludge. In case the return flow rate is too low, the addition to the return stream is inappropriate and the best location for toxicant addition is in the aeration tank.

Table 5.6 Oxidizing agent dosage criteria to face filamentous bulking for: (a) chlorine; (b) ozone (c) hydrogen peroxide.

(a)	**Chlorine (g kg^{-1} SS d^{-1})**	**References**
	1–15	Jenkins *et al.* (1982)
	5–15	Neethling *et al.* (1985)
	0.43–1	Adey and McClintock (1996)
	1.42–3.07	Saayman *et al.* (1996)
	4–10	Marten and Daigger (1997)
	6–8	Ramirez *et al.* (2000)
	4–8	Ramirez *et al.* (2001)
(b)	**Ozone (g kg^{-1} SS d^{-1})**	**References**
	2	van Leeuwen (1992)
	12–48	Collignon *et al.* (1993)
	6.7–16.6	Goi *et al.* (1994)
	0.36–1.42	Saayman *et al.* (1996)
	16.5–27.6	Kim and Somiya (1998)
(c)	**Hydrogen Peroxide (g kg^{-1} SS d^{-1})**	**References**
	1.0–6.0	Jenkins *et al.* (2004)
	8.0	van Leeuwen (1992)
	1.5–9.5	Saayman *et al.* (1996)

In addition to the daily dosage of the biocide reported in Table 5.6, other important parameters in the dosage strategy are the concentration of the biocide at the dose point (defined as the ratio of the mass of biocide dosed per day and the flow rate past dose point) and the frequency of exposure of activated sludge to chlorine dose (Wanner, 1994a). Recommended values are, for the first parameter, 15–20 mg Cl_2 L^{-1} in order to avoid floc-former damage, and for the second one a value higher than 2.5–3 d^{-1}.

The most widely used biocide agent is Cl_2, provided as chlorine solution, or sodium hypochlorite solution. It is also the oldest control measure against bulking (Smith & Purdy, 1936). Table 5.6 shows that the amount of chlorine fed to activated sludge systems varies in the range 1–15 g Cl_2 kg^{-1} VSS d^{-1} (roughly maintenance-strong action).

Table 5.7 Reported criteria for weighting or flocculating agents dosage: (a) Iron salts or lime; (b) polymers; (c) talc.

(a) **Substance**	**Amount**	**References**
$FeSO_4$ $7H_2O$	7 mg Fe L^{-1}	ATV (1989)
$FeSO_4$ $7H_2O$	35 mg Fe L^{-1}	Wanner (1994a)
$Fe_2(SO_4)_3$ (43–45%)	7.5 mg Fe L^{-1}	Echeverria *et al.* (1993)
Lime	10–15 mg Ca kg^{-1} SS d^{-1}	Wanner (1994a)
(b) **Type of Polymer (Commercial Name)**	**Amount**	**References**
Cationic	<1 mg L^{-1}	WEF Manual of Practice (1992)
Cationic charged polyacrylamide	0.5 mg L^{-1}	Shao *et al.* (1989)
Cationic charged polyacrylamide, MW 10^7 (Zetag 88 N)	8 mg L^{-1}	Vanderhasselt and Verstraete (1999)
Cationic charged polymer, MW 6 10^6	3–12 mg L^{-1}	Seka *et al.* (2001)
(Keystone KF 4249), MW $2.5–3.5 \times 10^6$	1 mL polymer/ 10 L wastewater	Der-Fong (2005)
(c) **Substance (Commercial Name)**	**Amount**	**References**
Talc (PE8418)	0.6–1 kg kg^{-1} SS	Eikelboom and Grovenstein (1998)
Multi-component additive: 1% biocide, 98.5% talc, 0.5% polymer	0.35–1 kg kg^{-1} SS	Seka *et al.* (2001), Seka *et al.* (2003)
Talc (Aquatal)	0.6–1 kg kg^{-1} SS	Bideault *et al.* (1997)
Talc (Aquatal)	0.7 kg kg^{-1} SS	Clauss *et al.* (1999)
Talc (Aquatal)	0.5–8.5 g L^{-1}	Vanderhasselt and Verstraete (1999)

In a pure culture study of *Sphaerotilus natans* and *Acinetobacter anitratus* (Caravelli *et al.* 2003), the authors showed that the amount of Cl_2 needed to obtain a more than 80% inhibition of respirometric activity of both microorganisms was about 12 mg Cl_2 g^{-1} VSS. In another study on a phosphorus removal sludge Chang *et al.* (2004) reported a significant improvement of settling properties (both SVI and zone settling velocity) of a sludge with high abundance of *Thiothrix* with chlorine dosages up to 40 mg NaOCl g^{-1} SS; on the other hand, a worsening of substrate uptake rate and phosphorus removal at chlorine dosages higher than 2.5 mg NaOCl g^{-1} SS was observed.

In order to choose the optimum chlorine dosage with the aim of killing filaments without affecting the floc-forming population, some control methods have been developed, based on a commercial stain mixture (LIVE/DEAD BacLight, Ramirez *et al.* 2000) that allow us to evaluate microscopically the living and dead microorganisms. In this way the effect of chlorine on the biocenosis can be easily monitored and chlorine can be added on the basis of filament and floc damage. Alternatively, chlorination effects can be assessed by traditional microscopic observation; in particular, the presence of small flocs and the small amounts of microfauna are frequently linked to chlorine overdosage. Information about the amount of chlorine that can be added without affecting nutrient removal can be found in Saayman *et al.* (1996, see below).

M. parvicella seems to be particularly difficult to control even with chlorine dosage. Hwang and Tanaka (1998) reported, that amounts of chlorine up to 20 g Cl_2 kg^{-1} SS had no effect on M. parvicella filaments, whilst this amount was highly detrimental to floc-forming microorganisms. An amount of 200 g Cl_2 kg^{-1} SS was needed to break down more than half of the coils of the filament. As the disinfecting effect of chlorine is a function of penetration rate through the cell wall (White, 1982), the poor success in M. parvicella bulking control with chlorine dosage could be due to the hydrophobic cell wall of this filament, which prevents the biocide agent from penetrating into the cell.

The addition of chlorine to activated sludge systems has caused some concerns about possible effluent quality deterioration due to the formation of halogenated organic compounds. Van Leeuwen and van Rossum (1990) showed in batch tests that the chlorination of activated sludge leads to the formation of halogenated organics such as $CHCl_3$ and $CHBrCl_2$, but in pilot continuous flow activated sludge systems these compounds were never detected in the final effluent owing to air stripping.

The first reports of the use of hydrogen peroxide (H_2O_2) are dated more than 40 years ago (Caropreso *et al.* 1974) and its use has been limited by its high cost (van Leeuwen, 1992). It has the advantage of not leaving toxic residuals since it is ultimately converted to water and oxygen:

$$2H_2O_2 \rightarrow 2H_2O + O_2$$

As far as the ozonation process is concerned, its dosage is more costly than chlorination (van Leeuwen, 1988, 1992): ozone has to be generated on site and requires costly equipment. Similarly to H_2O_2, its addition does not leave toxic residuals.

A comparative study was made by Saayman *et al.* (1998) who compared the effects of the three biocides on a large BNR plant (13,000 m^3 d^{-1}). Their conclusions were that: (1) chlorine was the best treatment and at doses of 3.07 g Cl_2 kg^{-1} MLSS d^{-1}, bulking control and target DSVI were obtained in 4 days, (2) the hydrogen peroxide had only a temporary beneficial effect; (3) ozone even at low doses (1.45 g O_3 kg^{-1} MLSS d^{-1}) gave relatively small but consistent improvement. A more recent study (Nilsson *et al.* 2014) has focused on the action of ozone on either the SVI, the filamentous bacteria populations (ascertained by FISH) or the nitrifying and poly-P bacteria activities. In the conditions adopted (5% of the return sludge treated, O_3 dosage 2.8–5.0 g O^3 kg^{-1} MLSS, hydraulic retention time in the reaction vessel

11 min) a benefit was observed for the SVI, without observing any negative effect on the N and P removal. In the period of the test, 43 days prolonged to 71 without O_3 addition, the filamentous bacteria percentage surprisingly did not show any substantial difference with the reference line (no O_3 addition). The authors conclude that further research is necessary, including detailed analysis of the costs associated with the process: ozone utilization is expensive and precise cost/benefit analysis has to be done for each specific WWTP.

5.3.2 Weighting or flocculating agents

Bulking can be controlled without the addition of potentially toxic oxidizing agents by adding weighting or flocculating agents (Table 5.7). Traditionally used flocculating agents are chemicals such as lime or Fe and Al salts. These compounds are also used in the chemical P precipitation and overcome the bridging or diffuse floc structure associated with high filament levels, yielding an immediate improvement in settling. As a further beneficial effect, the addition of iron salts can precipitate sulfides and thus depress the growth of sulfide-utilizing filamentous microorganisms such as *Thiothrix* or *Beggiatoa* (Echeverria *et al.* 1993). The usage of these products may have anyhow a negative consequence: the increase of the amount of sludge (especially with lime) in the systems, with additional remarkable costs for wasting it. As alternative flocculating agents, polymers (usually cationic) can be used.

Divalent cations such as Ca^{2+} and Mg^{2+} can also be used to favor the formation of biological flocs. According to the divalent cation bridging theory (DCBT) (Sobeck & Higgins, 2002), divalent cations bridge the negatively charged functional groups which are present on exocellular biopolymers and therefore promote good floc formation. When the ratio of monovalent/divalent cations in the wastewater is high (more than approximately 2, expressing the ions as meq L^{-1}), divalent cations are displaced from the biomass by monovalent cations and settling properties (SVI and effluent solids) deteriorate. Therefore, in order to control bulking, the ratio of monovalent/divalent cations in mixed liquor should be lowered to values less than 2, either by increasing the divalent cation concentrations or by decreasing the monovalent cation concentrations. Successful case studies have been reported (Higgins *et al.* 2004a, b).

Talc-based additives have also been proven successful (Bidault *et al.* 1997; Claus *et al.* 1999). Talc is a hydrated magnesium silicate often combined with other minerals. Talc is insoluble, it is not toxic to bacteria and it is compatible with other chemicals commonly used in wastewater treatment plants. For floc weighting, talc minerals are usually used as water suspensions and are added directly into the aeration tank. The action of talc is to modify flocculation in spite of the filaments by creating larger flocs. These larger flocs usually entrap the filaments and prevent them from decreasing sludge settleability. Experimental studies show that the average floc dimensions and the settling velocities increase with higher talc additions. It is evident that when talc-based additives are added, the amount of TSS and ash in the system increases, but the total mass of sludge after dewatering on the filter-press remains unchanged, because more water is expelled from excess sludge due to the better dewatering properties of the sludge to which the talc was added (Bideault *et al.* 1997).

The positive effect on settling properties of biocides and of weighting agents can be combined in one single multi-component additive (Seka *et al.* 2001, 2003): it has been shown that an additive in powder form composed of (w/w) 1% biocide (a quaternary ammonium compound), 98.5% talc and 0.5% polyelectrolytes could simultaneously both burden the flocs (due to the action of the polymer and of talc) and reduce the filaments (due to the action of the biocide). Through the addition of the multi-component additive, both immediate and long-term effects (for several months) were obtained.

5.3.3 Specific biocide

After an unsuccessful chlorination action against Type 021N filamentous bacteria, in a continuous activated sludge system, bulking was controlled by a biocidal agent, cetyltrimethyl ammonium bromide (CTAB), which exhibited a stronger biocidal effect on these filaments, and which significantly improved sludge settleability (Guo *et al.* 2012). CTAB was utilized in the range 20–50 g kg^{-1}, with an optimal dosage individuated in 30 g kg^{-1}. Type 021N, properly identified by FISH probe, with filamentous index of 5 was selectively killed; floc-formers recovered quickly their activity after CTAB treatment. The study highlights the importance of the biocide penetration capacity to the cell wall of chlorine-resistant Type 021N bacteria.

5.4 CONTROL OF NON-FILAMENTOUS BULKING ('VISCOUS BULKING')

The problem of non-filamentous (viscous) bulking has not been as much studied as filamentous bulking. However, two main reasons seem to be evident for this type of bulking:

(a) A high amount of fatty acids and oleic compounds in the wastewater;
(b) An insufficient amount of nutrients, such as nitrogen and phosphorus.

Based on this evidence, specific control methods for non-filamentous bulking can be adopted.

In a laboratory scale study (Novak *et al.* 1993), the decrease in the oleic acid proportion in the influent wastewater (from 33 to 19% of the total influent COD) was supposed to be the reason for a shift from bad-settling to well-settling sludge in about 3 weeks. In an SBR fed with brewing wastewater (Peng *et al.* 2003), non-filamentous bulking was observed due to nitrogen limitation (influent BOD:N ratio 100:0.94) whereas a well-settling sludge was obtained when the influent composition was more balanced (BOD:N 100:4). At the Detroit Wastewater Treatment Plant (Franklin *et al.* 2001) the problem of non-filamentous bulking, which was apparently due to low concentrations of soluble phosphate (<0.2 mg L^{-1}), was corrected by increasing the soluble phosphate feed to the plant. A part separation problems, even an influence on nitrifying bacteria presence has been shown (van den Akker *et al.* 2010). A negative relationship between the AOB/total Bacteria ratio and SVI was found, showing that a poor settling sludge generally coincided with periods of relatively low AOB abundance.

As far as nonspecific control methods are concerned, they seem usually ineffective in the control of non-filamentous bulking. However, a successful case study was reported (van Leeuwen, 1989) where an O_3 dosage of 1 g kg^{-1} MLSS d^{-1}) decreased the SVI from 125 to 70 ml g^{-1}, mainly by preventing excessive Zoogloeal growth.

5.5 AVOIDING POOR SETTLING PROPERTIES: ALTERNATIVE SEPARATIONS OF ACTIVATED SLUDGE

The activated sludge separation problems described above are all related to low settling rates, which cause problems in the secondary sedimentation. One control that solves the problem at its origin is to use a different device for the sludge–supernatant separation, such as the flotation or membrane process. Flotation is normally utilized for sludge thickening and also for sludge separation (Zlokarnik, 1998), even though secondary clarification by gravity is the traditional (and cheaper) approach. Severe bulking cases might be solved by flotation.

A technology that has had more application for bulking control is the use of membrane for avoiding poor activated sludge/effluent separation (Stephenson *et al.* 2000; Viviani *et al.* 2016).

The scheme of a bioreactor equipped with a membrane for solid–liquid separation (a membrane bioreactor, MBR) is shown in Figure 5.10. The use of membranes avoids the effect of filamentous micro-organisms on settling because the filterability of the sludge is independent of the number of filaments present. The solids concentration in an MBR can be much higher than a secondary clarifier can handle: thus, in addition to the advantages of not suffering from bulking problems, MBR activated sludge also has the advantage of a smaller footprint.

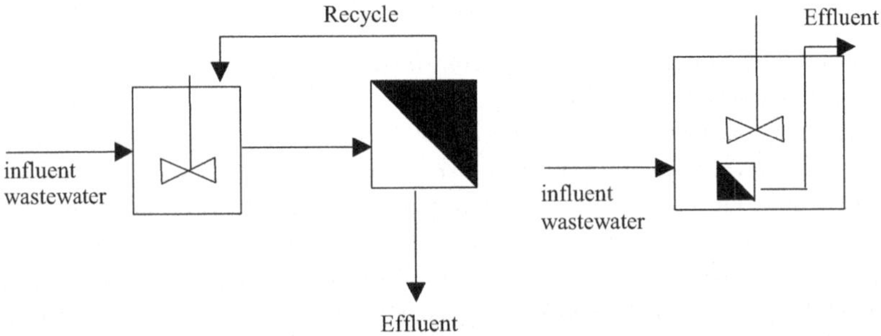

Figure 5.10 Alternative separation for activated sludge: membrane bioreactors. Sidestream (left) and submerged (right). *Source*: Stephenson *et al.* (2000).

The disadvantage of MBRs is their high cost. Membrane costs are approximately proportional to plant size, whereas in conventional plants there is an economy of scale. Additional disadvantages are membrane fouling that requires periodic cleaning, and high aeration requirements for preventing membrane fouling.

However, also in MBRs, as in conventional activated sludge processes with secondary sedimentation, the operating conditions of the activated sludge plant are important in determining the feasibility of the membrane separation process (Liao *et al.* 2004; Fenu *et al.* 2010). As an example, it was observed that the sustainable flux across membranes increased by increasing the sludge age from 3–8 to 33 d, likely because of a lower membrane fouling due to the lower concentration of extracellular polymeric substances (EPS) at high sludge age (Chang & Lee, 1998). Membrane fouling was also lower because of less EPS content at low organic loading content (Nagaoka *et al.* 2000) and with an increase in COD/N ratio (Hao *et al.* 2016).

Therefore, it cannot be expected that, in a hypothetical future where membranes will replace secondary clarifiers, we will not deal with activated sludge separation problems any more. Simply, activated sludge processes will have to be designed in order to select for sludges with 'good filterability properties', instead of 'good settling properties'.

REFERENCES

Adey A. H. and McClintock S. A. (1996). Control of filamentous bulking in a complete-mix activated sludge process treating high-strength industrial wastewater. *Proc. WEFTEC '96 Annu. Conf. Expo. 69th*, **3**, 395–404.

Albertson O. E. (1991). Bulking sludge control – progress practice and problems. *Water Science and Technology*, **23**(4–5), 835–846.

Albuquerque M. G. E., Carvalho G., Kragelund C., Silva A. F., Barreto Crespo M. T., Reis M. A. M. and Nielsen P. H. (2013). Link between microbial composition and carbon substrate-uptake preferences in a PHA-storing community. *ISME Journal*, **7**, 1–12.

Andreasen K., Agertved J., Petersen J. O. and Skaarup H. (1999). Improvement of sludge settleability in activated sludge plants treating effluents from pulp and paper industries. *Water Science and Technology*, **40**(11–12), 215–221.

Appeldoorn K. J., Boom A. J., Kortstee G. J. J. and Zehnder A. J. B. (1992). Contribution of precipitated phosphates and acid-soluble polyphosphate to enhanced biological phosphate removal. *Water Research*, **26**, 937–943.

Arvin E. (1985). Biological removal of phosphorus from wastewater. *CRC Crit. Rev. Env. Cont.*, **15**, 25–64.

Aulenta F., Dionisi D., Majone M., Parisi A., Ramadori R. and Tandoi V. (2003). Effect of periodic feeding in sequencing batch reactor on substrate uptake and storage rates by a pure culture of *Amaricoccus kaplicensis*. *Water Research*, **37**, 2764–2772.

ATV Working Group 2.6.1 (1989). Prevention and control of bulking sludge and scum. *Korrespondenz Abwasser*, **36**(13), 165–177.

Badoer S., Cuomo M., Miana P., Fabris M., Rossetti S. and Tandoi V. (2003). Bological scum by Microthrix parvicella in a large wastewater treatment plant in Northern

Italy: causes and solutions. IWA Regional Conference "Water key to sustainable development in Africa", Cape Town, 14–17 September.

Badoer S., Miana P., Della Sala S., Marchiori G., Tandoi V. and Di Pippo F. (2015). Nitrifying bacterial biomass and nitrification activity evaluated by FISH and an automatic on-line instrument at full-scale Fusina (Venice, Italy) WWTP. *Environmental Science and Pollution Research*, **22**(24), 19786–19792.

Beccari M., Majone M., Massanisso P. and Ramadori R. (1998). A bulking sludge with high storage response selected under intermittent feeding. *Water Research*, **32**(11), 3403–3413.

Bertanza G., Canato M., Laera G. and Tomei M. C. (2015). Methodology for technical and economic assessment of advanced routes for sludge processing and disposal. *Environmental Science and Pollution Research*, **22**(10), 7190–7202.

Bidault A., Clauss F., Helaine D. and Balavoine C. (1997). Floc agglomeration and structuration by a specific talc mineral composition. *Water Science and Technology*, **36**(4), 57–68.

Blackall L. L., Tandoi V. and Jenkins D. (1991). Continuous culture studies with *Nocardia amarae* from activated sludge and their implications for *Nocardia* foaming control. *Research Journal of the Water Pollution Control Federation*, **63**, 44–50.

Bortone G., Cech J. S., Bianchi R. and Tilche A. (1995). Effects of an anaerobic zone in a textile wastewater treatment plant. *Water Science and Technology*, **32**(9–10), 133–140.

Bucci V., Majed N., Hellweger F. L. and Gu A. Z. (2012). Heterogeneity of intracellular polymer storage states in Enhanced Biological Phosphorus Removal (EBPR) – observation and modeling. *Environmental Science Technology*, **46**, 3244–3252.

Çığgın A. S., Rossetti S., Majone M. and Orhon D. (2011a). Effect of feeding and sludge age on acclimated bacterial community and fate of slowly biodegradable substrate. *Bioresource Technology*, **102**, 7794–7801.

Çığgın A. S., Rossetti S., Majone M. and Orhon D. (2011b). Short-term and long-term effects on carbon storage of pulse feeding on acclimated or unacclimated activated sludge. *Water Research*, **45**, 3119–3128.

Caravelli A. H., Contreras E. M., Giannuzzi L. and Zaritzky N. E. (2003). Modeling of chlorine effect on floc-forming and filamentous microorgnaisms of activated sludges. *Water Research*, **37**, 2097–2105.

Caropreso F. E., Raleigh C. W. and Brown J. C. (1974). Attack bulking sludge with H_2O_2 a microscope. *Bull California Water Poll Control Assoc*, p. 44.

Chang I. S. and Lee C. H. (1998). Membrane filtration characteristics in membrane coupled activated sludge systems-effect of physiological states of activated sludge on membrane fouling. *Desalination*, **120**, 221–233.

Chang W. C., Jou S. J., Chien C. C. and He J. A. (2004). Effect of chlorination bulking control on water quality and phosphate release/uptake in an anaerobic-oxic activated sludge system. *Water Science and Technology*, **50**(8), 177–183.

Chevakidagarn P., Orth H. and Ratanachai C. (2001). The use of a selector when improving the efficiency of the conventional activated sludge process under tropical temperature conditions. In: Poster 48 Proc. of the 3rd IWA Specialised Conference on Microorganisms in Activated Sludge and Biofilm Processes, V. Tandoi, R. Passino and M. C. Blundo (eds), Rome, 13–15 June.

Chudoba J. and Wanner J. (1989). Notes on oxic anoxic and anaerobic selectors. *Newsletters of the IAWPRC Specialist Group on Activated Sludge Population Dynamics*, **1**(2), 46–49.

Chudoba J., Grau P. and Ottova V. (1973). Control of activated sludge filamentous bulking-II Selection of microorganisms by means of a selector. *Water Research*, **7**, 1389–1406.

Clauss F., Helaine D., Balavoine C. and Bidault A. (1998). Improving activated sludge floc structure and aggregation for enhanced settling and thickening performances. *Water Science and Technology*, **38**(8–9), 35–44.

Collignon A., Martin G., Martin N. and Laplanche A. (1993). Treatment of bulking by ozonation. Mechanisms of ozone on microorganisms. *Tribune de l'Eau*, **46**, 46–57.

Daigger G. T. and Nicholson G. A. (1990). Performance of four full-scale nitrifying wastewater treatment plants incorporating selectors. *Research Journal of the Water Pollution Control Federation*, **62**, 676–683.

Daigger G. T., Robbins M. H. and Marshall B. R. (1985). The design of a selector to control low F/M filamentous bulking. *Journal Water Pollution Control Federation*, **16**, 35–53.

Davoli D., Madoni P., Gugliemi L. and Pergetti M. (2001). Testing the effect of selectors in the control of bulking and foaming in small activated-sludge plants. *Water Science and Technology*, **46**(1–2), 495–498.

Der-Fong J. (2005). Effects of synthetic polymer on the filamentous bacteria in activated sludge. *Bioresource Technology*, **96**(1), 3140.

Dionisi D., Majone M., Tandoi V. and Beccari M. (2001). Sequencing batch reactor: influence of periodic operation on physiological state microbial composition and performance of activated sludge in biological wastewater treatment. *Industrial & Engineering Chemistry Research*, **40**, 5110–5119.

Dionisi D., Majone M., Vallini G., Di Gregorio S. and Beccari M. (2006). Effect of the applied organic load rate on biodegradable polymer production by mixed microbial cultures in a sequencing batch reactor. *Biotechnology and Bioengineering*, **93**, 76–88.

Dueholm M. S., Marques I. G., Karst S. M., D'Imperio S., Tale V. P., Lewis D., Nielsen P. H. and Nielsen J. L. (2015). Survival and activity of individual bioaugmentation strains. *Bioresource Technology*, **186**, 192–199.

Duine A. and Kunst S. (2001). Control of bulking sludge caused by type 021N and type 0961 in an industrial wastewater treatment plant with an aerobic selector. *Water Science and Technology*, **46**(1–2), 29–33.

Dyson Z. A., Tucci J., Seviour R. J. and Petrovski S. (2016). Isolation and characterization of bacteriophage SPI1, which infects the activated-sludge-foaming bacterium *Skermania piniformis*. *Archives of Virology*, **161**, 149–158.

Echeverria E., Seco A. and Ferrer J. (1993). Control of activated sludge settleability using preaeration and preprecipitation. *Water Research*, **27**, 293–296.

Eilkelboom D. H. (1991). The role of competition between floc-forming and filamentous bacteria in bulking activated sludge. In: Biological Approach to Sewage Treatment Process: Current Status and Perspectives, P. Madoni (ed.), Luigi Bazzucchi Center, Perugia, Italy, pp. 143–149.

Eilkelboom D. H. (1994). The *Microthrix parvicella* puzzle. *Water Science and Technology*, **29**(7), 271–279.

Eilkelboom D. H. (1997). Control of Microthrix parvicella by addition of PAX 14. TNO Report R 97/305.

Eikelboom D. H. and Grovenstein J. (1998). Control of bulking in a full scale plant by addition of talc (PE, 8418). *Water Science and Technology*, **37**(4–5), 297–301.

Eikelboom D. H. and van Buijsen H. J. J. (1983). Microscopic Sludge Investigation Manual, Report A 94. tno Research Institute for Environmental Hygiene, Delft, The Netherlands.

Fenu, A., Guglielmi G., Jimenez J., Sperandio M., Saroj D., Lesjean, B., Brepols C., Thoeye C. and Nopens I. (2010). Activated sludge model (ASM) based modelling of membrane bioreactor (MBR) processes: A critical review with special regard to MBR specificities. *Water Research*, **44**, 4272–4294.

Fialkowska E. and Pajdak-Stos A. (2008). The role of Lecane rotifers in activated sludge bulking control. *Water Research*, **42**(10–11), 2483–2490.

Franklin C., Hanson C., Verma K. and Packman J. (2001). Case study; viscous bulking-importance of nutrient monitoring. *Conference Proceedings – Water Env Fed Annual Conference and Exposition 74th Atlanta*, USA, 13–17 Oct, 4794–4822.

Frigon D., Muyzer G., van Loosdrecht M. C. M. and Raskin L. (2006). rRNA and Poly-{beta}-hydroxybutyrate dynamics in bioreactors subjected to feast and famine cycles. *Applied Environmental Microbiology*, **72**, 2322–2330.

Goi M., Nishimura T., Kuribayashi S., Okouchi T. and Murakami T. (1994). An experimental study to suppress scum formation accompanying the abnormal growth of *Nocardia* by adding ozone in the aeration tank. *Water Science and Technology*, **30**(11), 231–234.

Grau P. (1991). Criteria for nutrient-balanced operation of activated sludge process. *Water Science and Technology*, **24**(3–4), 251–258.

Guida M., Cesaro G., Lipardi I. L. and Melluso G. (2001). A full scale application in the control of the filamentous bulking generated by type 021N and *Thiothrix* sp. *Water Science and Technology*, **46**(1–2), 507–510.

Gujer W., Henze M., Mino T. and van Loosdrecht M. C. M. (1999). Activated sludge model No 3. *Water Science & Technology*, **39**(1), 183–193.

Guo J., Peng Y., Peng C. Y., Wang Z., Chen Y., Yuan Z. and Yang X. (2012). Control filamentous bulking caused by chlorine-resistant Type 021N bacteria through adding a biocide CTAB. *Water Research*, **46**, 6531–6542.

Hao L., Liss S. N. and Liao B. Q. (2016). Influence of COD:N ratio on sludge properties and their role in membrane fouling of a submerged membrane bioreactor. *Water Research*, **89**, 132–141.

Higgins M. J., Tom L. A. and Sobeck D. C. (2004a). Case study I: application of the divalent cation bridging theory to improve biofloc properties and industrial acytivated sludge system perfomance-direct addition of divalent cations. *Water Engineering Resources*, **76**, 344–352.

Higgins M. J., Sobeck D. C., Owens S. J. and Szabo L. M. (2004b). Case study II: application of the divalent cation bridging theory to improve biofloc properties and industrial acytivated sludge system perfomance-using alternatives to sodium-based chemicals. *Water Engineering Resources*, **76**, 353–359.

Hsu M. and Wilson T. E. (1992). Activated sludge treatment of municipal wastewater- USA practice. In: Activated Sludge Process Design: Theory and Practice, W. W. Eckenfelder and Grau P. (eds), Technomic Publishing Co., Inc., Lancaster, PA, pp. 37–68.

Hwang Y. and Tanaka T. (1998). Control of *Microthrix parvicella* foaming in activated sludge. *Water Research*, **32**, 1678–1686.

Inamori Y., Kuniyasu Y., Sudo R. and Koga M. (1991). Control of the growth of filamentous microorganisms using predacious ciliated protozoa. *Water Science and Technology*, **23**(4–6), 963–971.

Jiang Y., Marang L., Kleerebezem R., Muyzer G. and van Loosdrecht M. C. M. (2011). Polyhydroxybutyrate production from lactate using a mixed microbial culture. *Biotechnology and Bioengineering*, **108**, 2022–2035.

Jiang Y., Marang L., Tamis J., van Loosdrecht M. C. M., Dijkman H. and Kleerebezem R. (2012). Waste to resource: Converting paper mill wastewater to bioplastic. *Water Research*, **46**, 5517–5530.

Jenkins D., Richard M. G. and Daigger G. T. (1993). Manual on the Causes and Control of Activated Sludge Bulking and Foaming. 2nd edn., Lewis Publishers Inc., Chelsea, USA, p. 193.

Jenkins D., Richard M. G. and Daigger G. T. (2004). Manual on the Causes and Control of Activated Sludge Bulking and Foaming and Other Solids Separation Problems, 3nd edn. Lewis Publishers, New York.

Kappeler J. and Gujer W. (1994a). Development of a mathematical model for aerobic bulking. *Water Research*, **28**, 303–310.

Kappeler J. and Gujer W. (1994b). Verification and application of a mathematical model for "aerobic bulking". *Water Research*, **28**, 311–322.

Kim Y. R. and Somiya I. (1998). Effect of ozonization on filamentous bulking control of activated sludge. *Journal of Environmental Conservation Engineering*, **27**, 590–597.

Kragelund C., Nielsen J. L., Thomsen T. R. and Nielsen P. H. (2005). Ecophysiology of the filamentous alphaproteobacterium *Meganema perideroedes* in activated sludge. *FEMS Microbiology Ecology*, **54**, 111–122.

Kruit J., Hulsbeek J. and Visser A. (2002). Bulking sludge solved? *Water Science and Technology*, **46**(1–2), 457–464.

Lee Y. and Oleszkiewicz J. A. (2004). Bench-scale assessment of the effectiveness of an anaerobic selector in controlling filementous bulking. *Environmental Technology*, **25**, 751–755.

Lee S. E., Koopman B. L., Jenkins D. and Lewis R. F. (1982). The effect of aeration basin configuration on activated sludge bulking at low organic loading. *Water Science and Technology*, **14**(6–7), 407–427.

Lemos P. C., Levantesi C., Serafim L. S., Rossetti S., Reis M. A. M. and Tandoi V. (2008). Microbial characterization of polyhydroxyalkanoate storing populations selected under different operating conditions using a cell-sorting RT-PCR approach. *Applied Microbiology and Biotechnology*, **78**, 351–360.

Levantesi C., Beimfohr C., Geurkink B., Rossetti S., Thelen K., Krooneman J., Snaidr J., van der Waarde J. and Tandoi V. (2004). Filamentous alphaproteobacteria associated with bulking in industrial wastewater treatment plants. *Systematic and Applied Microbiology*, **27**, 716–727.

Levantesi C., Rossetti S., Thelen K., Kragelund C., Krooneman J., Eikelboom D., Nielsen P. H. and Tandoi V. (2006). Phylogeny, physiology and distribution of *Candidatus* Microthrix calida, a new Microthrix species isolated from industrial activated sludge wastewater treatment plants. *Environmental Microbiology*, **8**, 1552–1563.

Liao B. Q., Bagley D. M., Kraemer H. E., Leppard G. G. and Liss S. N. (2004). A review of biofouling and its control in membrane separation bioreactors. *Water Environment Research*, **76**, 425–436.

Linne S. R., Chiesa S. C., Rieth M. G. and Polta R. C. (1989). The impact of selector operation on activated sludge settleablity and nitrification: pilot scale results. *Journal of the Water Pollution Control Federation*, **61**, 66–72.

Majone M., Dircks K. and Beun J. J. (1999). Aerobic storage in the activated sludge processes. *The state of the art. Water Science and Technology*, **39**(1), 49–60.

Majone M., Beccari M., Dionisi D., Levantesi C., Ramadori R. and Tandoi V. (2007). Effect of periodic feeding on substrate uptake and storage rates by a pure culture of *Thiothrix* (CT3 strain). *Water Research*, **41**, 177–87.

Mamais D., Andreadakis A., Noutsopoulos C. and Kalergis C. (1998). Causes of and control strategies for *Microthrix parvicella* bulking and foaming in nutrient removal activated sludge systems. *Water Science and Technology*, **37**(4–5), 9–17.

Marten W. L. and Daigger G. T. (1997). Full scale evaluation of factors affecting the performance of anoxic selectors. *Water Engineering Resources*, **69**, 1272–1281.

Martins A. M. P., Heijen J. J. and van Loosdrecht M. C. M. (2003). Effect of feeding patterns and storage on the sludge settleability under aerobic conditions. *Water Research*, **37**, 2555–2570.

Martins A. M. P., Pagilla K., Heijnen J. J. and van Loosdrecht M. C. M. (2004). Filamentous bulking sludge – a critical review. *Water Research*, **38**, 793–817.

McIlroy S. J., Nittami T., Seviour E. M. and Seviour R. J. (2010). Filamentous members of cluster III *Defluviicoccus* have the in situ phenotype expected of a glycogen-accumulating organism in activated sludge. *FEMS Microbiology Ecology*, **74**(1), 248–256.

McIlroy S. J., Kristiansen R., Albertsen M., Karst S. M., Rossetti S., Nielsen J. L., Tandoi V., Seviour R. J. and Nielsen P. H. (2013). Metabolic model for the filamentous "Candidatus 'Microthrix parvicella'" based on genomic and metagenomic analyses. *The ISME Journal*, **7**(6), 1161–1172.

McIlroy S. J., Albertsen M., Andresen E. K., Saunders A.M., Kristiansen R., Stokholm-Bjerregaard M., Nielsen K. L. and Nielsen. P. H. (2014). 'Candidatus Competibacter'-lineage genomes retrieved from metagenomes reveal functional metabolic diversity. *ISME Journal*, **8**, 613–624.

Mininni G., Laera G., Bertanza G., Canato M. and Sbrilli A. (2015). Mass and energy balances of sludge processing in reference and upgraded wastewater treatment plants. *Environmental Science Pollution Research Interantional*, **22**(10), 7203–7215.

Nagaoka H., Kono S., Yamanishi S. and Miya A. (2000). Influence of organic loading on membrane fouling in membrane separation activated sludge process. *Water Science and Technology*, **41**(10–11), 355–362.

Nakhla G. and Lugowski A. (2003). Control of filamentous organisms in food-processing wastewater by intermittent aeration and selectors. *Journal of Chemical Technology and Biotechnology*, **78**, 420–430.

Neethling J. B., Jenkins D. and Johnson K. M. (1985). Chemistry microbiology and modelling of chlorination for activated sludge bulking control. *Journal of the Water Pollution Control Federation*, **57**, 882–889.

Nielsen P. H., Roslev P., Dueholm T. E. and Lund Nielsen J. (2002). *Microthrix parvicella* a specialized lipid consumer in anaerobic-aerobic activated sludge plants. *Water Science and Technology*, **46**(1–2), 81–90.

Nilsson F., Hagman M., Mielczarek A. T., Nielsen P. H. and Jönsson K. (2014). Application of Ozone in Full-Scale to Reduce Filamentous Bulking Sludge at Öresundsverket WWTP. *Ozone: Science & Engineering*, **36**(3), 238–243.

Novak L., Larrea L., Wanner J. and Garcia-Heras J. L. (1993). Non-filamentous activated sludge bulking in a laboratory scale system. *Water Research*, **27**, 1339–1346.

Pagni M., Beffa T., Isch C. D. and Aragno M. (1992). Linear growth and poly(β-hydroxybutyrate) synthesis in response to pulse-wise addition of the growth-limiting substrate to steady-state heterotrophic continuous cultures of *Aquaspirillum autotrophicum*. *Journal of General Microbiology*, **138**, 429–436.

Palm J. C., Jenkins D. and Parker D. S. (1980). Relationship between organic loading dissolved oxygen concentration and sludge settleability in the completely mixed activated sludge process. *Journal of the Water Pollution Control Federation*, **52**, 223–232.

Peng Y., Gao C., Wang S., Ozaki M. and Tagikawa A. (2003). Non-filamentous sludge bulking caused by a deficiency of nitrogen in industrial wastewater treatment plant. *Water Science and Technology*, **47**(11), 289–295.

Prendl L. and Kroiss H. (1998). Bulking sludge prevention by an aerobic selector. *Water Science and Technology*, **38**(8–9), 19–27.

Pujol R. (1992). Control of filamentous bulking by the contact zone: design and results. *Newsletters of the IAWPRC Specialist Group on Activated Sludge Population Dynamics*, **4**(1), 24–26.

Queiròs D., Rossetti S. and Serafim L. S. (2014). PHA production by mixed cultures: a way to valorise wastes from pulp industry. *Bioresource and Technology*, **157**, 197–205.

Queiròs D., Lemos P. C., Rossetti S. and Serafim L. S. (2015). Unveiling PHA-storing populations using molecular methods. *Applied Microbiology and Biotechnology*, **99**, 10433–10446.

Ramirez G. W., Alonso J. L., Villanueva A., Guardino R., Basiero J. A., Bernecer I. and Morenilla J. J. (2000). A rapid direct method for assessing chlorine effect on filamentous bacteria in activated sludge. *Water Research*, **34**, 3894–3898.

Ramirez G. W., Alonso J. L., Basiero J. A., Benecer I. and Morenilla J. J. (2001). Survey of the filamentous populations using a rapid method for assessing chlorine effect in activated sludge. In: Proceedings of the 3rd IWA specialised conference on Microorganisms in Activated Sludge and Biofilm Processes, V. Tandoi, R. Passino and M. C. Blundo (eds), Rome, 13–15 June, paper 145.

Rensink J. H. and Donker H. J. G. W. (1991). The effect of contact tank operation on bulking sludge and biosorption processes. *Water Science and Technology*, **23**(4–6), 857–866.

Richard M. G., Shimizu G. P. and Jenkins D. (1985). The growth physiology of the filamentous organism Type 021N and its significance to activated sludge bulking. *Journal of the Water Pollution Control Federation*, **57**, 1152–1162.

Roels T., Dauwe F., Van Damme S., De Wilde K. and Roelandt F. (2002). The influence of PAX-14 on activated sludge systems and in particular on *Microthrix parvicella*. *Water Science and Technology*, **46**(1–2), 487–490.

Rossetti S., Tomei M. C., Nielsen P. H. and Tandoi V. (2005). "*M. parvicella*" a filamentous bacterium causing bulking and foaming in activated sludge systems: a review of current knowledge. *FEMS Microbiology Reviews*, **29**, 49–64.

Rossetti S., Qi R., Fan N., Levantesi C., Tandoi V. and Yang M. (2015). Causes and bulking control strategies in sewage treatment plants located at high altitude in China. NEWE Workshop, The 4th International Conference on Environmental Simulation and Pollution Control, ESPC & CSES, Beijing, 3–4 November 2015.

Saayman G. B., Schutte C. F. and van Leeuwen J. (1996). The effect of chemical bulking control on biological nutrient removal in a full scale activated sludge plant. *Water Science and Technology*, **34**(3–4), 275–282.

Saayman G. B., Schutte C. F. and van Leeuwen J. (1998). Chemical control of filamentous sludge bulking in a full-scale biological nutrient removal activated sludge plant. *Ozone: Science & Engineering*, **20**(1), 1–15.

Seka A. M., van de Wiele T. and Verstraete W. (2001). Feasibility of a multi-component additive for efficient control of activated sludge filamentous bulking. *Water Research*, **35**, 2995–3003.

Seka A. M., van de Wiele T. and Verstraete W. (2003). Full-scale evaluation of a multi-component additive for efficient control of activated sludge filamentous bulking. *Environmental Technology*, **23**, 67–72.

Serafim L. S., Lemos P. C., Rossetti S., Levantesi C., Tandoi V. and Reis M. A. M. (2006). Microbial community analysis with a high PHA storage capacity. *Water Science and Technology*, **54**, 183–188.

Seviour R. and Nielsen P. H. (2010). Microbial Ecology of Activated Sludge. IWA Publishing, London.

Shao Y. J., Starr M., Kaporis K., Kim H. S. and Jenkins D. (1989). Polymer addition as a solution to Nocardia foaming problems. *Water Environment Research*, **69**, 1059–1060.

Shoji T., Satoh H. and Mino T. (2003). Quantitative estimation of the role of denitrifying phosphate accumulating organisms in nutrient removal. *Water Science and Technology*, **47**(11), 23–29.

Simpson J. R., List E. and Dunbar J. M. (1991). Bulking sludge: a theory and successful case studies. *Water and Environmanagement Journal*, **5**, 302–311.

Smith R. S. and Purdy W. C. (1936). The use of chlorine for correction of sludge bulking in the activated sludge process. *Public Health Report*, **51**, 1–617.

Sobeck D. C. and Higgins M. J. (2002). Examination of three theories of cation-induced bioflocculation. *Water Research*, **36**, 527–536.

Soddell J. (1999). Foaming, In: The Microbiology of Activated Sludge, R. J. Seviour and L. L. Blackall (eds), Kluwer Academic, London, 1–422.

Stephenson T., Judd S., Jefferson B. and Brindle K. (2000). Membrane Bioreactors for Wastewater Treatment, IWA Publishing, London UK.

Still D. A., Ekama G. A., Wentzel M. C., Casey T. G. and Marais G.v.R. (1996). Filamentous organism bulking in nutrient removal activated sludge systems. Paper 2: Stimulation of the selector effect under aerobic conditions. *Water SA*, **22**, 97–118.

Switzenbaum M. S., Plate T. R. and Woodworth B. K. (1992). Filamentous bulking in Massachusetts: extent of the problem and case studies. *Water Science and Technology*, **25**(4–5), 265–271.

Tandoi V., Majone M, May J. and Ramadori R. (1998a). The behaviour of phosphate accumulating Acinetobacter isolates in an anaerobic aerobic chemostat. *Water Research*, **32**, 2903–2912.

Tandoi V., Rossetti S., Blackall L. L. and Majone M. (1998b). Some physiological properties of an Italian isolate of *Microthrix parvicella. Water Science and Technology*, **37**(4–5), 1–8.

Thomsen T. R., Blackall L. L., de Muro M. A., Nielsen J. L. and Nielsen P. H. (2006). Meganema perideroedes gen. nov., sp. nov., a filamentous alphaproteobacterium from activated sludge. *International Journal of Systematic and Evolutionary Microbiology*, **56**, 1865–1868.

Vanderhasselt A. and Verstraete W. (1999). Short term effects of additives on sludge sedimentation characteristics. *Water Research*, **33**, 381–390.

van den Akker B., Beard H., Kaeding U., Giglio S. and Short M. D. (2010). Exploring the relationship between viscous bulking and ammonia-oxidiser abundance in activated sludge: a comparison of conventional and IFAS systems. *Water Research*, **44**, 2919–2929.

van Leeuwen J. (1988). Bulking control with ozonation in a nutrient removal activated sludge system. *Water SA*, **14**, 119–124.

van Leeuwen J. (1989). Ozonation for non-filamentous bulking control in an activated sludge plant treating fuel synthesys wastewater. *Water SA*, **15**, 127–132.

van Leeuwen J. (1992). A review of the potential application of non-specific activated sludge bulking control. *Water SA*, **18**, 101–106.

van Leeuwen J. and van Rossum P. G. (1990). Trihalomethane formation during sludge bulking control with chlorine. *Water and Environmanagement Journal*, **4**, 530–534.

van Niekerk A., Jenkins D. and Richard M. G. (1987). The competitive growth of *Zoogloea ramigera* and Type 021N in activated sludge and pure culture – a model for low F:M bulking. *Research Journal of the Water Pollution Control Federation*, **59**, 262–273.

van Niekerk A., Jenkins D. and Richard M. G. (1988). A mathematical model of the carbon-limited growth of filamentous and flocforming organisms in low F/M sludge. *Journal of the Water Pollution Control Federation*, **60**, 100–106.

Viviani G., Torregrossa M. and Mannina G. (2016). Biomac 2016: Bioreattori a membrane (MBR) e trattamenti avanzati per la depurazione delle acque. Caracol Palermo, pp. 1–373, ISBN: 978-88-98546-58-9.

Wagner F. (1982). Study of the causes and prevention of sludge bulking in Germany. In: Bulking of Activated Sludge-Preventative and Remedial Methods, B. Chambers and E. J. Tomlinson (eds), Ellis Horwood Ltd., Chichester, pp. 29–40.

Wang J., Qi R., Liu M., Li Q., Bao H., Li Y., Wang S., Tandoi V. and Yang M. (2014). The potential role of 'Candidatus Microthrix parvicella' in phosphorus removal during sludge bulking in two full-scale enhanced biological phosphorus removal plants. *Water Science and Technology*, **70**(2), 367–375.

Wanner J. (1993). Kinetic and metabolic selection in controlling the filamentous organisms in activated sludge systems. In: Prevention and Control of Bulking Activated Sludge, D. Jenkins, R. Ramadori and L. Cingolani (eds), Luigi Bazzucchi Center, Perugia, Italy, pp. 17–42.

Wanner J. (1994a). Activated Sludge Bulking and Foaming Control. Technomic Publishing Co Inc., Lancaster, USA.

Wanner J. (1994b). The implementation of bulking control in the design of activated sludge systems. *Water Science and Technology*, **29**(7), 193–202.

Wanner J. and Grau P. (1989). Identification of filamentous microorganisms from activated sludge: a compromise between wishes needs and possibilities. *Water Research*, **23**, 883–891.

WEF Manual of Practice No 8 and ASCE Manual and Report on Engineering Practice No 76 (1992). Population dynamics and control. *Design of Municipal Wastewater Treatment Plants*, **1**, 538.

White G. C. (1982). Handbook of Chlorination. Van Nostrand Reinhold, New York.

Wilderer P. A., Irvine R. L. and Goronszy M. C. (2001). Sequencing Batch Reactor Technology. Scientific and Technical Report no 10, IWA Publishing, London, UK.

Yaguchi J., Chigusa K. and Ohkubo Y. (1991). Isolation of microorganisms capable of lysing the filamentous bacterium "Type 021N" (Studies on lytic enzyme against

the filamentous bacterium "Type 021N" Screening Studies). *Water Science and Technology*, **23**(4–6), 955–962.
Zeng R. J., Yuan Z. and Keller J. (2003). Enrichment of denitrifying glycogen-accumulating organisms in anaerobic/anoxic activated sludge system. *Biotechnology Bioengineering*, **81**(4), 397–404.
Zlokarnik M. (1998). Separation of activated sludge from purified waste water by induced air flotation (IAF). *Water Research*, **32**, 1095–1102.

Chapter 6

Experiences in various countries

6.1 INTRODUCTION

This chapter reports experiences in different countries when faced with bulking and foaming problems. This chapter is an attempt to give a whole picture of the approaches used worldwide. It cannot be homogenous in style, content and descriptions, because in different countries different scientific approaches and experimental research have been developed and carried out; consequently several control methods and solutions as have been adopted.

The general overview of the activities carried out in 13 different countries shows a dramatic difference from the depiction ten years ago in the first edition of this book. The traditional activated sludge microscopic description is applied worldwide, constituting the basis, obtained with minimal effort, for a systematic approach to effectively facing plant dysfunctions. The necessity to give a common basis for a proper description of activated sludge microbial components is faced in Austria via an NGO that examines large numbers of plants: in Italy qualification circuits are periodically organized to train plant operators to a common description of their activated sludge samples. In the Czech Republic, which has historical experience on these topics, considerable effort has been made to define plant configurations and mechanical devices to prevent bulking and foaming phenomena. In Belgium, China, France, Greece and the USA, large numbers of plants are monitored, intervining with specific and aspecific correction actions: several full-scale cases are considered and described. Some biomolecular tools, such as FISH and PCR, are beginning to be applied in full-scale plants (Australia, Denmark, Italy, Spain, South Africa), even though not yet finding systematic wide application at wastewater treatment plants. Finally, in Malaysia, new approaches are under investigation, such as magnetic field effects in controlling bulking and foaming phenomena.

doi: 10.2166/9781780408637_139

6.2 AUSTRALIA

H. Stratton, M. Christie, P. Griffiths and R.J. Seviour

6.2.1 General situation

The difficulties of making generalized statements relating to bulking and foaming monitoring and control within Australia were discussed in the earlier edition of this book (Tandoi *et al.* 2006). This situation still largely remains. On the other hand, comments made then in relation to which filaments seemed to occur more commonly in Australia than in other countries and *vice versa* may not necessarily be the case now. So what if anything has changed? Almost all activated sludge plants constructed in the past 20 years in Australia are designed to remove nitrogen and many also operate to achieve biological phosphorus reduction (Enhanced Biological Phosphorus Removal, EBPR, plants). Globally, anecdotal evidence indicates that frequently observed filaments, such as *Candidatus* 'Promineofilum breve' (Type, 0092), are more commonly encountered in EBPR systems than in conventional activated sludge plants. The Gram negative, Neisser negative Eikelboom *Nostocoida limicola* II morphotype, shown quite recently to belong to the genus *Defluviicoccus* (see below), is a dominant filament in EBPR plants in eastern Australia (Nittami *et al.* 2009), but from the little literature available, is only encountered rarely elsewhere. While *Gordonia amarae*-like organisms (GALOs) are still the most commonly seen foam forming bacteria, in Australia (de los Reyes, 2010), anecdotal evidence suggests *Skermania piniformis* is less frequently encountered now. No convincing explanation is forthcoming to explain such a trend. Plants in many other countries produce foams dominated by *S. piniformis*-like organisms (Eales *et al.* 2005, 2006). In our experience, the dominance of '*Candidatus* P. breve' in foams seems to be more common than it once was, but this may reflect its increased relative abundance in mixed liquors, since de los Reyes (2010) argues that it should not be considered a true foam former. A number of authors (Eales *et al.* 2005, 2006; Petrovski *et al.* 2011; Dyson *et al.* 2016) suggest that foaming should be viewed as a flotation event, and any highly hydrophobic organism has the potential to stabilise foams in activated sludge plants. Yet McIlroy *et al.* (2016) suggest that '*Candidatus* P. breve' is not hydrophobic.

Over the past ten years, little has improved in the application of microscopic and molecular methods like FISH for identifying the filamentous bacteria responsible. Very few water authorities in Australia, in our experience, possess suitable microscopes or employ appropriately trained microbiologists for this purpose. Consequently, in most cases, no routine examination of their activated sludge biomass is performed, where changes in the relative abundances of particular filaments may be seen to relate to changes in plant operational conditions or influent characteristics. Generally, it is only when a severe bulking or foaming incident has occurred that single samples are sent for analyses (only three laboratories, two on the Australian East coast and the other in South Australia, offer such a service, and only one carries out FISH based identifications), and more often than not, no examination is undertaken subsequently to assess whether any of the selected or proposed control strategies have been effective. This situation is despite the availability of (well attended) courses run through Griffith University, Brisbane, to train operators and laboratory staff to 'identify' these filamentous bacteria and relate their presence/proliferation to operational parameters. Unfortunately, very few appear subsequently to continue to apply and refine these skills and knowledge.

There are rare exceptions. A small number of the larger water corporations (e.g. SA Water) employ microbiologists to carry out FISH based analyses on a regular basis on site, and attempt then to relate filament dominance profiles to plant operational data (Keegan *et al.* 2012). How successful they are in predicting and controlling the problems the filaments cause is uncertain. Therefore, while considerable progress has been made in the past 25 years to develop FISH based molecular *in situ* methods for filament identification, and the eco-physiological understanding of many of the filaments, most water utilities have not yet taken advantage of this new knowledge. Commercial FISH kits (e.g. from Vermicon), which have been designed for several of the more common filaments (e.g. '*Candidatus* M. parvicella') were available in Australia. However, presumably because of low interest from the industry, these kits are no longer readily available and without a commercial distributor. With recent developments in amplicon sequencing and its ability to fingerprint activated sludge communities (van Loosdrecht *et al.* 2016), operational conditions that provide the environment for one filament to dominate over another are expected to be better measured than microscopy or FISH can achieve alone.

The existing situation reflects an inherent cultural problem, which is probably not restricted only to Australia. Engineers, who, in the vast majority, remain unconvinced of the value of having such microbiological information, dominate the wastewater treatment industry. Here, every operational problem is considered in engineering terms and so every solution implemented is an engineering one, with no consideration generally given to the role of the bacterial community.

Although it is difficult to convince water authorities of the value of research on bulking and foaming solutions, South-East Water and Melbourne Water have recently contributed to work at La Trobe University to investigate foaming control strategies based on bacteriophage biocontrol therapy (Dyson *et al.* 2016). This approach certainly works at lab-scale levels, although whether phage therapy is suitable for large-scale treatment plants is still not clear. Due to the high host cell specificities exhibited by many of these phages (Petrovski *et al.* 2012), this control strategy approach relies on an initial unequivocal identification of the community of foaming bacteria in each individual plant by using FISH. Yet FISH probes are currently available for relatively few of the foam causing filamentous bacteria, only, the mycolata (de los Reyes, 2010), and the microbial ecology of foam forming communities continue to be poorly understood globally (Petrovski *et al.* 2011).

6.2.2 What do we know of these filamentous bacteria?

Since the first edition of this book was published, the systematics of many more of the bulking and foaming unculturable Eikelboom filamentous bacterial morphotypes have been resolved, and FISH probes designed and validated for their *in situ* identification. In all cases, these probes, developed mainly from 16S rRNA sequence data obtained using culture independent methods, have been shown to perform well with samples from plants around the world. From the work primarily carried out in Australia and Denmark, many of the Eikelboom morphotypes are all now known to be phylogenetically diverse members of the phylum Chloroflexi. These include the morphotype Type 0914 which is now named '*Candidatus* Sarcinathrix' spp. (Speirs *et al.* 2011), Type 0092, '*Candidatus* P. breve' (McIlroy *et al.* 2016) and Type 1851 which belongs to the genus *Kouleothrix* also in the phylum Chloroflexi (Beer *et al.* 2002; Kohno *et al.* 2002).

Work at La Trobe University has revealed that the Australian Type 0803 morphotype differs phylogenetically from the Danish Type 0803 Chloroflexi member (Speirs *et al.* 2015), with the Danish phylotype more commonly seen in the few FISH based surveys carried out in Australia (Speirs *et al.* 2015). This finding adds further weight to the view that Eikelboom's microscopic identification method underestimates considerably the diversity of these troublesome filaments (Nielsen *et al.* 2009; Seviour & Nielsen, 2010).

The availability of FISH probes has also allowed insight into the in situ physiology of these filaments using FISH/microradiography (FISH/MAR) (Nielsen & Nielsen, 2005; Wagner *et al.* 2006). While most of this work was developed and performed in Prof Per Nielsen's laboratory in Aalborg, Denmark, it has also been used widely at La Trobe University for both foaming and bulking filaments (e.g. Eales *et al.* 2006), and provides *in situ* information not readily accessible by any other current technique. For example, the Gram-negative filamentous *Nostocoida limicola* II morphotype, seen frequently at high abundance in EBPR plants in eastern Australia was shown by FISH/MAR to possess the phenotype of a glycogen accumulating *Defluviicoccus* sp. (McIlroy *et al.* 2010); it was thus capable not only of causing bulking, but also of competing with the polyphosphate accumulating organisms (PAO) for substrates in anaerobic reactors of EBPR plants.

However, FISH/MAR only provides a 'snapshot' of an organism's *in situ* physiological potential. Whole genome sequence data, which reveal an organism's total metabolic potential, have now been generated in collaboration with Prof. Per Nielsen's group for the Australian *Tetrasphaera jenkinsii* ('*Nostocoida limicola*' II) (Kristiansen *et al.* 2013) '*Candidatus* M. parvicella' (McIlroy *et al.* 2013), and '*Candidatus* P. breve' (McIlroy *et al.* 2016). Metabolic models for these filaments have then been constructed revealing their physiology under different reactor conditions, providing clues as to how their population densities might be controlled. The number of filament whole genome sequences is expected to increase as DNA sequencing technology and genome annotation methods continue to become cheaper and more user friendly.

The following survey and case study presented are based on microscopy only for 'identifying' the problematic filamentous bacteria. The data collected during the surveys were largely in response to requests by treatment plants experiencing settling problems and/or foaming incidents, and plant personnel who initiated the investigations. While the correlation of operational parameters with the proliferation of filamentous bacteria was not possible for the survey work, due to the lack of useful operational information, some anecdotal observations are offered throughout the case study, which may lead to ameliorating operational control for bulking and foaming problems.

6.2.3 Bulking and foaming filamentous bacteria in Australian activated-sludge plants

There are over 1300 wastewater treatment plants in Australia with over half of these being activated sludge systems. Between 2002 and 2015, twenty-two different filamentous morphotypes were identified in 46 wastewater treatment plants within Australia. The microscopic evaluations were carried out in the Griffith University laboratory with the conventional techniques described in Seviour and Nielsen (2010). The relative abundance of each filament (filament index) was assessed qualitatively using a score of 0–6 according to (Jenkins *et al.* 2004), where the numbers represent the numbers of filaments per floc: 1 = incidental, 2 = one in the occasional floc; 3 = 1–5, 4 = >5–20, 5 = >20 and 6 = too many to count. Bulking was assessed by the operational staff at the treatment plants using a settled sludge volume index (SSVI), SVI or diluted SVI (DSVI) and >100 mL g^{-1}, 250 mL g^{-1} or 150 mL g^{-1} respectively. When these values were exceeded, the sludge was said to be bulking. However, the samples usually arrived without any accompanying operational data, and the little analytical data provided by plant personnel was usually unhelpful, reflecting their lack of understanding of the process. This situation has not improved since the last edition of this book was published.

With microscope assessments, we decided that the biomass sample was from a bulking sludge when the filament index was 3 or above and the filaments had an adverse effect on the floc structure, by being involved either in interfloc bridging or by creating open flocs. The filament with the highest index was considered dominant although other filaments were often present in high numbers. Thus, we have reported the top three highest rating filaments according to the 0–6 rating as dominant/co-dominant, high numbers and moderate to low.

Filamentous bacteria, like all bacteria, have an optimal temperature range over which they grow and their proliferation often relates predictably to seasonal change. According to climatic zoning from the Bureau of Meteorology, 22 of the surveyed treatment plants fall within the temperate climatic zone, with average maximum temperatures ranging from approximately 6 to 14 in winter and 16 to

26 in summer. Twenty four (24) of the surveyed treatment plants fell within the sub-tropical climate zone which is defined as having warm (average maximum 17 in winter) or hot (average maximum 30 in summer), humid conditions.

From the 22 treatment plants in the temperate climate zone, 64 mixed liquor and 19 foam samples were analysed. The 24 plants in sub-tropical climatic conditions provided 88 mixed liquor and 48 foam samples. Approximately 23% of these plants were considered to be both bulking and foaming at the time of sampling, based on the information provided by the plant personnel and the filament abundances observed in the samples.

6.2.3.1 The main foaming organisms

The foam samples examined contained well-documented foamers like *Gordonia amarae*-like organisms (GALOs), *S. piniformis* (seen only in plants in the colder temperate zone) and '*Candidatus* M. parvicella'. Other non-filamentous bacteria commonly seen in high numbers in one foam sample from an MBR treatment plant were thought to be related to *Rhodococcus*, a known foamer with a life cycle involving filament fragmentation (Goodfellow & Williams, 1983) or *Nocardia* spp. (Stratton *et al.* 1996). They are referred to here as Gram positive cocci/diptheroid Rods (GPCDR).

From the 88 mixed liquor (ML) samples from sub-tropical treatment plants, 48 (approximately 55%) of those provided accompanying foam samples. From temperate treatment plants, 19 foam samples were collected together with the 64 mixed liquor samples. Biological foaming did not correlate with any particular plant design, as all types of plants surveyed, MBR, SBR, EBPR, Oxidation ditches and MLE configured plants were prone to foaming.

The survey results for the foam-causing filaments showed the culprit varied between foams from temperate and sub-tropical zone plants. Those from temperate zone plants were dominated invariably by GALOs and '*Candidatus* P. breve', with both present as dominant or co-dominant in 37% of samples, followed by '*Candidatus* M. parvicella' and *S. piniformis* (both found to be the culprit in 26% of samples). Of the 19 samples, 11 were almost pure cultures of GALOs, *S. piniformis* or '*Candidatus* M. parvicella'. The presence of '*Candidatus* P. breve' in foams has been reported previously but whether its presence stabilizes foams or it arrives there accidentally has been discussed by de los Reyes (2010).

In sub-tropical foam samples, GALOs were present in 67% of the samples, '*Candidatus* P. breve' in 29%, and Type 0914, (not seen in temperate foams) in 13%, while *S. piniformis* were present in only 10%. A higher frequency of dominance of '*Candidatus* M. parvicella' and *S. piniformis* was seen in temperate zone foams than in those from the sub-tropical, consistent with reports that both of these filaments are often seen in samples from cooler climates (Levantesi *et al.* 2006; Seviour *et al.* 1990). Comparisons of the main filaments in the foam samples (qualitative analysis) are given in Fig. 6.1.

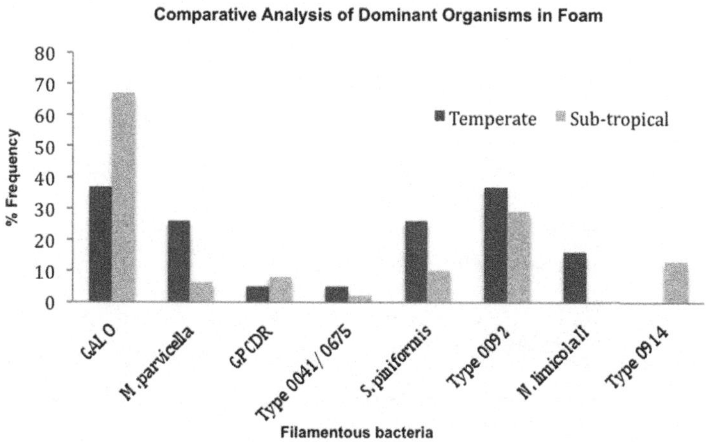

Figure 6.1 Comparison of dominant filamentous bacteria in foam samples from treatment plants in temperate and sub-tropical climate zones in Australia, shown as % frequency (percentage of samples in which the organism was dominant).

6.2.3.2 The main bulking filamentous bacteria

Several filaments have been identified as putative 'bulking' organisms based on being observed in high numbers and the changes they cause to floc morphology. The data from mixed liquor (152) samples from 46 treatment plants of various configurations, which were collected intermittently over a 14-year period in temperate and sub-tropical climate zones within Australia are summarized in Figs 6.2(a, b). From all 152 samples, 19 main filamentous morphotypes were identified.

All mixed liquor (ML) samples were collected to identify the probable organisms responsible for bulking at the time of sampling. The most frequent and abundant filamentous bacteria in ML sampled within the sub-tropical zone were Type 0041/0675 and '*Candidatus* P. breve' respectively. These were followed in decreasing abundances by '*Candidatus* Sarcinathrix' spp., *Kouleothrix* spp., *T. jenkinsii* and Type 021N. '*Candidatus* P. breve' was the most frequently seen dominant filament, followed by Type 0041/0675 (unrelated to Chloroflexi), '*Candidatus* Sarcinathrix' spp. and *Kouleothrix* spp. The frequency of abundance of the 19 main morphotypes in sub-tropical plants is summarized in Fig. 6.2(a).

The most abundant filamentous bacteria in the 64 mixed liquor samples from plants in the temperate zone were Type 0041/0675 and '*Candidatus* P. breve'. However, unlike sub-tropical plants, Type 0041/0675 dominated in more samples than '*Candidatus* P. breve'. Frequency of occurrence was then Type O21N, *Kouleothrix* sp. and T. jenkinsii. The filaments *Kouleothrix* spp., '*Candidatus* M. parvicella' and Type 0803 were more frequently observed as dominant, in that order. '*Candidatus* M. parvicella' was dominant in 11% of bulking sludge samples from temperate plants but only 2% from sub-tropical plants.

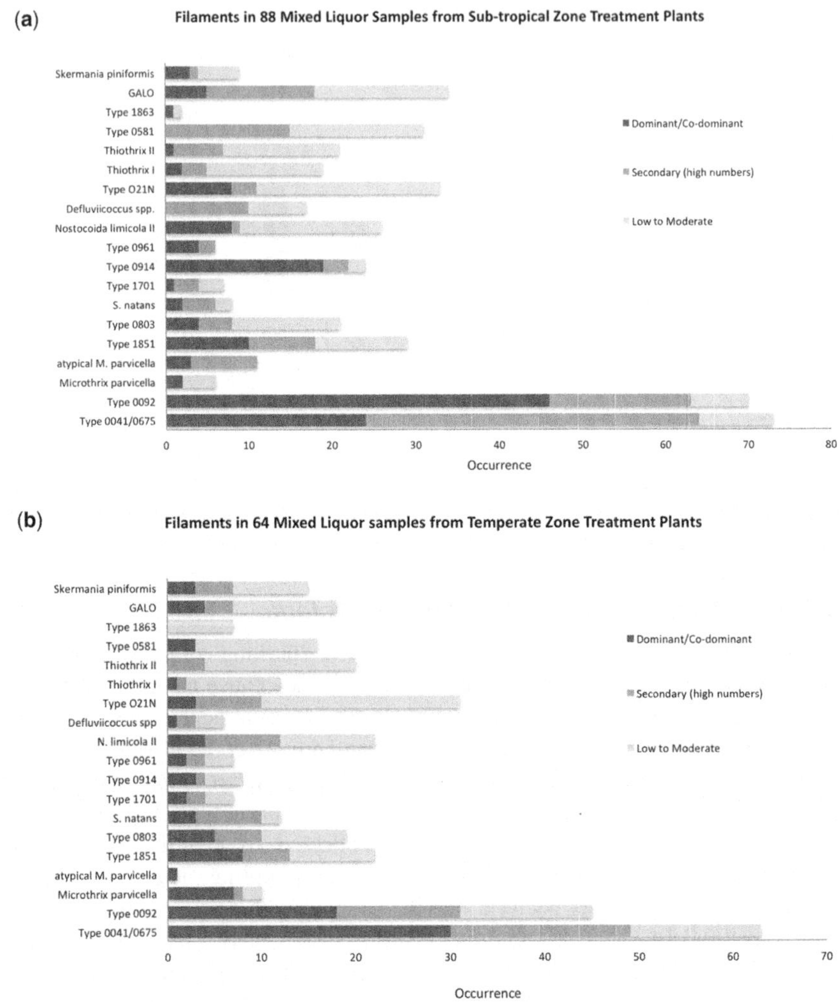

Figure 6.2 (**a**) Survey data from a variety of plant configurations showing major filamentous bacteria in bulking treatment plants sampled within sub-tropical climate zones (QLD and WA) in Australia, $n = 88$; (**b**) Survey data from a variety of plant configurations showing filamentous bacteria in bulking treatment plants sampled within temperate climate zones (VIC, NSW, SA and WA) in Australia, $n = 64$.

'*Candidatus* Sarcinathrix' spp. was more often seen in sub-tropical plants than temperate, particularly in those operating as oxidation ditches, while morphotype 0803 was more frequently dominant in temperate plants. A high frequency of occurrence of Type 0041/0675 was observed in all the plants and across all plant design types.

Based on microscopy assessment alone, '*Candidatus* P. breve' was abundant both within and between flocs in many of the samples from plants in both climate zones, corroborating results in other surveys (Blackbeard *et al.* 1986; dos Santos *et al.* 2015). Another survey that used FISH for treatment plants in eastern Australia by Speirs *et al.* (2009) suggested that three morphological variants of the '*Candidatus* P. breve' filament exist, and that they appeared simultaneously in most of the plants. Although '*Candidatus* P. breve' is common in Australian plants, more than one organism may be involved and so caution in interpreting its presence is recommended. Along with these common filaments, an unidentified morphotype resembling '*Candidatus* M. parvicella' was dominant in a number of samples. It was first observed in 2011 as dominant in mixed liquor of a bulking plant in South Australia and in foams of three plants in Queensland. Although FISH is required to see if it may be morphotype 0581 (rarely if ever seen in Australia) or '*Candidatus* Microthrix calida', a thinner version of '*Candidatus* M. parvicella', it is referred to in Figs 6.2 (a, b) as atypical '*Candidatus* M. parvicella'.

Resampling plants infrequently, in some cases years apart, suggested the filament community appeared to change little. As with the foaming data, no clear relationship was established between dominant filament morphotypes and particular plant configurations. The only correlation appeared to be between *S. piniformis* and '*Candidatus* M. parvicella' and '*Candidatus* P. breve' and ambient temperature, as these were more often dominant in the biomass in cooler, temperate zone plants. While '*Candidatus* M. parvicella' has been reported to prefer cooler climates (Seviour *et al.* 1990), *S. piniformis* has been shown to grow between 15°C and 31°C (Blackall *et al.* 1989). In warmer sub-tropical conditions, '*Candidatus* Sarcinathrix' spp. and '*Candidatus* P. breve' were more dominant in mixed liquors.

6.2.4 Do filamentous bacteria populations in the same treatment plant change over time and can we control them?

Our understanding of why filamentous bacteria grow excessively would be enhanced if full plant operating information together with precise information on which filamentous bacterial populations were present under each condition. With the emergence of 16S rRNA amplicon sequencing (McIlroy *et al.* 2015), the correlation of plant configuration and operation with whole community analyses and ecophysiology, this is now possible. However, this method of community fingerprinting has not been used widely for this purpose in Australia. Only rarely are full plant data available. Surrogate measures such as solids settling (SSVI, DSVI or SVI) are used to indicate the 'potential' for filamentous bulking, although this does not indicate which are the causative organism(s). Those extending from the floc surface (i.e. the best way to see if they are potential bulkers); can now be more definitively identified by FISH.

Once a plant is designed, built and commissioned, only a few operating variables lend themselves to manipulation. They include, dissolved oxygen concentration, sludge age, return activated sludge pumping rate and internal recirculation rates (mixed liquor or A-Recycle). Some parameters are not controllable, such as mixed liquor temperature, influent seasonal changes, surge loading under wet weather flows and to a large extent the influent characteristics.

Detecting increases in problematic filament populations before problems arise can be achieved by regular analyses and determining operational parameters and what a normal microbial community looks like as a baseline during periods of satisfactory biomass separation in the clarifiers. A baseline filament survey was conducted on a treatment plant in Queensland to establish their 'normal' filament communities and how changes to it may relate to environmental and operational parameters. The purpose of the study was driven by the utility wanting to control ongoing excess foaming problems. While operational data were sparse during the long-term study, we have attempted to suggest possible control methods.

6.2.4.1 Case Study – Excessive biological foam accumulation trouble shooting

This plant, located near Brisbane, was an oxidation ditch comprised of two equivalent treatment trains comprising bioreactors, reactor 1 (R1) and reactor 2 (R2). The plant is designed for 100,000 P.E. with an average flow of mainly domestic waste of 20 ML d^{-1}. The two treatment trains receive the same influent. It was designed for nitrogen reduction only, but some incidental EBPR occurs from the low nitrate effluent achieved, suggesting anaerobic conditions for part of the day. Filamentous bacteria from both reactors were surveyed between October 2003 and April 2005, and subsequently intermittent sampling between 2005 and 2014.

Mixed liquor samples ($n = 71$) were taken at weekly or fortnightly intervals. The dominant filaments in both R1 and R2, '*Candidatus* P. breve' and *T. jenkinsii*, were co-dominant initially but then *T. jenkinsii* became less abundant, as shown in Figs 6.3 and 6.4, which demonstrate the similarity between the filamentous communities in both tanks. The data also show regular increases in abundances of other filaments, GALOs, Type 0803 and '*Candidatus*, Sarcinathrix' spp. between July and November 2004, at which point poor settling occurred. Detailed plant operational data, including SVI were not available. From the accessible plant reports, the P-removal values were consistent and around the licensing limit, but effluent suspended solids exceeded license requirements more than once between July and November 2004.

The plant was surveyed again in 2009 during an upgrade for capacity and performance. This upgrade included replacing mechanical surface aerators with diffused air aeration divided into four aeration cells, and four mechanical mixers provided as two pairs within the ditch for recirculation of the MLSS. Dedicated aerobic, de-aeration and anoxic zones were thus provided and the sludge age reduced. Mechanical foam harvesters were used to manage foams caused by

G. amarae-like organisms. These foams had persisted since the earlier survey. The dominant mixed liquor filament was again '*Candidatus* P. breve'. Several other filaments were present in both the mixed liquor and foam at moderate (Type, 0581; Type 0041/0675), low or incidental (Type, 0803; *Kouleothrix* sp., Type O21N and GPCDR) numbers.

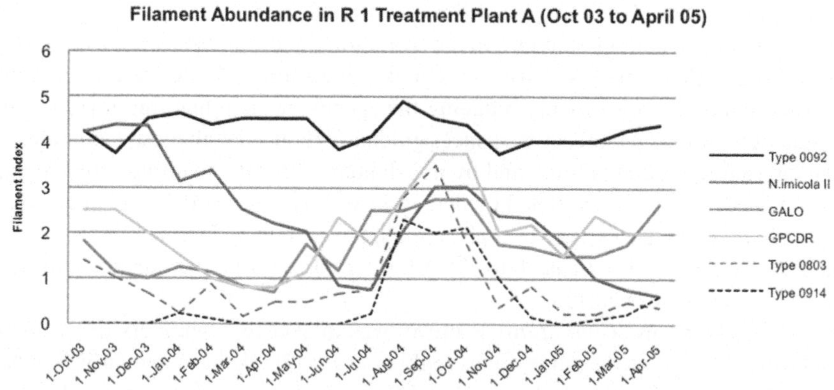

Figure 6.3 The filamentous bacterial population abundances rated from 1 to 6 in samples of mixed liquor from the aeration tank R1, Plant A, between October 2003 and April 2005 ($n = 71$).

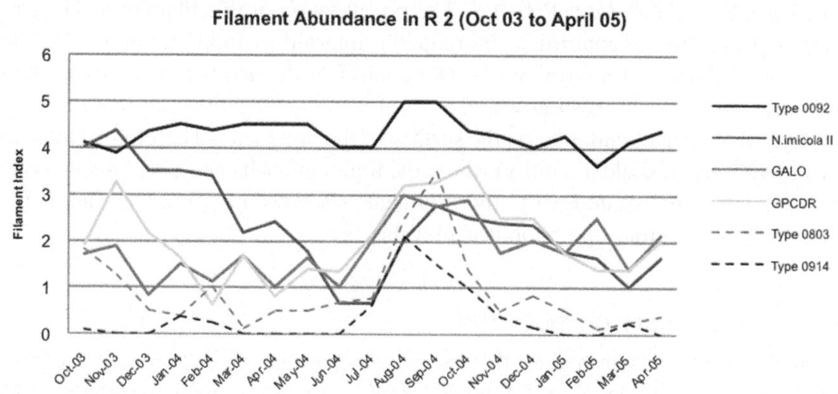

Figure 6.4 The filamentous bacterial population abundances rated 1–6 in samples of mixed liquor taken from the aeration tank 2, Plant A, between October 2003 and April 2005 ($n = 71$).

During commissioning, following the upgrade, foam accumulated on the bioreactor surface, despite mechanical foam harvesters operating 24 hr d^{-1}. The system was bulking with a SVI of 279–290 mL g^{-1}. Initially, it was suspected that Type 0041/0675, not a known foamer, was becoming entrained with the foam forming bacteria, as it was in high numbers (4/6) in foam and mixed liquor samples. DO concentration set points in the four aeration zones in each oxidation ditch were increased from 1 mg L^{-1} to 1.3 mg L^{-1} on 4 September 2009, but subsequent SVI values increased to 300 mL g^{-1} and above. Both mixed liquor and foam were dominated by '*Candidatus* P. breve' and Type 0041/0675 had fallen to low numbers (2/6). GALOs were still present in the foam.

The internal recirculation rate in the oxidation ditch was reduced progressively over the next three months. This was achieved by turning off internal recirculation mixers and then progressively reducing the speeds of the remaining mixers. The extent of foaming on bioreactor surfaces gradually reduced with these reductions of the internal recirculation rate, and by mid-January 2010 the foaming problem had diminished. This was paralleled by decreases in '*Candidatus* P. breve' abundance in the foam from 'excessive' to 'high', although it remained dominant in the mixed liquor. The SVI also decreased (to 175 mL g^{-1}) and major improvements to overall plant performance occurred.

It is unclear why '*Candidatus* P. breve' was in excessive numbers in the foam and mixed liquor at the same time. Jenkins *et al.* (2004) have shown it is more often dominant in plants removing N and P than in fully aerobic systems. It is not clear whether '*Candidatus* P. breve' can denitrify, but McIlroy *et al.* (2016) have shown from whole genome sequencing that it is a facultative anaerobic organism, able to generate energy by both fermentation and aerobic respiration. Given these observations a fully aerobic system (i.e. an aerobic selector) may be one approach to controlling '*Candidatus* P. breve'. The proliferation of this filament in the case here appeared to be 'controlled' by reducing internal recirculation rates to about 20 times ADWF and improving the DO control in the aerobic zone of the plant. However, its total elimination was not essential to manage its contribution to the biological foam accumulation on the surface of the bioreactor. Therefore, it appears that there is a threshold (quantity) where the filamentous bacteria become an issue, as has been demonstrated with other filaments (Petrovski *et al.* 2011). Below this threshold, satisfactory operation is achievable.

6.2.4.2 Summary of case study

From this study, we can suggest that the two key parameters that seemed to control filamentous bulking by '*Candidatus* P. breve' and foaming involving the GALOs and '*Candidatus* P. breve' were:

- The replacement of mechanical surface aerators with diffused subsurface aerators with better DO control and
- Decrease in the internal MLSS recirculation (i.e. reduction in sludge age)

Little change in the filament populations was observed throughout the study period, including the extended studies described in 6.2.4.1.

6.2.5 The future

In the past twenty years, enormous advances in molecular microbiological techniques have enabled ecosystems to be better understood and processes to be better controlled in other industries such as food, beverage production and medicine. While biological wastewater treatment systems have been operating for more than a century, largely as black boxes, it is now time to apply these tools in a similar way that other sectors do.

With the additional metabolic information that NGS DNA sequencing provides, it seems certain that the Eikelboom numerical nomenclature system will be replaced by providing all these filaments with valid microbiological names. This in turn will ensure that communication between microbiologists around the world will be more efficient, since ambiguity over which organism/s are being discussed will be replaced with certainty. This is beginning to happen. Recent suggestions of a universal, expandable database for these and other activated sludge bacteria, MIDAS (McIlroy *et al.* 2015) is an important initial step in this direction. By the time the next edition of this book is published, MIDAS should provide the common source of taxonomic/phenotypic information upon which all of these articles will be based. This online database (http://www.midasfieldguide.org) attempts to link the identity of the abundant organisms, as determined by 16S rRNA amplicon sequencing of full-scale plant communities, to their functional role there, based on shared taxonomic principles. Already Prof. Per Nielsen's group in Denmark are using amplicon sequencing to fingerprint full-scale plants (see http://dnasense.dk), and correlating them with plant operating parameters (Albertsen *et al.* 2013). The technology is cheap and rapid, and allows plant operators the power of profiling their communities at very regular intervals and seeing how they change in response to changes in plant operation. This approach is a game changer and a safe prediction is that it will become common practice in the next few years in plants all around the world.

6.3 AUSTRIA

N. Kreuzinger and N. Matsché

6.3.1 Intention of the investigation

Starting with the later 1990s, in Austria, operators of municipal wastewater treatment plants increasingly reported on bulking sludge phenomena occurring at their activated sludge plants. The increase in observations seemed to be linked with the increased construction of low loaded plants as well as the upgrade of existing plants to full nitrification and denitrification as required by Austrian law. Initially the observed cases were dealt with individually, but as reported cases increased, a decision on a comprehensive assessment of the bulking sludge situation in Austria was taken in order to provide a sound base for the development of possible, nationwide management strategies.

The goals of the assessments were to answer the following:

- How many municipal wastewater treatment plants are facing bulking problems?
- Is it a seasonal or a continuous problem?
- What are the main responsible filamentous bacteria responsible for bulking?

After a preparation phase of some months, the assessment phase could be started at the beginning of 2006.

6.3.2 Organization of the assessment

6.3.2.1 General

Due to the usual F/M ratio applied as a central design criterion for activated sludge plants in Austria and previous experience, it was expected that *M. parvicella* would be the dominant organism responsible for the bulking phenomenon and therefore a higher percentage of plants were expected to show bulking during the cold season. Another reason for choosing the cold season as the sampling period was a pragmatic approach linked to the shipment of sludge samples to the institution carrying out the microscopic investigations. As preliminary investigations showed, the typical duration of sample transport from abstraction until delivery to the laboratory was 2 to 3 days and neither sludge morphology nor staining behaviour was influenced by a potential degeneration of the sludge sample. Therefore, it was decided to go forward without special sample cooling during transport, making the handling easier for the operators and allowing them to conduct the survey during the first month of the year 2006.

Due to very limited resources available for the investigation and in order to maximize the feedback (sending back the supplied questionnaire as well as sludge samples) in regard to representative answers for the research questions, it was decided to minimize the effort for the operators necessary to deal with the assessment not only in regard to sample handling but in regard to the information required for filling the questionnaire too. On the other hand, that approach restricted the scientific potential of the assessment but was considered to sustain the core goals of the investigation.

6.3.2.2 Carrying out the investigation

In Austria, wastewater treatment plants are organized on an NGO level by so-called 'treatment plants neighbourhoods'. These structures are intended to provide a joint base for information exchange, discussion of common challenges and mutual support in their daily routine in case of necessity. Participation is on a voluntary base and there is no limit for the plant size in order to be part of the organization. So especially for smaller plants with only limited personal resources, the neighbourhoods are central and frequently the only means of professional training and information exchange. Typically, around 10 plants in closer vicinity are organized in a joint ‘neighbourhood’ headed by an elected ‘speaker’. All neighbourhoods of a province are coordinated by an ‘advisor’ usually coming from a larger treatment plant or administration. Together with representatives from research, administration and other relevant institutions, all advisors again are part of a steering board that is responsible for elaboration of supporting material, such as, training and education material and courses, provision of relevant technical and legal documents, organization of laboratory comparison tests, etc.

The infrastructure of the wastewater treatment plant neighbourhoods was used for the distribution of the survey material. During their regular autumn meetings, the detailed logistic procedure was presented and sampling scheduled for February the following year. Questionnaire, sampling containers and envelops for postal transport were distributed to the ‘speakers’ via the ‘advisors’ for final distribution to the treatment plants in January. The material was distributed to all members of the neighbourhoods regardless of their treatment system in order to keep the logistics simple. Operators with trickling filters or other than activated sludge municipal plants were instructed by an accompanying letter to ignore the survey.

6.3.2.3 Questionnaire

The questionnaire was deliberately kept simple, so even small plants without computer aided data documentation and only simple measurements of operational parameters were able to provide all the information requested. This of course put a limitation to a more elaborate and scientific data processing but increased the feedback, which was expected to be around 20% of the approached plants.

Beside basic contact information and the design capacity, the following parameters were requested:

- Volume of all aeration basins operated
- Mean value of daily wastewater flow during the month prior to sampling
- Mean BOD_5 concentration in the inflow to the plant for the month prior to sampling
- Mean temperature in the aeration tank for the month prior to sampling
- Mean value for TSS in the aeration tank for the month prior to sampling
- Most recent value for mixed liquor suspended solids (MLSS) in the aeration tank
- Most recent value for SVI measured
- Most recent calculation of the sludge retention time (SRT)
- Month of the year with an SVI > 150 mL g^{-1}

The content of the questionnaire was discussed with selected operators representing smaller treatment plants in a preliminary study in order to make sure that the information could be provided without effort. Though initially considered, more elaborate parameters were not sought as it was not clear that plausible information could be expected.

6.3.2.4 Participation of WWTPs

At the time of the survey, 951 plants of all size are participating in the wastewater treatment plant neighbourhoods and were provided with sampling containers and questionnaires. Starting with February 2006 till April, 645 questionnaires and 625 sludge samples were sent back, equalling to 68% of all plants and >80% of the activated sludge plants. As some of the plants represented others then activated sludge plants, data from a total number of 582 WWTPs finally was used for further processing.

The unexpectedly high participation led to the situation that not all sludge samples could be investigated by microscopy and staining. It was decided to use all the questionnaires for statistical and descriptive considerations, but only sludges with an SVI > 120 mL g^{-1} were investigated by staining and microscopy for detailed examination of the filamentous bacteria content; 225 mixed liquor samples were accessed by microscopy and staining. The SVI criteria of 120 mL g^{-1} was applied because it is the typical design criteria for secondary clarifiers in Austria and represents a solid distance to the criteria of 150 mL g^{-1} for the definition of bulking sludge.

6.3.2.5 Categorization of plants and sludges

For statistical analysis and practical handling, the information obtained from the questionnaires was clustered, for example, according to plant size, SVI and filament classes.

Table 6.1 gives an overview of the number of participating WWTPs grouped for size and SVI. Size classes follow the classification used for the definition of treatment requirements in the Austrian legislation, SVI classes are based on equidistant numbers of about 30 mL g^{-1} around the 120 mL g^{-1} and 150 mL g^{-1} criteria discussed above.

Table 6.1 Number of WWTPs participating in the survey according to size classes (P.E. used and SVI classes chosen).

			P.E.(COD_{120})				
SVI [mL g^{-1}]	**≤500**	**≤1000**	**≤5000**	**≤50,000**	**≤50,000**	**total**	**%**
≤75	2	2	35	59	12	110	19
≤100	3	6	53	67	10	139	24
≤120	4	6	48	52	10	120	21
≤150	3	8	42	37	15	105	18
≤150	5	14	39	43	7	108	19
	17	36	217	258	54	582	

The level of participation above 1000 P.E. reflects the size distribution of Austrian WWTPs and is typical for ambitious operators. Participation of plants >1000 P.E. is higher than 95%.

6.3.3 Methodology

6.3.3.1 Microscopy

The samples of mixed liquor sent to the Institute together with the questionnaire were examined microscopically as fast as possible. In a total number of 225 samples with an SVI above 120 mL g^{-1}, the following aspects were examined:

- Floc structure
- Filament index (Eikelboom)
- Identification of dominating and accompanying filaments
  - Phase contrast microscopy
  - Gram staining
  - Neisser staining
  - Typification according to Eikelboom
- Documentation with digital microscopic photographs
  - Floc structure and filament index
  - Dominant filaments (non-dominant filaments)
- Samples were fixed and frozen (for further DNA extraction)

A detailed description of the applied microscopic methods for the determination of the filaments can be found in Eikelboom (2000, 2002).

For the determination of the dominating filaments (DF), one filament type out of the observed species considered the most frequent and/or most responsible for the filament index or SVI respectively in the corresponding sample was identified by subjective judgement. A maximum of two dominant filament types was documented. As non-dominant or accompanying filaments (AF) all those filament types were characterized, which appeared in minor abundance and/or had a lower influence on the SVI.

Bias due to false characterization by microscopy and staining cannot be excluded but due to the high number of samples and the typical morphology that is rather easy to determine for the most abundant filaments, the final results should not be influenced significantly by wrong phenotypic classifications.

6.3.3.2 Parameters for analysis

As mentioned above, information on the following parameters was obtained directly from the questionnaire:

- Volume of all aeration basins operated
- Mean value of daily wastewater flow during the month prior to sampling
- Mean BOD_5 concentration in the inflow to the plant for the month prior to sampling
- Mean temperature in the aeration tank for the month prior to sampling
- Mean value for TSS in the aeration tank for the month prior to sampling
- Most recent value for mixed liquor suspended solids (MLSS) in the aeration tank

- Most recent value for SVI measured
- Most recent calculation of the sludge retention time (SRT)
- Month of the year with an SVI > 150 mL g^{-1}

These parameters allowed the calculation of additional parameters as:

- Mean volumetric load in the month before sampling (kg BOD_5 m^{-3} V_{AT} d^{-1})
- Mean sludge load = F/M ratio in the month before sampling (kg BOD_5 kg^{-1} MLSS d^{-1})
- Actual loading of plant (based on 60 g BOD_5 PE^{-1})

6.3.3.3 Data plausibility

Possibilities for the verification of the received data material were rather limited. A visual evaluation of the questionnaire data was applied for all parameters. By this means, rough mistakes and errors in the questionnaires, such as decimal point faults could be corrected. If not obvious, the identified implausible entries from the questionnaire were checked by directly contacting the operator and discussing the data provided.

As one example of a few routines applied, data from the questionnaire was used to estimate the actual loading of the plant based on 60 g BOD_5 PE^{-1} (=PE_{60}) and compared with the design capacity. Deviation of the calculated PE_{60} from the corresponding PE_{design} higher than the standard deviation of all correlations was used as a criterion to get in contact with the operator and discuss the data provided. This simple comparison could give at least some information, if other data, such as sludge age (SRT) or MLSS concentration that cannot directly be validated could be trusted. Additionally, the MLSS, SRT and F/M ratio provided for the last month prior to sampling was compared to data from the annual reporting within the neighbourhoods.

For a rough verification for the rather subjective aspects in microscopy, derived filament index classes were compared to the SVI provided, as shown in Fig. 6.5.

6.3.3.4 Data processing and statistical analysis

The information from all questionnaires ($n = 645$) was transferred to MS-Excel in a first step, evaluated and corrected and finally the results from the microscopic analysis (occurrence of filaments, filament index, dominant filaments, etc.) was added.

Particular datasets were flagged and categorized (e.g. all activated sludge plants) for further consideration. The final data sheet was transferred to SPSS (now: IBM SPSS statistics) and descriptive and more elaborated statistics was done by this software. After a basic data analysis for normal distribution, spikes and outliers, analysis of correlations and component analysis were done for the following basic hypothesis and questions:

- Is there a correlation between occurrence of bulking and plant size?
- What is the most dominant filament responsible for bulking?

- Is there a seasonal fluctuation in the bulking occurrence?
- Is there a correlation between F/M ratio and bulking/SVI?
- Can a correlation between M. parvicella bulking and F/M ratio be seen?
- Do particular filaments typically lead to more severe bulking (higher SVI)?

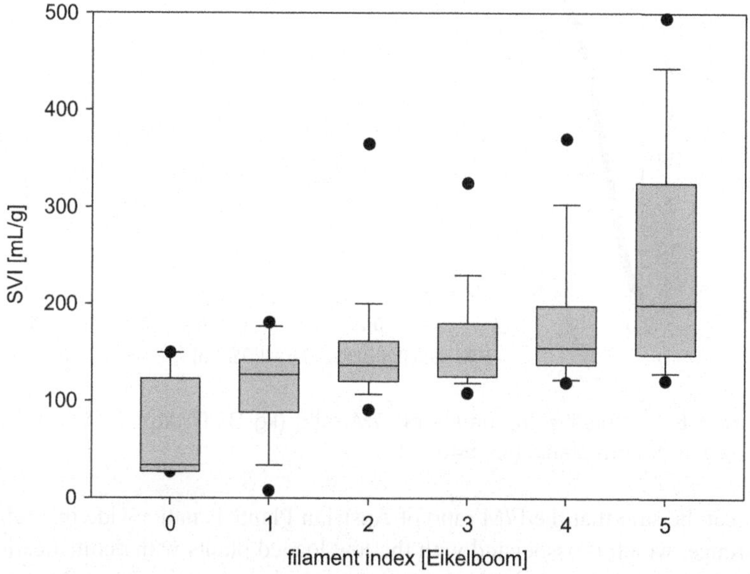

Figure 6.5 Plausibility of subjective grouping of the filament index and the SVI ($n = 225$).

6.3.4 Results

6.3.4.1 F/M ratio of the treatment plants

The actual loading of the plants based on 60 g BOD_5 P.E. $^{-1}$ d^{-1} at the time of sampling was calculated from the mean BOD loading of the month prior to sampling and was used for the calculation of the F/M ratio together with the MLSS during sampling in order to get an as close as possible value that can be provided by all plants without scientific elaborations. The results of that calculation are shown in Fig. 6.6 and represent a nationwide overview of the loading situation of treatment plants in Austria. Results above 1 kg BOD kg^{-1} MLSS d^{-1} are considered as not plausible due to wrong data, but were only appearing for singular plants ($n = 6$).

The 50% percentile of the F/M rations is calculated to be about 0.04 kg BOD kg^{-1} MLSS d^{-1}, the 75% percentile 0.075 and the 90% percentile was 0.137 respectively. The most frequent F/M ratio obtained was around the median value of 0.04 kg BOD kg^{-1} MLSS d^{-1}.

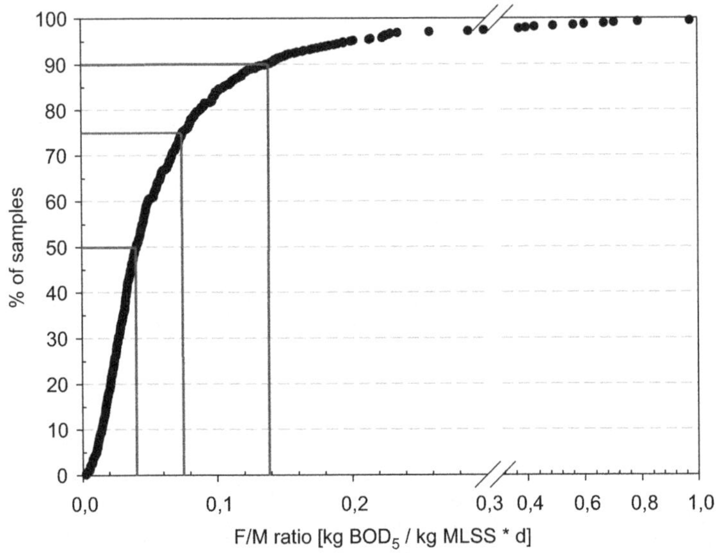

Figure 6.6 Cumulative frequency of F/M ratio (kg BOD_5 kg^{-1} MLSS d^{-1}) in 540 Austrian treatment plants ($n = 540$).

It can be said that the F/M ratio of Austrian Plants is nationwide in a relatively low range, which is associated with the low loaded plants with denitrification and simultaneous aerobic sludge stabilization typical in Austria. The F/M data were used for further investigations on, for example, a correlation with SVI.

6.3.4.2 Sludge volume index (SVI)

The SVI considered for this investigation was the actual value measured on the day when the samples for the microscopic investigations were taken. In Fig. 6.7 the cumulative curve of the collected SVI data of all plants is shown.

The 50% percentile of the SVI was 106 mL g^{-1}, the mean value 124 mL g^{-1} respectively. The SVI was below the value of 120 mL g^{-1} (used for design purposes of secondary clarifiers) in about 65% of the cases. Bulking sludge with an index above 150 mL g^{-1} occurred in approximately 20% of the plants. This corresponds well with the information provided for the seasonal occurrence of bulking (see paragraph 6.3.4.5). As described above, the general conditions for the sampling were intentionally chosen for expected higher SVI values during spring time in February and March. The results of this investigation in respect to the SVI therefore reflect a 'worst case' situation for Austria.

Samples showing an SVI > 120 mL g^{-1} further used for microscopic analysis are indicated by a dotted line in Fig. 6.7. The number corresponds to the 225 samples that were stained and investigated by microscope.

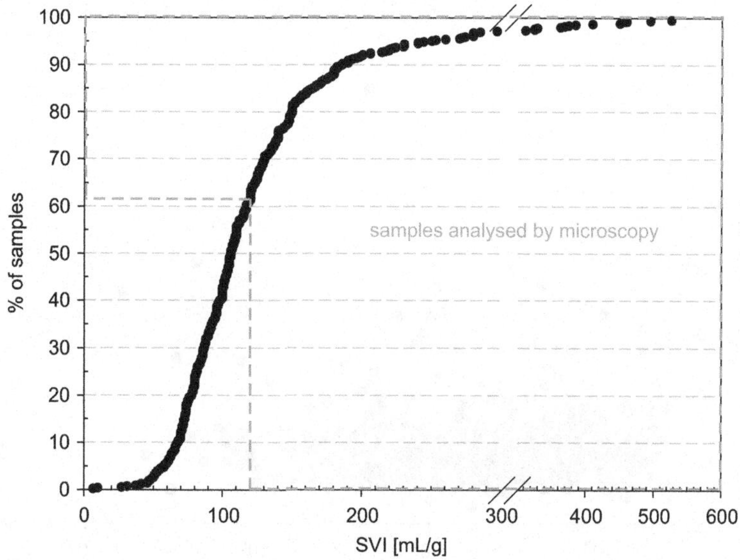

Figure 6.7 Cumulative frequency of the SVI of Austrian treatment plants (*n* = 582).

6.3.4.3 Sludge volume index and plant size

A general question of interest for the survey was, whether the bulking phenomenon is accumulating at specific plant sizes, as WWTPs of similar size often have realized very similar process engineering and plant configurations, such as pre-flotation units, pre-sedimentation; similar nitrification/denitrification concepts, control strategies, etc., which are different to other sized plants.

Figure 6.8 shows the results obtained for the correlation of plant size and observed SVI. It can be clearly seen that there is no correlation between the occurrence of bulking and the SVI respectively with the plant size. Detailed statistical analyses revealed that the occurrence in size groups (1000 and 2500 P.E. used as cluster size) showed no significant difference in the frequency of bulking. So, bulking occurs in all plants sizes.

6.3.4.4 Occurrence of filament types

Mixed liquor (225 samples) was analysed microscopically as described above. The presence of filaments in the investigated samples was documented by microphotography and the type of filamentous bacteria determined according to Eikelboom (2000, 2002). The results of this assessment are summarized in Table 6.2, however, only the most frequently occurring organisms in regard to the index dominating filament (DF) and mayor accompanying filaments (AF) were accessed. Single filament types with an abundance of below 10% are summarized to 'others'.

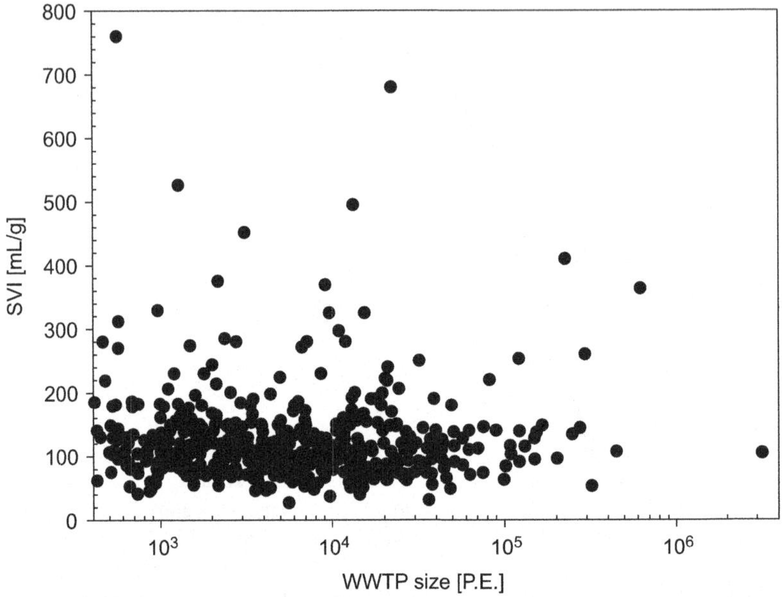

Figure 6.8 SVI at Austrian WWTPs sorted for the size of the plants ($n = 582$).

Table 6.2 Occurrence of filamentous bacteria in Austrian WWTP (base for Figs 6.5 and 6.8).

	SVI >120(n)	SVI >120 (%)	DF (%)	AF (%)
M. parvicella	186	84	73	10
Type 0041	179	81	23	58
N. limicola	103	46	12	34
NALO	88	40	15	25
Type 1851	66	30	8	22
Type 021N	57	26	9	17
Type 1701	45	20	4	17
Type 0092	44	20	8	12
Others	48	22	12	10

Note: DF = characterized as dominant filament AF = identified as accompanying side filament; NALO = *Nocardia amarae*-like organisms. Data in % of $n = 225$

In Fig. 6.9, all filamentous organisms observed as dominant and accompanying side filaments in Austrian WWTPs with an SVI above 120 mL g^{-1} are shown as percentage occurring in the 225 samples. The information corresponds to Table 6.2, but the category 'others' is split up into the individual filament types.

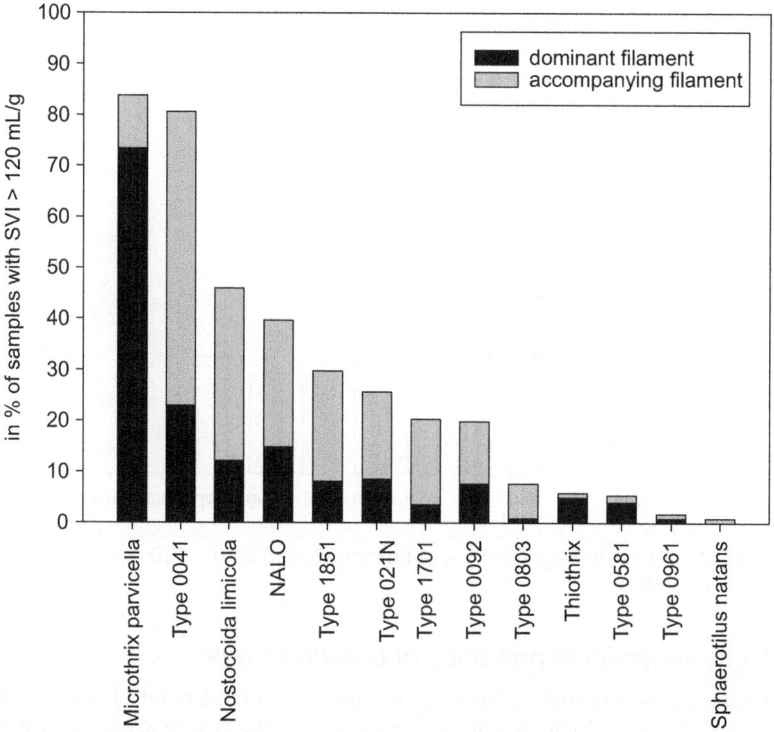

Figure 6.9 Filamentous bacteria in Austrian WWTP (main- and side-filaments).

The most frequently observed filament types are M. parvicella and Type 0041. They were found in about 85% of all plants with an index above 120 mL g^{-1}. However, the significant difference between these two organisms is their different significance and influence on the SVI. The investigations showed that only 23% of the sludges had Type 0041 as dominant filament; M. parvicella on the other hand was responsible for bulking in 73% of the cases (Table 6.2 and Fig. 6.9). This is a clear indication of the importance of M. parvicella for bulking problems in Austria. In 45% of the cases M. parvicella was the only filament followed by 16% with Type 0041 and 10% with NALOs. The rest splits itself into *Nostocoida limicola* (8%), Type 1851 (6%), Type 021N (6%), Type 0092 (5%), Type 1701 (3%) and Type 0803 (1%). The remaining Types could only be observed in a few cases.

The result of the investigations with the dominating M. parvicella in the examined samples from February and March coincides quite well with the seasonal appearance of bulking in the majority of the WWTPs (see Fig. 6.10). Even possible errors in the identification of the filaments are of minor importance for the results, since the dominating filaments could be detected very well with the help of staining and the typical morphology of these filaments.

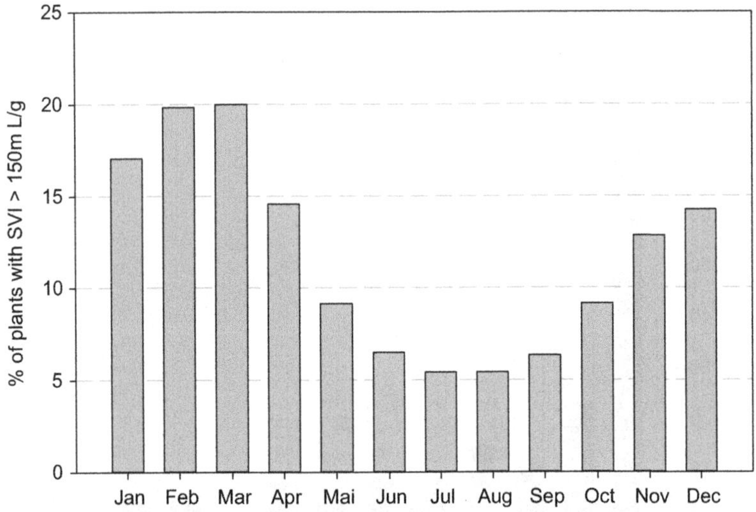

Figure 6.10 Seasonal appearance of bulking sludge (SVI > 150 mL g^{-1}) in Austrian WWTPs ($n = 582$).

6.3.4.5 Seasonal appearance of bulking sludge

The seasonal appearance of bulking was another central point of interest of the whole investigation. In order to get this information, the treatment plants were asked in the questionnaire for those months when the SVI exceeds values of 150 mL g^{-1}. The results of this investigation are presented in Fig. 6.10 as a percentage of 582 evaluable answers.

The results indicate a significant trend in the occurrence of bulking in Austria. In the months of February and March, bulking occurs in 20% of WWTPs. This 'spring maximum' also reflects the results of the SVI determination given in Fig. 6.7 which were collected in February and March. During summer, the number of affected plants is reduced to about 5%. This seasonal trend and the general situation in Austrian WWTPs suggest that bulking is most probably caused by M. parvicella which is the dominant filamentous organism in low loaded plants during spring.

6.3.4.6 F/M ratio and SVI

Based on the hypothesis and confirmed by the investigations that M. parvicella is the predominant filament causing bulking in Austrian plants, a possible correlation between F/M ratio and the occurrence of bulking (via SVI) was postulated because M. parvicella is known to prevail under low loading conditions. For that reason, F/M ratio and SVI were correlated for 582 plants and compared to a subset of data representing samples with an SVI > 120 mL g^{-1} and M. parvicella as the predominant filament (graphical result in Fig. 6.11).

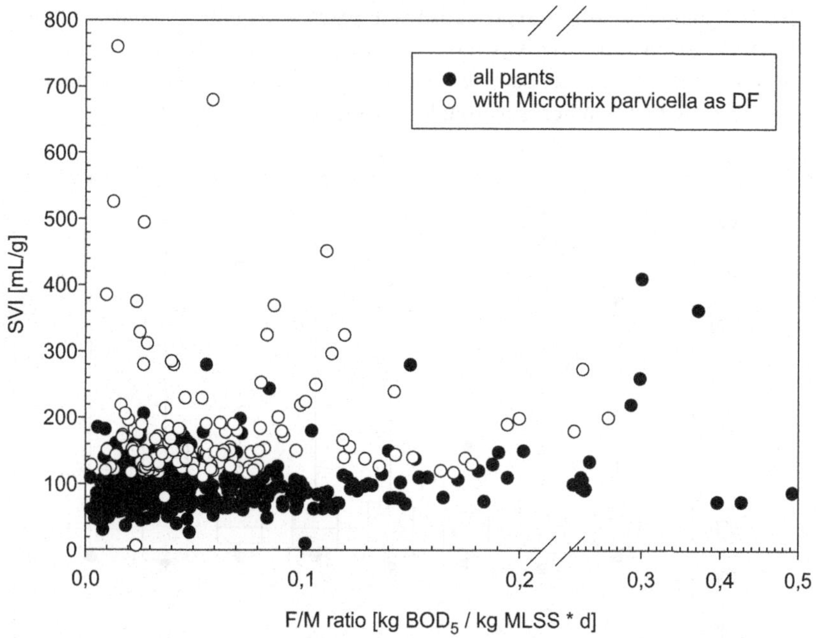

Figure 6.11 All data: correlation between F/M ratio and SVI ($n = 582$); empty dots: subset of data with SVI > 120 mL g^{-1} and M. parvicella as dominant filament ($n = 163$).

No statistically significant correlation could be found between the F/M ratio and the SVI nor between the occurrence of M. parvicella and F/M ratio. This indicates, that under the same F/M ratios and – as sampling took place at the same time – under the same lower than average temperature conditions (mean value over all samples $T = 7.9°C$), M. parvicella not mandatory dominates the biocenosis. F/M ratio and temperature definitively are relevant factors favouring the growth of M. parvicella, but obviously are not the only significant factors and WWTPs can be operated with a low SVI without M. parvicella during lower temperature conditions. There was even no significant correlation between occurrence of M. parvicella and the phosphorus precipitation agent applied (Fe vs. Al; additional information collected *a posteriori* for 25 randomly selected plants).

6.3.4.7 Influence of filament types on SVI

Due to the different morphology of the various filaments it could be expected that their effect on the SVI could be different. For that reason, measured SVI data was linked with the dominating filaments in the respective samples.

Statistical analysis showed no significant difference in the SVI measured for the different filament types. This is indicated too by Fig. 6.12. As a cut-off criterion of SVI 120 mL g^{-1} was chosen (see above), no information on the occurrence in samples below that number is available that influences the results.

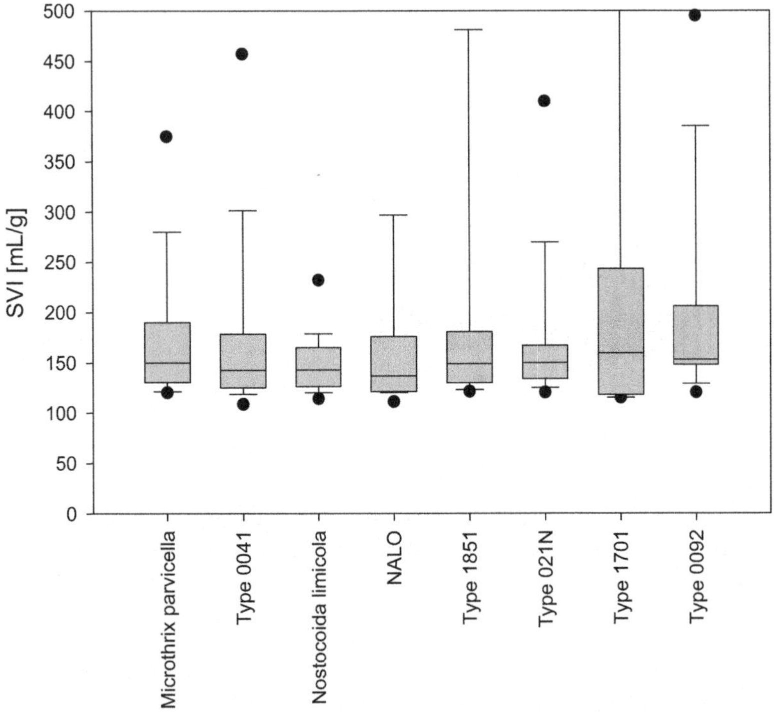

Figure 6.12 Dominant filaments in samples with an SVI > 120 mL g^{-1} and span of associated SVI observed ($n = 225$).

6.3.5 Summary and conclusion

The presented investigation was carried out in order to answer the following core questions on a nationwide scale in Austria:

- Does bulking occur in all activated sludge plants independent of their size?
- What is the most dominant filament responsible for bulking?
- Is there a seasonal fluctuation in the bulking occurrence?

For logistical support of the investigations, the structure of the 'WWTP – neighbourhoods', a NGO association of WWTP operators was used. Out of 951 plants addressed, 645 filled in questionnaires and 625 sludge samples were sent back. 225 samples with an SVI > 120 mL g^{-1} were analysed by microscopy.

Results identified M. parvicella as the most relevant filament responsible for bulking phenomena in Austria. Bulking predominantly occurs in February and March during times with lower wastewater inflow temperatures and 20% of Austrian WWTPs were affected. In about 80% of those cases M. parvicella is the dominating filament. The lowest abundance of bulking is observed in summer (July, August).

Basically, all plants independent of size and F/M ratio (and SRT) can be affected by bulking and no significant correlation between those factors and the SVI could be found.

6.4 BELGIUM

A. Fenu, J. Deurinck and S. Van Damme

6.4.1 General situation

Aquafin is responsible for expanding, operating and pre-financing the wastewater treatment infrastructure in Flanders. It operates more than 300 WWTPs, of which nearly 250 have an activated sludge system, 1500 pumping stations and the related sewer network. In total, a population of about 6 million people equivalents is connected to Aquafin infrastructure. The WWTPs have been built in different periods and use different technologies and process schemes, known in their times to be the state of the art of WWTP technology. Most WWTPs are equipped with screw pumps and a pre-treatment compartment, whose complexity depends on the WWTPs size. In our experience, influent screw pumps provide a considerable oxygenation of influent flow.

All activated sludge systems in Flanders (98.9% advanced and 1.1% secondary treatment) are very low loaded (design load of 0.05 g BOD kg^{-1} MLSS d^{-1}) to ensure nutrient removal of at least 80%. Due to dilution in the sewer system, the actual influent loads are even lower. Total influent nitrogen concentration is typically in the range of 30–45 mg TN L^{-1} and the BOD/TN ratio is generally in the range of 2 to 3. A combination of more factors raises a challenge in meeting the effluent quality standards imposed by the actual legislation: (i) urban waters and drainage waters are collected in combined systems, (ii) influent loads dilution is strongly due to rain events in central Europe, (iii) Aquafin WWTPs are designed to biologically treat 6 times the dry weather flow, (iv) stringent legislation prohibits the discharge of vast amounts of – mostly concentrated – industrial wastewater, (v) groundwater infiltration is common in winter time causing further dilution and inflow of iron salts.

The vast majority of process schemes make use of oxidation ditches with intermittent aeration. The length of the aerobic and anoxic period is in most cases online controlled by NO^{3-} and/or NH^{4+} sensors. About 75% of the activated sludge systems is also equipped with an anoxic selector, designed at a hydraulic retention time of 25 min at dry weather flow. Recycle flow to these selectors can be controlled by the use of by-pass valves. A fine-tuning of this flow would allow the establishment of a full anoxic compartment, controlling at the same time the hydraulic retention time in the selector. Settleability in WWTPs is proven to slightly improve when anoxic selectors are placed in Aquafin WWTPs (Callens, 2007). However, due to the initial sewage oxygenation provided by the influent screw pumps, dissolved oxygen (DO) is regularly measured at the entrance of the selector tanks. Some filamentous bacteria can be selectively favoured by the low DO levels present in the selectors but, as mentioned by Jenkins *et al.* (2004) and Casey *et al.* (1993), anaerobic, anoxic and aerobic selectors have been proven unable to prevent low food/micro-organisms filament proliferation. Therefore, despite the broad use of selectors within Aquafin, bulking sludge is still a major issue in the daily operation of activated sludge systems within Aquafin.

6.4.2 The M. parvicella problem

Aquafin started in 1994 with activated sludge examination by conventional microscopy. The laboratory is accredited by the Belgian Accreditation Body as a testing laboratory (1999), according to the requirements of NBN ENISO/IEC 17025:2005 for the Standard Microscopic Research of Activated Sludge.

Filamentous bacteria are identified by use of the Eikelboom and van Buijsen dichotomous key (Eikelboom & van Buijsen, 1983). Type 0041 and Type 0675 are combined into one group Type 0041/Type 0675; *Nostocoida limicola* I, II and III into group *Nostocoïda limicola* and *Thiothrix* I and II into group *Thiothrix spp.* Filaments that do not match morphological or staining filament characteristics described in Eikelboom and van Buijsen (1983) or Jenkins *et al.* (2004) are labeled *unidentified type*. Relative abundance of a specific type is indicated with the terms dominant (i.e. most numerous) and subdominant (i.e. two times less than the most numerous type). In some cases, more than one (sub-) dominating type occurs. Absolute abundance is assessed by reference pictures resulting in five filament scores (from *few* to *excessive*). The terms *lower limit* and *upper limit* are used to better indicate the evolution.

In order to characterize the filamentous bacteria present in Flemish WWTPs, the results of all analyzed samples (3933 activated sludge samples and 431 scum samples) during a period of 10 years (2003–2013) are discussed below. No distinction is made in terms of motivation for microscopic analysis (*inter alia* follow up, filamentous problems, problems with toxicity).

6.4.2.1 The filamentous types in the activated sludge

Microthrix parvicella is the most frequent filamentous type in the activated sludge of Flemish WWTPs (Fig. 6.13). This is mainly due to the use of low-loaded activated sludge systems with intermittent aeration and extended anoxic conditions, at moderate to low temperatures (Knoop & Kunst, 1998). Other low F/M bacteria are also common: Type 0041/Type 0675 is the the second most and Type 0092 the third most detected filamentous type. The predominance of Type 0041/Type 0675 over Type 0092 can be explained by the broad implementation of selector tanks (or *pre-mixing zones*) which favors the former (Eikelboom, 1999). Type 0041/Type 0675 is known to bridge smaller flocs, increasing the floc size. Type 0092 creates open areas inside the flocs, 'spreading' the flocs in the water phase. Therefore, the impact of both types on sludge settleability is limited, in contrast to M. parvicella.

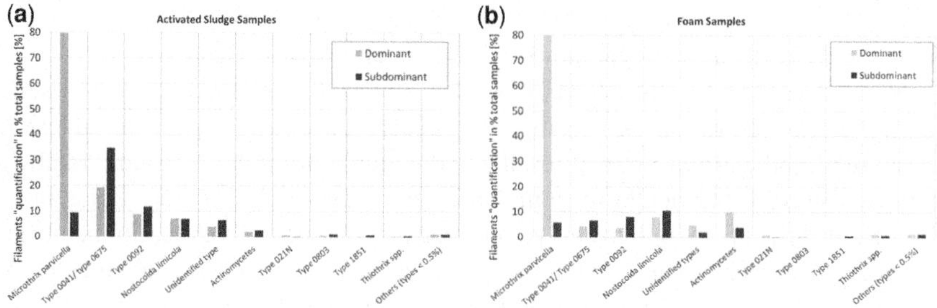

Figure 6.13 Filaments per type and quantification, and reported as % of the total amount of samples. (**a**) Activated sludge samples dataset. (**b**) Scum samples dataset.

Figure 6.14 shows a seasonal effect in the competition between M. parvicella, Type 0041/Type 0675 and Type 0092. A proliferation of M. parvicella at water temperatures below 15°C and a switch to Type 0092 as the temperature rises above this value is common to other European cases (Eikelboom, 1999).

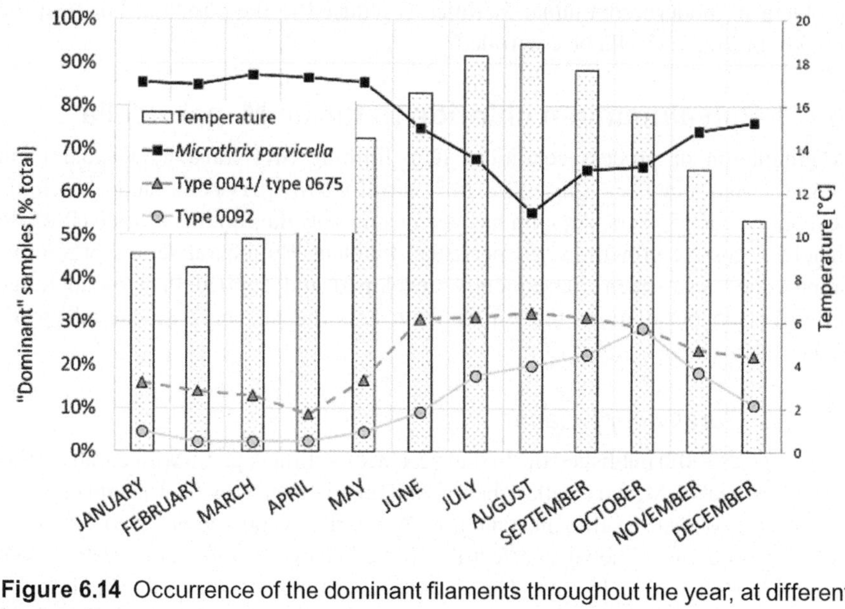

Figure 6.14 Occurrence of the dominant filaments throughout the year, at different temperatures.

6.4.2.2 The filamentous types in the foam

As was the case with the activated sludge, M. parvicella is by far the most abundant filamentous bacteria in sludge foam (Fig. 6.13). Contrary to the activated sludge, *Actinomycetes* is detected second most followed by *N. limicola*. Indeed, the Gram-positive characteristics of these three types are linked with hydrophobicity, which is known to cause foaming. This is in contrast to non-foaming Gram-negative filamentous bacteria like Type 0041/Type 0675 and Type 0092.

In most cases, the growth of *N. limicola* can be explained by the presence of higher concentrations of readily degradable substrate (e.g. from industrial discharges and carbon source for denitrification). *Actinomycetes* are likely to compete with M. parvicella for long chain fatty acids (LFCA) from lipids in the municipal influent, as both profit from high hydrophobicity. In Aquafin membrane bioreactors, which receive extended aeration, known to be detrimental to M. parvicella, *Actinomycetes* is detected to be dominant in the foam. Because low concentrations of *Actinomycetes* in the activated sludge can cause severe foaming, microscopic foam analysis is performed to implement effective process measures.

6.4.2.3 Conclusion

A combination of low loaded activated sludge systems, low temperature (<15°C) during the winter months and intermittent aeration for nutrient removal makes M. parvicella the most abundant filamentous bacteria in both activated sludge and foam in the Flemish WWTPs. This confirms Eikelbooms theory that the bulking problem in modern communal facilities is reduced to the question: how can the growth of M. parvicella be controlled?

6.4.3 Polyaluminium chloride to tackle M. parvicella

Manipulation of process conditions (e.g. limiting SRT, raising DO and foam removal) has not been successful in controlling M. parvicella sludge bulking. Therefore, for 15 years, Aquafin has been using polyaluminium chloride (PAC) to handle excessive growth of this persistent filament. Both curative and preventive approaches have been consequently developed and applied in a broad range of full-scale installations that suffer from foaming and bulking caused by M. parvicella.

6.4.3.1 Curative approach

Roels *et al.* (2002) published the first experiences within Aquafin with curative PAC dosing to tackle M. parvicella when abundant filament growth was endangering suspended solids (SS) effluent standards. A continuous dosage of 1.5 to 4.5 g Al^{3+} kg^{-1} MLSS d^{-1} was shown to affect the diluted sludge volume index (DSVI) and scum layer after 10 to 15 days. The abundance of filaments was reduced after 3 to 3.5 weeks.

As most municipal activated sludge systems in Flanders are operated at a sludge age between 15 and 25 days, a dosage of 2.0 g Al^{3+} kg^{-1} MLSS d^{-1} is applied whenever microscopic analysis suggests that a problematic dSVI value is caused by M. parvicella, as in Roels *et al.* (2002).

Figure 6.15 shows the effect of PAC on dSVI, filament score and effluent phosphate in the case study of WWTP Wommelgem (2007–8). A first period of 2.0 g Al^{3+} kg^{-1} MLSS d^{-1} dosing led to a decrease of DSVI from 140 mL g^{-1} to 75 mL g^{-1} in three weeks. The M. parvicella filament score dropped from 'very much' (> 2.5) to 'much' (1.5 < score < 2.5). However, due to a sudden switch to Fe-dosing, problems reoccurred two months later. The same dosage was applied, but after the initial phase of high dosage, dosing was prolonged at a lower level (~1.0 g Al^{3+} kg^{-1} MLSS d^{-1}) until a steady decline of filament score to the value 'moderate' (0.5 < score < 1.5) was observed. Complete phosphate elimination is observed at high dosage (i.e. PO_4-P < 0.3 mg L^{-1}), but as soon as Al^{3+}-addition is lowered, the phosphate values increase again.

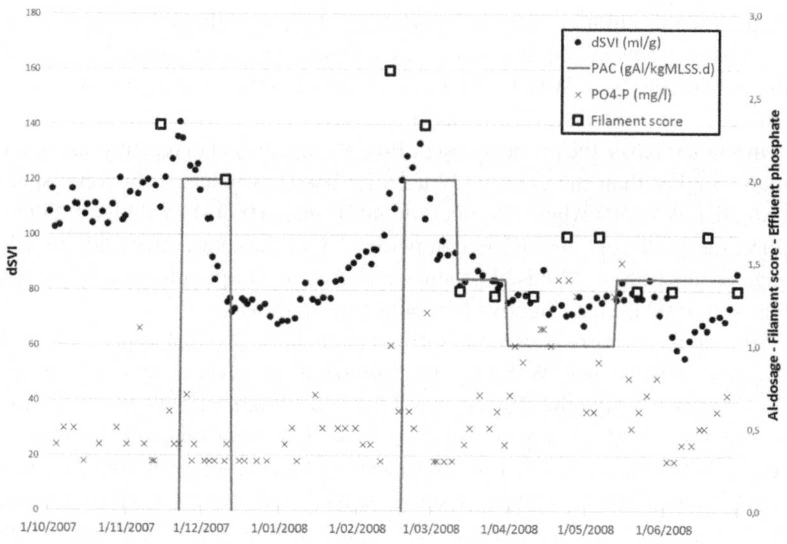

Figure 6.15 Effect of curative dosing on DSVI and filament score at WWTP Wommelgem.

Our experiences with this strategy can be summarized as follows:

- PAC dosage results in an effective decrease in DSVI and a significant reduction of the risk for sludge wash-out, provided that a sufficiently high dosage is used at the start. Dosage can be lowered (e.g. halved) as soon as DSVI drops to acceptable values. A sudden stop of dosing might result in rapid re-growth of M. parvicella.

- There is no different effect of PAC dosing if we compare (1) products of different suppliers or (2) products with a different Al-concentration.
- M. parvicella destruction can result in a slight increase of effluent suspended solids. In the case of sludge with a weak floc structure or phosphorus limitation it is to be evaluated whether additional dosing of polymer to entrap pinpoint flocs or even (temporary) ending of the dosage can be necessary. Therefore, microscopic follow-up is crucial during each treatment with PAC.
- In some cases, PAC is not effective. Firstly, in a Unitank system the above approach does not result in decreasing DSVI and/or deterioration of the condition of M. parvicella. Little improvement could be achieved by variation in dosage location, dosage regime or dosage amounts. Secondly, PAC turns out to be not always effective at WWTPs partially treating industrial wastewater containing long-chain fatty acids.

6.4.3.2 Preventive approach

The use of the curative approach indicates the possibility of using continuous PAC dosing at low concentrations as a preventive approach against M. parvicella bulking. Therefore, at WWTPs with chemical phosphorus precipitation, a switch from Fe-based products to PAC was considered.

On a molar basis, the product cost of PAC has decreased in recent years but it still remains higher than the commonly used Fe-based products. However, especially for smaller WWTPs where the operational (transport) cost of sludge treatment is higher, the total cost of PAC is competitive with the alternatives due to a lower sludge production of Al-based products. Therefore, if operational advantages can be proven, PAC is cost-effective for chemical P-removal.

Aiming for a correct comparison between both product types and to filter individual effects per WWTP, the following procedure was followed. All WWTPs within Aquafin that switched from a variable Fe-dosing to a variable but continuous PAC-dosing in 2012, 2013 or 2014 were selected. In none of the selected WWTPs was a curative approach applied. After excluding one WWTP because of a significant inflow of sludge washed out from a connected industrial WWTP (brewery), this led to a list of 11 representative WWTPs. M. parvicella has been observed as a (co-)dominant filament in these WWTPs in the past. For this selection the decrease in average DSVI between the first period (from 1/1/2011 until the switch) and second period (from the switch until 31/12/2015) are compared. Based on 3775 DSVI analyses over 5 years a drop of 14.9 ± 11.3 mL g^{-1} is noticed. With an average dosing of 0.80 ± 0.36 g Al^{3+} kg^{-1}MLSS d^{-1}, chemical precipitation was always well below the recommended dosing for curative action.

At first sight this improvement is rather small. However, the improvement turns out to be the strongest during high DSVI periods, when settling capacity of the secondary clarifiers is most critical. As illustrated in Fig. 6.16, DSVI values in winter time are mostly 20 to 30 mL g^{-1} lower in the case of continuous PAC-dosing.

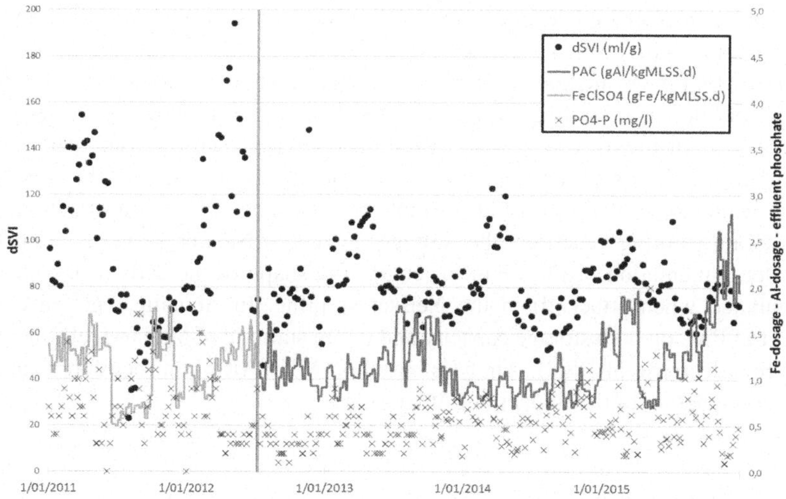

Figure 6.16 Effect of preventive dosing on DSVI at WWTP Zichem.

6.4.3.3 Conclusions

Although a rapid decrease in DSVI can be realised with a curative PAC dosing and a significant improvement is noticed with a preventive PAC dosing, this approach (1) still implicates a considerable operational cost and (2) is not always effective. To tackle these drawbacks and for a more constant follow-up of the M. parvicella evolution, for example at WWTPs with preventive PAC dosing, faster microscopic characterization techniques are desirable.

6.4.4 *Microthrix parvicella* monitoring: a revised methodology

6.4.4.1 The MICPARV method

When operational experience indicates activated sludge bulking or foaming at a specific WWTP, standard filamentous identification (Eikelboom & van Buijsen, 1983) is performed at first. If this comprehensive microscopic technique determines M. parvicella to be the dominant and problem-causing filament, a method called MICPARV is used for follow-up. Being less time-consuming, this approach enables us to increase the follow-up frequency, despite our growing number of WWTP's.

MICPARV is a microscopic analysis method focusing on the abundance and condition of M. parvicella. As this novel approach still uses both phase and bright-field microscopy techniques, a general overview of the biological population (other filaments, flocs, protozoa and metazoa, etc.) is guaranteed. In the case of disruption in microscopic image or turbid supernatant, a thorough microscopic examination is performed.

The MICPARV method is based on a quantitative and qualitative assessment of M. parvicella according to a well-defined reference framework. Assessment of the amount of M. parvicella is based on reference pictures combined with a detailed description of the impact on floc structure. Assessment of the condition uses evaluation criteria enlisted in Table 6.3. Aspects under consideration are: the Gram-staining reaction; the storage capacity (Neisser positive granules); the defragmentation of the filament length; the shape of the cells (cells become granulated when attacked) in the Neisser staining. In Aquafin experience, the effect of PAC on both storage capacity and Gram staining of M. parvicella appears before a decrease of DSVI. Analysis based on MICPARV can therefore enable a fast(er) detection of the impact of PAC dosing.

Table 6.3 Description of the criteria and terms used to evaluate M. parvicella condition.

Condition	Evaluation Criteria
Very good	• Strongly Gram-positive • Much Neisser-positive granules • Intact shape
Good	• Mostly Gram-positive • Frequently Neisser-positive granules • Intact shape
Moderate	• Mostly Gram-positive and frequent Gram-variable • Occasionally Neisser-positive granules • Granulated shape (Neisser staining)
Poor	• Mostly Gram-variable • Occasionally Neisser-positive granules • Granulated shape (Neisser staining)
Bad	• Mostly Gram-negative • Rarely Neisser-positive granules • Granulated shape

6.4.4.2 The real-time PCR method

The microscopic evaluation of M. parvicella by MICPARV, is a semi-quantitative method that requires manual interpretation and skilled microscopists. The method is – for example – sensitive to variations in activated sludge concentration. Moreover, facing a growing number of WWTPs, a quicker monitoring method is desired for continuous follow-up of M. parvicella growth, for example during preventive PAC dosing.

In the literature, Kaetzke *et al.* (2005) and Kumari *et al.* (2009) presented a method to quantify M. parvicella by real-time Polymerase Chain Reaction (qPCR). In 2012, Aquafin introduced a real-time PCR procedure (Vanysacker *et al.* 2014) consisting of the SYBR green assay and primers targeting 16S rDNA genes used to quantify the amount of M. parvicella.

After some optimizations aiming for automatization of the process and a more reliable sample homogenization, the following procedure was applied:

(i) a DNA-extraction consisting of a sample pre-treatment and the Magmax Total NA isolation extraction method, guaranteeing robust and semi-automatic extraction of the sludge DNA
(ii) a real-time PCR method, improved by using a custom designed plasmid as standard, to ensure the stability of the assay

Further automatization is implemented by use of a pipetting robot (Tecan Evoware). In 2014, the method was validated as conforming to ISO 3534–1.

The method results in a percentage of M. parvicella (% MP), based on the amount of M. parvicella DNA gained from the PCR and the total DNA extracted from the sludge sample. The measurement thus takes into consideration the activated sludge concentration.

Since PCR only targets one specific (filamentous) strain/type, this method does not provide an overview of the general activated sludge characteristics (flocs, other filaments, etc.). Despite its sensitivity for contamination and its significant operational cost, the technique is profitable when a large quantity of samples is to be analyzed.

6.4.4.3 Novel methods in practice

The real-time PCR information needs to be cross-linked with data deriving from microscopy methods (MICPARV) and more conventional monitoring tools, such as the DSVI. This is done with a dataset of 30 samples in Fig. 6.17. To be able to compare the semi-quantitative MICPARV method with these two quantitative approaches, a score (%) is calculated from the abundance and condition determined. Results obtained with both novel methods are comparable. The DSVI value by contrast shows a bad correlation with the two novel methods. Indeed, absolute DSVI values are strongly influenced by a broad range of parameters (e.g. inorganic fraction in the sludge, other codominant filamentous types, floc structure) differing for each WWTP. Compared to microscopy, this PCR-method requires totally different skills and equipment. The protocol is complex and has to be carried out very conscientiously to obtain reliable and reproducible results. Aquafin is currently upscaling this PCR-method in order to use it as a routine method instead of MICPARV. In addition, MICPARV will be used to check unexpected PCR-results.

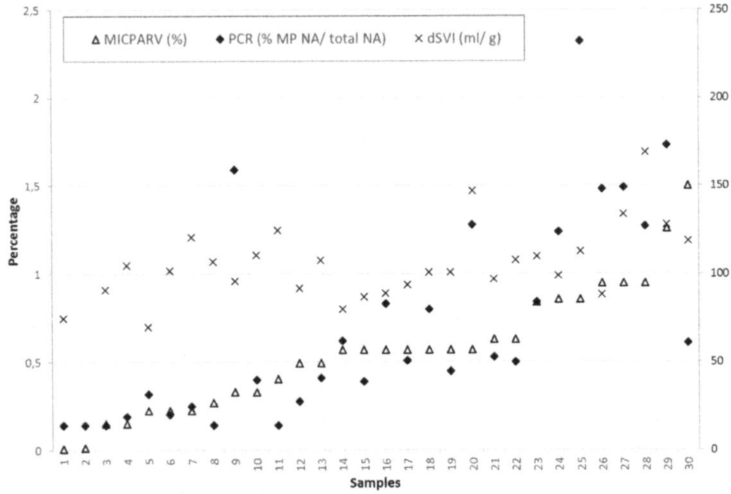

Figure 6.17 Comparison of the different follow-up methods; 30 samples are ranked on the MICPARV method score.

6.4.5 Conclusion

WWTPs in Flanders suffer periodically from severe sludge bulking caused by M. parvicella. More than 15 years of microscopic and operational data show that low temperature (<15°C) is the trigger for M. parvicella proliferation. Due to stringent nutrient removal standards, there is little possibility to affect two other decisive factors in the abundance of this filament: sludge retention time and variation in anoxic/aerobic conditions. Furthermore, selectors have not always proven to be effective tackling M. parvicella.

Aquafin developed successful strategies to both control and prevent sludge bulking with a combination of polyaluminium chloride (PAC) dosing and microscopic follow-up. To further reduce the operational cost of chemical dosing and standard microscopic filament procedures, two novel techniques are elaborated specifically aiming at M. parvicella scoring. Both the MICPARV method and the PCR technique allow us to analyse more samples.

6.5 CHINA

M. Yang, R. Qi and J. Wang

6.5.1 Introduction

By June 2015, 2149 municipal wastewater treatment plants (MWTP) have been constructed in Chinese cities with a total daily treatment capacity of 1.32×10^8 m^3, most of which were constructed in the last decade. With the promulgation of the revised edition of wastewater discharge standard (GB18918–2002) in 2003, which requires strict control of nitrogen and phosphorus, more and more MWTPs employed an Enhanced Biological Phosphorus Removal (EBPR) process, which represented as the anaerobic–anoxic–aerobic (A^2/O), oxidation ditch (OD) and inverted A^2/O (anoxic–anaerobic–aerobic) (Fig. 6.18).

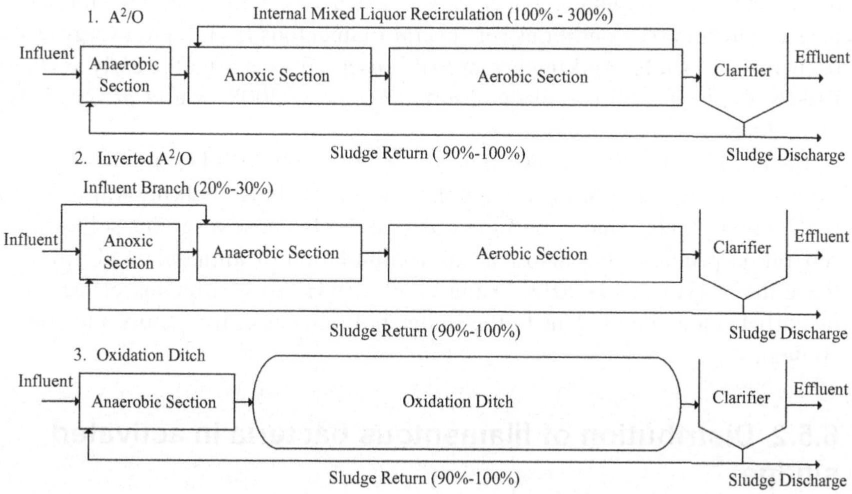

Figure 6.18 Diagram flowcharts of the three major EBPR processes in China.

It is estimated that these EBPRs accounted for over 60% of MWTPs in China, and such a proportion will be increased in the future due to the old ones upgrading (Jin *et al.* 2014; Ministry of Housing & Urban–Rural Development of the People's Republic of China, 2015). For instance, in Beijing, the treatment capacity of these three processes was 2.4×10^6 m^3, accounting for 71% of the total capacity. However, the various configurations resulted in different nutrients removal performance and related bacterial distribution. In comparison with the A^2/O system, the inverted one in the same MWTP exhibited better long-term phosphorus removal performance (0.2 ± 0.3 vs. 0.7 ± 0.7 mg L^{-1}), which should be attributed to the higher P release rate (0.79 vs. 0.60 kg P kg^{-1} MLSS d^{-1}) in the non-aerated compartments.

FISH analysis results showed that the abundances of the '*Candidatus* Accumulibacter phosphatis' population in the two systems were quite similar, with $19.1 \pm 3.27\%$ and $18.4 \pm 4.15\%$ of the total microbe population in the inverted and conventional systems, respectively. However the abundances of the whole polyphosphate accumulating organisms (PAOs) in the aerobic ends were quite different being 45 ± 4.18% and 35 ± 5.39%, respectively (Qi *et al.* 2012).

In addition, China is the biggest developing country across the widest longitude (73°33′E–135°05′E) and latitude (3°51′N–53°33′N), leading to various fluctuation ranges of influent flow rate, temperature and nutrient concentrations in MWTPs. Even with similar process configurations, the plants performed with different nutrient removal efficiencies and operational problems, particularly of sludge bulking phenomena. For instance, the big northern cities have a more drastic change of pollutant concentrations and a longer and colder winter, which may create suitable conditions for several filamentous bacteria to overgrow. So occurrence of sludge bulking began to be observed soon after the early northern EBPRs were put into use in the 2000s (Wu *et al.* 2008; Xie *et al.* 2007; Liu *et al.* 2008).

Recent investigation shows that all of the A^2/O MWTPs in Beijing have experienced sludge bulking, though the severity of sludge bulking was different in different plants. Sludge bulking could not only cause a serious solid–liquid separation problem, but also lead to accumulation of ammonia and nitrite in the effluent (Hu *et al.* 2013; Wang *et al.* 2014). So continuous efforts have recently been made to identify the sludge bulking causative factors and control strategies.

6.5.2 Distribution of filamentous bacteria in activated sludge

The authors conducted a survey on filamentous bacteria in 12 EBPR systems located in 4 cities of China over a period of 2011–2014. Based on the Eikelboom and Jenkins method identification as shown in Fig. 6.19, Type 0092, Type 0041, Type 1851, M. parvicella, Type 021N, *N. limicola* II & III, *H. hydrossis* and mycolata were found to be the abundant filamentous species in these systems during the 'Non-Bulking' period from June to October. During this period, activated sludge was at a filamentous abundant level of 0–2 with an SVI value varying between 32 and 86 mL g^{-1}. The floc size was in general > 150 μm. Type 0092 and Type 0041 were observed in most of the samples, showing that they might have formed the skeletons of activated sludge flocs. Microthrix parvicella was observed in the inside of flocs with very low abundancy. *Haliscomenobacter hydrossis* and mycolata could only be detected in 5 Southern China samples with a low abundance. Sludge bulking was observed in most of the Northern China MWTPs (SVI, 210–330 mL g^{-1}) over a period between November to the next March, when wastewater temperature was around 12–16°C. The filamentous abundance was 5–6. Microthrix parvicella instead of Type 0092 and Type 0041 became the dominant filamentous species in flocs. During the period, the population of filamentous species did not change much in activated sludge samples from South China where the wastewater temperature was in general above 20°C.

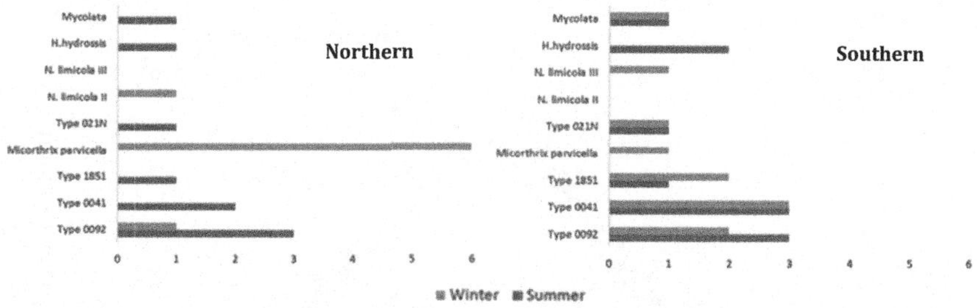

Figure 6.19 Filamentous bacteria distribution in southern and northern EBPR processes during summer and winter (filamentous abundance average value of the processes of A^2/O, inverted A^2/O and oxidation ditch).

6.5.3 Studies on sludge bulking processes and control strategy

6.5.3.1 Wastewater treatment performance and population dynamics over a sludge bulking cycle

Sludge bulking repeatedly occurred in a full-scale MWTP (4.0×10^5 m^3 d^{-1}) located in North China, which included the two EBPR systems (A^2/O, Inverted A^2/O) operated in parallel. As shown in Fig. 6.20, the SVI value in general increased to above 150 mL g^{-1} from December, reached the peak value in February, and then decreased back to below 150 mL g^{-1} from May.

Clone library analysis results indicated that the bacterial clones affiliated with the phylum Actinobacteria constituted the vast majority (60–66%), followed by the clones affiliated with the phyla *Firmicutes* (8–30.8%) and *Proteobacteria* (10–26%). Furthermore, 50 to 60% of clones affiliated with *Actinobacteria* in the bacterial libraries were related to filamentous '*Candidatus* M. parvicella'. It was clear that the excess growth of M. parvicella caused sludge bulking in North China MWTPs. Our previous study shows that nitrite and ammonia oxidations were affected by sludge bulking. It was speculated that washout of nitrite oxidizing bacteria (NOB) and ammonia oxidizing bacteria (AOB) might be responsible for the significant accumulation of nitrite and ammonia, as shown in Fig. 6.21. In spite of washout of '*Candidatus* Accumulibacter phosphatis', in the major PAOs, phosphorus removal was not decreased. The presence of a large amount of filamentous bacteria M. parvicella was supposed to be responsible for the removal of phosphorus on behalf of '*Candidatus* Accumulibacter phosphatis' during sludge bulking.

Previous research reported that M. parvicella had the ability to produce polyphosphate inclusion, and a polyphosphate ATP NAD+ (nicotinamide adenine dinucleotide) kinase has been found in strain RN1 (Rossetti *et al.* 2005; McIlroy *et al.* 2013).

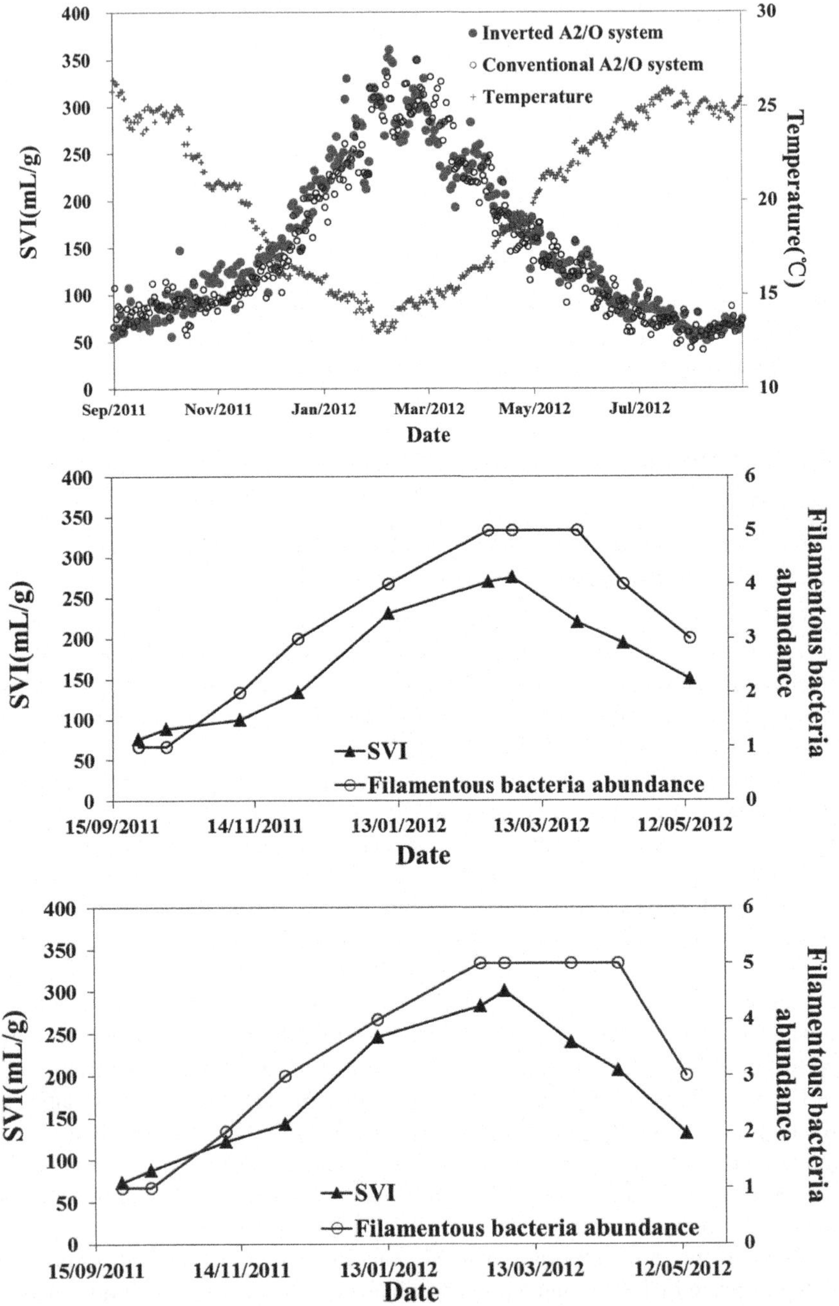

Figure 6.20 Variations of SVI, temperature and filamentous abundance in both EBPRs.

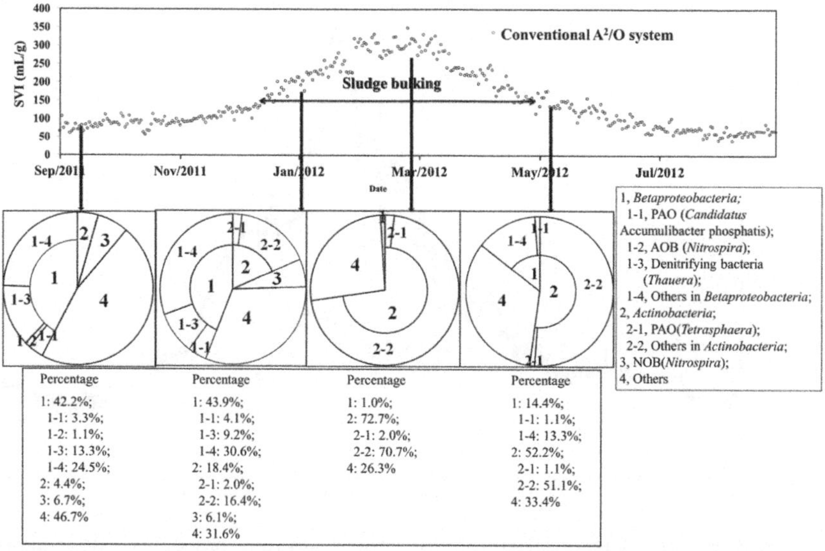

Figure 6.21 Population dynamics of functional groups during the bulking period.

6.5.3.2 Control strategy for M. parvicella overgrowth

It is known that the presence of the anaerobic-aerobic environment favoured the growth of M. parvicella (Martins *et al.* 2004c). This could well explain why the occurrence of sludge bulking occurred mainly after the adoption of the A^2/O and inverted A^2/O processes. Since the inverted A^2/O process exhibited the most serious sludge bulking, our first target was to control sludge bulking in the inverted A^2/O system. While it is reported that the addition of inorganic coagulants could effectively control sludge bulking caused by M. parvicella, the relatively high cost (0.10–0.20 RMB m^{-3}), makes it difficult to implement this strategy in China at the present time. So we decided to test the possibility of cutting the anaerobic-aerobic cycle by temporarily cancelling the anaerobic operation. Thus the inverted A^2/O process will be operated as an A/O one for several months during the low temperature period (at temperatures < 18°C). As shown in Fig. 6.22, this strategy has successfully realized the control of sludge bulking since 2013. Of course, ferric reagent will be needed for the removal of phosphorus with an added cost of 0.05–0.1RMB m^{-3}.

However, we still lack a strategy to control sludge bulking in the conventional A^2/O system. Better understanding of the reason for the excess growth of M. parvicella is important for the establishment of an economically favourable strategy for the control of sludge bulking. The difficulty in the isolation and cultivation of M. parvicella is still the big challenge for the bulking study. On the other hand, optimization of the control strategy for the inverted A^2/O process is also required, in order to reduce the running cost. So we need to continue our efforts in sludge bulking studies.

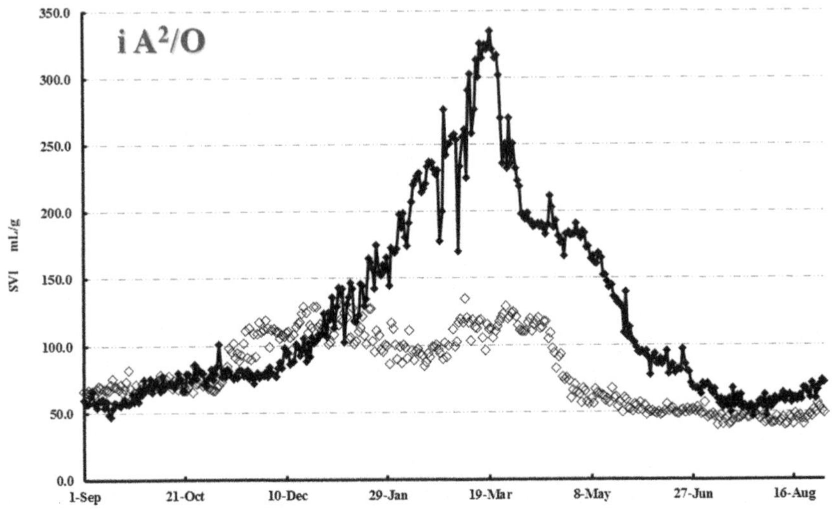

Figure 6.22 SVI variations before/after control strategy practicing under similar temperature conditions in the inverted A²/O process.

6.6 CZECH REPUBLIC

I. Ruzickova, O. Krhutkova and J. Wanner

6.6.1 Separation problems – situation up to the 1980s

Although some of the first papers on identification of activated sludge filamentous microorganisms came from (former) Czechoslovakia (Cyrus & Sladka, 1970; Sladka & Ottova, 1973), a systematic monitoring was not performed in this country at that time. There were several reasons for this:

- activated sludge plants built and operated up to the mid-1980s were either conventional, rather high loaded plants ($\Theta_X = 4–6$ d) or extended aeration plants with aerobic sludge stabilization ($\Theta_X > 25$ d); in both extremes the activated sludge plants did not exhibit significant sludge separation problems which could attract the attention of either plant owners or researchers
- no team in the country was equipped, supported and skilled enough
- serious sludge separation problems started at the end of 1980s and beginning of 1990s when new biological nutrient removal activated sludge plants were introduced over a short period of time.

Although no official national survey existed, it can be concluded from examinations of numerous activated sludge that the commonly occuring types were *Sphaerotilus natans*, Type 021N, *Thiothrix* spp., Type 0041/0675, *Haliscomenobacter hydrossis*, *Nostocoida limicola*, M. parvicella, Nocardioforms and fungi. Severe bulking problems were not frequently reported in municipal plants and the bulking problems usually were connected with the seasonal food industries in many towns (e.g. fruit and vegetable canneries, sugar beet mills, wine grapes processing, distilleries). Wastewaters from breweries (in all major Czech towns), caused a permanent presence of filamentous microorganisms in activated sludges.

Foaming in activated sludge plants treating domestic wastewaters was virtually unknown until the late 1980s. The first serious case was in an anaerobic–oxic system (EBPR); the microorganisms found in the foam were *Nostocoida limicola* and Type 0041 (Wanner & Grau, 1989).

6.6.2 Separation problems – situation up to the mid-1990s

In the 1990s, a shift in filamentous microorganisms dominance was observed at many plants. The Prague central plant is a well-documented example because its activated sludge has been used for more than 15 years as an inoculum for experimental units.

The plant has a plug-flow system aeration basin and filamentous microorganisms were always present. In the 1980s *Sphaerotilus natans* and/or Type 021N were dominant (Wanner *et al.* 1987, 1988) however, in 1991 and 1992 *Nostocoida limicola* had become dominant and *Sphaerotilus natans* and Type 021N levels were usually negligible. Because no changes had occurred, the shift could be attributed to changes in the wastewater.

At this time, foaming was observed in conventional activated sludge systems treating domestic sewages. The causative microorganisms were nocardioforms. Perhaps the biological foaming was related to new detergents introduced into the Czechoslovak market after 1990 and to a shift in the household consumption of fats from animal to cheaper vegetable.

6.6.3 WWTPs screening – 1995–2000

Regular activated sludge examination started in the early 1990s at plants in the Prague area and the spread survey was then extended to the whole country. The aim of this study was to determine the most common filamentous microorganisms in Czech plants and to study their effects on activated sludge flocs and on their separation properties.

6.6.3.1 Methodology

In year 1995, a questionnaire was mailed to supervisors at most activated sludge plants in the country to collect the answers to the following questions:

- design and current flow of plant; sludge age; mixed liquor concentration; SVI and average summer and winter temperature
- process type and schematic configuration

- wastewater type (sewage, municipal wastewater, type of industrial wastewaters), average composition (COD, BOD, SS, N and P) and required effluent quality
- for upgraded WWTPs data on changes in process facility, settling properties, etc.
- *separation problems*: if any, SVI data, behaviour of sludge in secondary clarifier, microscopic picture of the sludge
- *for sludge bulking*: extent of problem, countermeasures taken (selection from a list)
- *for foaming*: description of observed problems, countermeasures taken (selection from a list)
- seasonal changes in separation problems.

From questionnaire responses, plants with persistent separation problems were selected, then visited to verify the data from the questionnaire response and to collect mixed liquor and foam samples for microscopic examination. Activated sludge samples were taken at the effluent end of the aeration basin, while foam samples were taken from the surface of the aeration basin.

6.6.3.2 Microscopic analysis and sedimentation properties evaluation

Microscopic analyses of samples were performed by the method of Eikelboom and van Buijsen (1981) and Jenkins *et al.* (1993) for:

- floc characteristics and effect of filaments on floc structure
- protozoa and zoogloeal colonies
- overall and individual filamentous microorganism abundance
- photographic documentation and videorecording of characteristic features of observed filaments for training and educational purposes.

Filamentous microorganism abundance was rated according to the scale 0–6; the filament occurrence was rated dominant when the individual abundance was 4 (more than 5 filaments per floc – very common) or higher.

Sludge sedimentation properties were measured by SVI. Activated sludge was classified by SVI value using the following: 'well settling' with SVI up to 100 mL g^{-1}, 'light' with SVI 100–150 mL g^{-1} and 'bulking' with SVIs higher than 150 mL g^{-1}.

6.6.4 Results of separation problems and filamentous microorganisms screening

In the years 1996 and 1997, 78 and 72 activated sludge plants were visited and sampled, respectively (Wanner *et al.* 1998, 2000).

Bulking was not such a critical problem as in the past perhaps because most contemporary plants were designed using bulking control recommendations (Wanner, 1994a).

These designs have been efficient in suppressing filaments from Groups S and C (Wanner & Grau, 1989) and in this survey it was difficult to find filaments like

Sphaerotilus natans or Type 021N in municipal WWTPs. The survey demonstrated that in the past 10 years, filaments from Group A (M. parvicella and *Nostocoida limicola*) have become dominant. The elimination of Group S and C filamentous microorganisms from BNR systems can be explained by metabolic selection to which Group A filaments are less susceptible.

6.6.4.1 Filamentous microorganism abundance

The abundance of dominant filaments exhibits an increasing tendency, as illustrated by Fig. 6.23: in 1996 abundances 4 and 5 were almost equal while in 1997 abundance 5 clearly dominated (and to a smaller extent similar for abundance 6). The quality of activated sludge flocs affected the settling properties at medium and low abundance of filaments, whereas high filament abundance always resulted in bulking or foaming regardless of the floc characteristics.

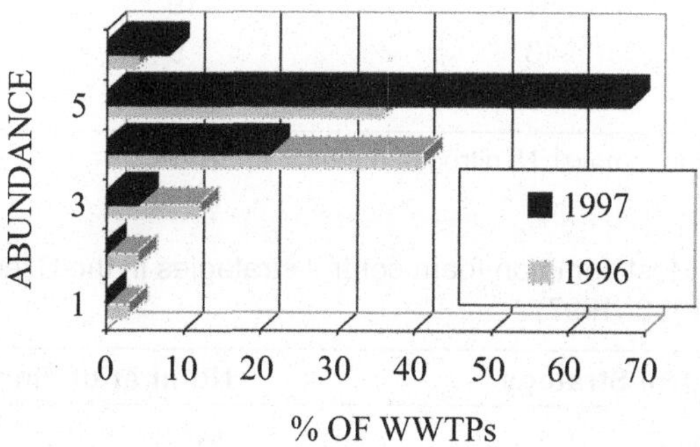

Figure 6.23 The difference in abundance of dominant filaments at 78 (1996)/72 (1997) activated sludge plants in the Czech Republic between 1996 and 1997.

6.6.4.2 Dominant filamentous microorganisms in sludge and foam

The commonest dominant filaments in conventional and BNR plants were M. parvicella, *Nostocoida limicola* and Types 0092 and 0041/0675.

The national survey showed that foaming by filamentous microorganisms (Group F) (Wanner & Grau, 1989) is becoming the most important solid separation problem. Table 6.4 shows the most common foam-forming filaments in plants with persistent foaming. Foams were dominated by M. parvicella, actinomycetes GALOs (previously known as NALOs – *Nocardia amarae*-like organisms) and by *Nostocoida limicola*. Table 6.5 shows that foaming was common in all activated sludge modifications and was most pronounced in plants with C and N removal (denitrification–nitrification). Fewer foam problems existed in conventional plants. The survey determined foam control strategies (Table 6.6).

Table 6.4 Most common filamentous microorganisms in biological foams (1996 and 1997).

1996			1997		
Filamentous Microorganism	**Number of Plants**	**%**	**Filamentous Microorganism**	**Number of Plants**	**%**
M. parvicella	16	73	M. parvicella	18	55
GALOs	6	27	GALOs	12	36
N. limicola	5	23	*N. limicola*	4	12
No. of foaming plants	22		No. of foaming plants	33	

Table 6.5 Correlation of activated sludge process type and frequency of biological foaming (1997).

Type of Process	Number of Plants	Foam Occurrence	%
C	51	19	37
C, N	13	9	69
C, N, P	8	5	63

C, organic carbon removal; N, nitrogen removal; P, EBPR.

Table 6.6 Most common foam control strategies in the Czech Republic (1996–1997).

Foam Control Strategy	Number of Plants	%
Water sprays	26	42
Skimming/mechanical foam removal	18	20
Return activated sludge manipulation	16	26
Sludge age reduction	15	24
Installation of baffles	13	21
Initial contact zones ('selectors')	10	16
Chlorination of foam	3	5

6.6.5 Development of filamentous population in Czech activated sludge plants between 1997 and 1998

Of the more than 70 activated sludge plants sampled in 1997, 33 WWTPs were selected for visitation in 1998.

6.6.5.1 SVI and total abundance of filaments

Table 6.7 shows that most sludges examined were 'light' in both 1997 and 1998. Bulking sludges increased from 21% in 1997 to 36% of bulking sludges in 1998.

High filament abundances did not necessarily mean high SVI (Wanner, 1994b). In some cases it was found that firm and compact floc morphology and an excessive quantity of filaments resulted in good settling, while in others, poor floc structure with low filament levels led to poor settling.

Table 6.7 Evaluation of total filament abundance and settling properties in 1997 and 1998 (33 WWTPs).

Year	SVI (mL g^{-1})	Total Abundance of Filaments				
		3	4	5	6	Σ**%
1997	<100	1	4	6	1	36
	100–200	1	2	11	–	43
	>200	–	–	7	–	21
	Σ*%	6	18	73	3	100
1998	<100	–	1	2	1	12
	100–200	1	3	11	2	52
	>200	–	–	11	1	36
	Σ*	3	12	73	12	100

* total % of activated sludge samples with overall filament abundance.
** % WWTPs in indicated SVI range.

During the two screening years, the mixed liquor samples could mostly be described as follows: small–medium flocs with a few compact centres, firm texture and irregular shape, filaments frequently present, affecting floc structure by bridging and by causing open floc structure.

6.6.5.2 Dominant filamentous microorganisms in activated sludge mixed liquor and foams

This survey covered both conventional and BNR plants, that is, plants, incorporating anaerobic and/or anoxic zones. Of the plants surveyed, 18 (55%) were conventional COD removal plants only, 8 WWTPs (24%) had a denitrification zone and 7 plants (21%) were BNR systems. In both years the four most frequent dominant filaments in mixed liquors were M. parvicella, *Nostocoida limicola*, Type 0092 and Type 0041/0675, and in 1998 *Haliscomenobacter hydrossis* was also seen (Fig. 6.24). The most frequently observed filaments were the same observed in all WWTP configurations.

Figure 6.25 shows the population distribution of secondary filaments. *Haliscomenobacter hydrossis*, Type 0041/0675 and Type 0803 were the commonest secondary microorganisms in 1997. In 1988 *Haliscomenobacter hydrossis* and Type 0041/0675 were the commonest secondary filaments. In addition, Types 0961, 1701 and *Thiothrix* spp. were common secondary filaments in 1998.

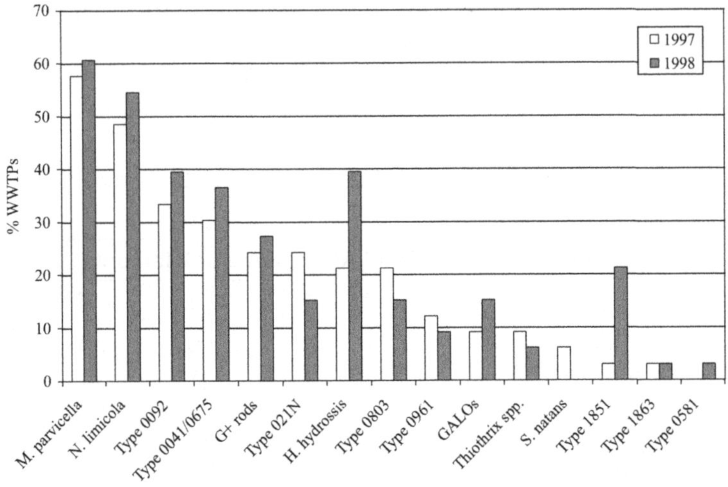

Figure 6.24 Comparison of the dominant filamentous bacteria distribution in activated sludges in 1997 and 1998.

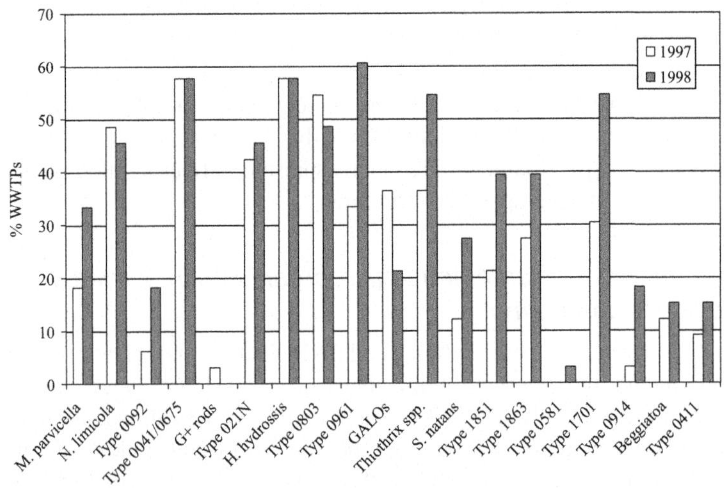

Figure 6.25 Comparison of the secondary bacteria distribution in activated sludge in 1997 and 1998.

In 1997 and 1998 13 biological foams (39% of all WWTPs) and 26 biological foams (79% of all WWTPs) were observed at the selected WWTPs. Thus foaming problems were more pronounced in the year 1998. Foams in aeration basin corners were most commonly observed. Figure 6.26 shows that M. parvicella,

Gram-positive rods (not evaluated in 1997) and GALOs were the most frequent foam-formers in both years. An increase of biological foam quantity containing Types 1863, 1851 and 0581 occurred. Other filaments were found in foams but they did not appear to support foam production because of their low abundance. Six types of filamentous bacteria were numerous in foam: M. parvicella, GALOs, *Nostocoida limicola*, Types 0092, 0041/0675 and 1851.

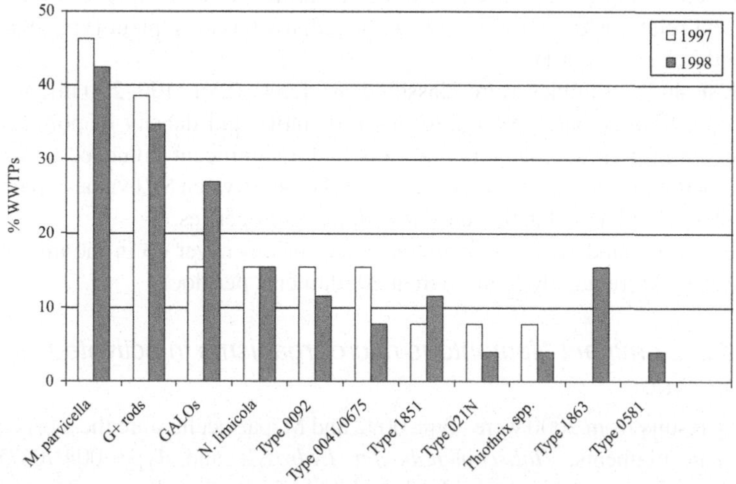

Figure 6.26 Comparison of distribution of filaments appearing in foams in 1997 and 1998 (some samples of foam were created by more than one dominant filamentous type).

Unidentified Gram-positive rods were common dominant microorganisms in mixed liquor and foam in both years. The rods were Gram-positive, Neisser-negative and most contained Neisser-positive granules. They grew as individual cells or very short chains (2–4) of cells of about 0.8 μm diameter. It is not known if these bacteria are a filamentous microorganism growth form (M. parvicella, GALOs) or an undescribed bacterial species. The Gram-positive rods were mostly found with both of the above mentioned filaments.

6.6.6 Screening of eight nutrient removal plants in 2000

6.6.6.1 Characterization of monitored WWTPs

All monitored WWTPs were BNR plants operated at low organic loads. Four were R-AN-D-N systems (regeneration–anaerobic–anoxic and oxic zones; Kos *et al.* 1992) with presettling and corridor-type basins, the other four were AN-D-N systems. The wastewaters were mostly of domestic sewage and wastewaters from food (milk, brewery, meat, preserving and starch industry) or tanning industries.

Each of the eight plants was microscopically examined at least four times a year (at least one sample of mixed liquor and one sample of foam per season).

6.6.6.2 Activated sludges and settling properties

Flocs in all 32 sludge samples had a firm texture and zoogloeal colonies did not affect their settling properties. Higher microorganisms were equally distibuted, among flagellates, amoeboid protozoa and stalked ciliates. No apparent relationship existed between floc characteristics, plant type and season (Krhutkova *et al.* 2001).

Most sludge samples were classified as 'light' (SVI: 100–200 mL g^{-1}). Floc structure, filamentous type and location, quantity and density of poly-P (PAOs) clusters, and presence of inorganic salts had a significant influence on activated sludge settleability. There was no clear correlation between SVI values, filamentous abundances and actual activated sludge settling properties.

The total abundance of filamentous bacteria was never <3 in the mixed liquor and was most frequently 5 (more than 20 filaments per floc).

6.6.6.3 Dominant filamentous microorganisms in activated sludge mixed liquor

Survey results from 2000 were: Type 0092 and M. parvicella were the most common dominant filaments, *Haliscomenobacter hydrossis* and Type 0041/0675 were secondary filaments always found at low abundance (1–3) at all operating conditions.

At some plants, P removal by chemical addition (Fe salts) was practiced. Three plants used ferric sulfate addition. The influence of chemical precipitation on filament types on settling properties is shown in Table 6.8.

Table 6.8 Correlation between SVI values, filamentous types and chemical precipitation.

	Addition of $Fe_2(SO_4)_3$			**No Addition of $Fe_2(SO_4)_3$**		
SVI (mL g^{-1})	<100	100–200	>200	<100	100–200	>200
Type 0092	80%	75%	33%	100%	100%	50%
M. parvicella	80%	62%	67%	25%	33%	67%
Type 1851	–	–	–	–	–	33%
N. limicola	–	–	33%	–	–	33%

At WWTPs without chemical precipitation there were 6 samples of bulking sludge containing: in descending order M. parvicella (67%), Type 0092, Type 0803 and *Nostocoida limicola* (33%). Type 0092 was identified in all 10 samples of well settling or light sludge. Microthrix parvicella was present in 30% of the samples. Type 0092 occurred in 16 samples where inorganic salts were added; its appearance was comparable to the samples without iron salt addition, about 80% in good settling sludges and 33% in bulking samples. An M. parvicella population was mostly found in 'well' settling sludges (80%). It was present 65% of the time in samples with higher SVI values. Thus the overall dominant presence of both Type 0092 and M. parvicella was in plants with and without chemical precipitation. The influence of M. parvicella on settling may have been partially eliminated by the addition of iron salts.

6.6.6.4 Biological foams

In 2000 there was no biological foam at 3 WWTPs and in 2 of them biological foams were always present. Seasonal shifts in foaming occurred. Most foams appeared in winter and spring, while biological foams were observed at 2 plants only after warm weather. Table 6.9 shows that the foam-forming filaments were mainly M. parvicella and Gram-positive rods, and some Types 0092, 0581 and GALOs. Well settling activated sludge formed foam less frequently than light bulking sludges.

Table 6.9 Occurrence of foam-forming filamentous types at eight WWTPs during year 2000.

	(R-AN-D-N)				(AN-D-N)			
	1	**2**	**3**	**4**	**5**	**6**	**7**	**8**
Winter	–	Type 0581	M. parvicella Type 0092	–	M. parvicella	–	M. parvicella	Gram + rods
Spring	–	Type 0581 Gram + rods	*	–	M. parvicella Gram + rods	–	M. parvicella	M. parvicella Gram + rods
Summer	–	–	M. parvicella Gram + rods	–	M. parvicella Gram + rods	–	GALOs	M. parvicella
Autumn	–	–	*	–	M. parvicella	–	GALOs	–

*unstable and sleazy foam was observed, sample was not taken.

6.6.7 FOAM CONTROL STRATEGIES

6.6.7.1 Water sprays

Biological foams can be destabilized by dilution. The amount of water sprayed on the foam needs to be much higher than in water sprays used for detergent foam control. Spraying with water does not eliminate the cause of foaming. The filaments from the collapsed foams are returned to the mixed liquor or RAS so that this method retains the filaments in the system. Thus, the use of water sprays should only be considered an immediate emergency measure, especially when a massive amount of floating biomass escapes the secondary settling tanks. The installation of water sprays in mixed liquor channels (Fig. 6.27) or on final clarifier bridges was an efficient method for reducing biomass losses at several WWTPs.

Figure 6.27 Water sprays in mixed liquor channels.

6.6.7.2 Skimming (mechanical removal) of foam

The mechanical removal of foam prevents foam-forming microorganisms from taking advantage of the prolonged biomass retention time in the foam. The latest secondary clarifier design recommendations is to equip both rectangular and circular settling units with skimmers which act over the whole water surface (Fig. 6.28) and with foam traps (Fig. 6.29). Biological foam can be trapped and mechanically removed in mixed liquor channels (Fig. 6.30). The foam removed from these traps or by the skimmers should be handled separately from WAS, especially when the WAS is stabilized in anaerobic digesters.

Figure 6.28 Skimming of secondary settling tank surface.

6.6.7.3 Return activated sludge manipulation

Increased RAS flow rate can reduce retention time in the thickened bottom sludge blanket preventing denitrification that helps float sludge. This countermeasure should not be used on a long-term basis.

Figure 6.29 Foam trap with final clarifier tank surface rotating bridge.

Figure 6.30 Example of biological foam mechanical removal.

6.6.7.4 Manipulation of SRT

The SRT required for reliable elimination of foam-forming microorganisms (especially GALOs) from activated sludges is low and may negatively affect the treatment efficiency. The low SRT consortium is dominated by fast-growing organotrophs, so that BNR is not possible.

6.6.7.5 Additional installation of baffles

Proper secondary clarifier design and construction is essential for foam control. Secondary clarifiers should be able to retain and collect floating biomass and prevent its escape into the effluent. Final clarifiers should be equipped with baffles that protect the effluent weirs and with efficient skimming devices; without these a low SS effluent cannot be produced (Figs 6.31 and 6.32).

Figure 6.31 Small patches of floating biomass.

Figure 6.32 Massive escape of biomass.

6.6.7.6 Initial contact zones ('selectors')

Methods such as kinetic or metabolic selection are less successful for foaming control than for bulking control (Table 6.6) because the selection pressure that can be achieved is not sufficient to suppress the growth of filaments like M. parvicella. Often, readily biodegradable organic substrate penetrates through anaerobic and/or anoxic zones into the oxic zone where it is used by filaments at a higher rate than in non-aerated zones. Extensive foaming problems were not found in activated sludge plants with well-established EBPR.

6.6.7.7 Chlorination of foam

Control of foam-forming filamentous microorganisms by chlorine is not widely used. When chlorination is used in a BNR plant, a temporary deterioration of nitrification and EBPR efficiency is to be expected.

6.6.8 Development in the last decade

Activated sludge separation problems connected with poor floc microstructures, that is, dispersed growth and microflocs, are no longer common in wastewater treatment practice. The explanation can be seen in:

- Improved design of activated sludge plants, especially as far as sludge age is concerned. New activated sludge plants are always designed for nitrification and denitrification which stresses the proper calculation of sludge age. IWA mathematical models ASM1 or ASM2d are commonly used for the calculation. The sludge age necessary for nitrification is usually in the range which excludes both dispersed growth and microfloc formation.
- Better protection of activated sludge plants which are potentially affected by industrial toxic wastewater, for instance, by on-line control of influent quality.
- The coagulation effect of ferric and aluminium salts dosed regularly for phosphorus removal in aeration tanks.

Viscous bulking was never considered to be an important activated sludge separation problem. On the other hand, floating sludge in secondary clarifiers (result of denitrification in settled and thickened sludge layer) is a much more common and serious problem, which can be, however, solved by:

(i) Installing baffles protecting the weirs of effluent launders (Fig. 6.33).
(ii) Using submerged perforated effluent pipes (Fig. 6.34).
(iii) Manipulation of activated sludge storage in the clarifier and by increasing the efficiency of denitrification in activated sludge system (increase in internal sludge recycle and/or addition of external organic substrate (Fig. 6.35).

Figure 6.33 Baffle protecting effluent launders.

Figure 6.34 Submerged perforated effluent pipes.

Figure 6.35 Stainless steel pipe for methanol dosing to anoxic zone.

6.6.8.1 Activated sludge bulking

The literature review of papers published in Czech national water journals or water utility reports has shown that the problem of activated sludge bulking is mentioned with much less frequency than in the previous decades and is usually connected with some specific conditions, especially with wastewater composition (food industry). In typical municipal wastewater treatment plants, the problems with bulking sludge were significantly reduced thanks to the following measures (Wanner & Jobbágy, 2014):

(i) Sludge age – modern activated sludge plants have been designed with sludge age suitable for nitrification, typically sludge age 12–18 days. In this region only certain types of filamentous microorganisms can proliferate. In addition, the observed biomass yield coefficient Y_{obs} is decreasing with increasing sludge age and thus the increased sludge age also reduces the amount of newly produced biomass of filaments.

(ii) Regeneration zone – in accordance with kinetic selection theory, especially with the principles of unbalanced growth (storage – regeneration), most of the newly built or upgraded activated sludge plants are equipped with separate activated sludge regeneration zones (Figs 6.36 and 6.37).

(iii) Common phosphorus precipitation by ferric salts – because of stricter limits for total phosphorus in final effluents as a result of the application of a combined approach to the formulation of effluent limits and because of the high risk of eutrophication in the whole country (a sensitive area according to EU legislation), most activated sludge plants in the country are equipped with simultaneous phosphorus precipitation. Continuous dosing of ferric salts in aeration basins helps to keep rather low values of SVI under normal circumstances (wastewater composition, sludge loading and age, proper configurations of activated sludge process).

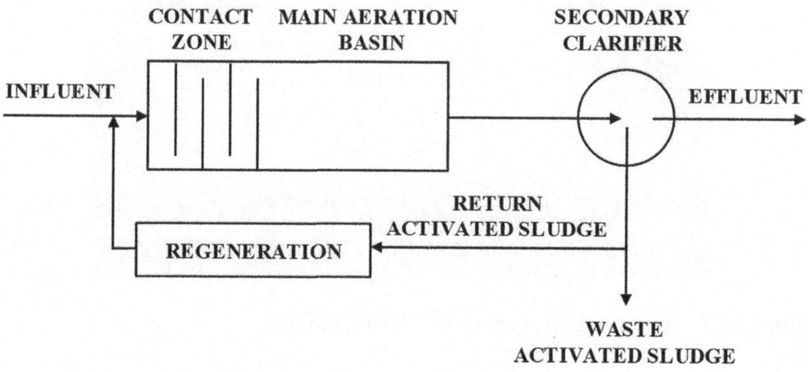

Figure 6.36 Scheme of activated sludge process with compartmentalized contact zone and sludge regeneration.

6.6.8.2 Activated sludge foaming

The previous surveys of foaming activated sludge in the Czech Republic showed that there are three principal filamentous microorganisms: '*Candidatus* M. parvicella', Gram-positive rods and *Gordonia amarae*-like organisms (GALO's).

Removal of Microthrix foams by aluminium salts. Although the problems caused by M. parvicella seemed initially very difficult to solve because of the very complex enzymatic equipment of this organism, the practical solution was finally surprisingly easy. Our group performed in the mid-1990s experiments at a wastewater treatment plant in Liberec. The Liberec plant was one of the first nitrification–denitrification plants with a sludge regeneration zone built in the Czech Republic. The plant suffered from extensive foaming problems caused by M. parvicella, which is one of the few exceptional filaments which cannot be controlled by common selection procedures. In the experiments performed in a pilot-scale model of the full-scale plant it was observed that the M. parvicella foam repeatedly disappeared when ferric sulfate was replaced with aluminium salts (Al^{3+}) for phosphorus precipitation (Váňa, 1996; Váňa & Wanner, 1996). However, at that time the company operating the plant was undergoing preparation for privatization and the experiments were stopped at the level of pilot-plant. This method has lately been verified in a full-scale plant in the Prague suburb wastewater treatment plant of Sedlec (Váňa & Wanner, 1998).

Figure 6.37 Regeneration zone in the stream of RAS.

The Liberec plant was upgraded at the beginning of the 21st century. Because the problems with M. parvicella foam persisted, the operator started dosage of aluminium salts instead of ferric salts, especially in winter conditions, when the foaming problems are most urgent. A similar strategy was later adopted also for controlling M. parvicella foam at the Central Wastewater Treatment Plant of Prague (Soukup *et al.* 2015).

Removal of other foams. The experience with the control of activated sludge biological foam resulted in commonly used design recommendations that all walls in activated sludge system should enable free movement of floating foam from individual zones to the final clarifier where it can be finally captured and removed in an efficient way (see also Figs 6.38 and 6.39). Today, the scrapers equipped also with water sprays are becoming more common. The sprayed water helps to dilute the floating foam, which finally settles together with the rest of the activated sludge. In this way, the foam-forming filaments can get into the waste activated sludge. However, the danger of foaming in anaerobic digesters can be significantly reduced by constructing with flat roofs, which proved to be more resistant to foaming problems in comparison with a conical shape of roof.

Figure 6.38 Secondary clarifier with scraper and water sprays.

Figure 6.39 Anaerobic digester with flat roof.

6.7 DENMARK

M. Nierychlo and P. H. Nielsen

6.7.1 General situation

Wastewater treatment plants (WWTPs) in Denmark are under continuous development in order to meet stringent effluent standards and minimize environmental impact. Furthermore, there is a strong interest in upgrading the larger plants for energy production (biogas), phosphate recovery (struvite) and general reduction in the use of chemicals and energy. During the last 25 years, the Danish plants have undergone an intense centralization process, whereby the number of plants with a capacity of over 30 person equivalents (PE) has decreased by 60%. Currently, there are

798 municipal WWTPs of over 30 PE, of which 250 plants treat 95% of incoming wastewater. These undertakings have resulted in reduction of nitrogen, phosphorus, and organic matter emission by 81%, 91%, and 94%, respectively (Ministry of Environment and Food of Denmark, 2015). The discharge limits for all WWTPs with a capacity over 5000 PE have been 1.5 mg P L^{-1} and 8 mg N L^{-1} (or lower for plant discharging to sensitive recipients). The WWTPs continuously meet these requirements with some rare exceptions. However, poor settling of the sludge due to filamentous microorganisms has always been a problem in the Danish WWTPs (Kristensen *et al.* 1994; Mielczarek *et al.* 2012). The problems have rarely been very serious, but, occasionally, plants experience severe bulking problems. Foaming, on the other hand, is nearly absent in Danish plants. In this chapter, we shall provide the status of the situation in Danish municipal WWTPs with nutrient removal regarding settling properties, identity, and abundance of filamentous microorganisms and introduce the new MiDAS database of microorganisms in activated sludge.

6.7.2 MiDAS: large-scale survey of the microbiology of Danish WWTPs

Since the year 2006, multiple Danish full-scale WWTPs with nutrient removal have been screened in relation to settling properties, identity, and abundance of various microorganisms in the framework of the Microbial Database for Activated Sludge (MiDAS) project, coordinated by the Center for Microbial Communities, Aalborg University. The activated sludge samples have been collected 2–4 times per year together with information about a number of operational parameters. Until now > 700 activated sludge samples from 56 Danish WWTPs have been registered in the MiDAS database. Each sample has been analyzed by light microscopy, and the bacterial community composition has been investigated by various methods. During the entire period, the characterization of activated sludge flocs and filaments has been carried out according to Eikelboom's manual (Eikelboom, 2000a, b). In addition, in the years 2008–2012, quantitative Fluorescent *in situ* Hybridization (q-FISH) with a range of oligonucleotide probes was applied to determine the abundance of probe-defined filamentous bacteria by fluorescence microscopy (Mielczarek *et al.* 2013). With the advent of 16S rRNA gene amplicon sequencing, the microbial diversity can now be sampled sufficiently to describe the composition and dynamics of most microorganisms (Caporaso *et al.* 2012), therefore, this technique is now routinely applied for the characterization of activated sludge communities within the MiDAS project. Since samples in the MiDAS project were frozen for potential later DNA analysis, we have sequenced all of them to get more detailed information about the microbial diversity, including filamentous bacteria, in the Danish plants (McIlroy *et al.* 2015).

The first large-scale 16S rRNA gene survey of the Danish plants showed that a substantial portion of the sequences obtained could not be classified at the genus level when using the available public taxonomic databases for bacterial classification, thus excluding a number of environmentally important organisms from the analysis.

To improve the classification for bacteria important in activated sludge systems, the MiDAS taxonomy was developed, for which the SILVA taxonomy (Quast *et al.* 2013) has been manually curated with annotations for all the abundant and important genera in full-scale activated sludge. Using the MiDAS taxonomy, we have collected key (abundant and/or important) organisms in activated sludge wastewater treatment systems, linked their identity with available information on their function, morphology, diversity, and distribution, and included this information in the MiDAS field guide (www.midasfieldguide.org) (McIlroy *et al.* 2015). The list of >150 genera currently described in the guide includes all bacteria known to possess filamentous morphology in activated sludge, thus making it possible to perform comprehensive analyses of filamentous populations in Danish WWTPs. Each genus contains a number of phylotypes called operational taxonomic units (OTU) that share >97% sequence identity and roughly represent one species. The online MiDAS field guide links the identity of annotated genera to the available information on their ecology. It is intended as a collaborative workspace, available to researchers and wastewater treatment practitioners, to facilitate a better understanding of the ecology of this biotechnologically important ecosystem.

6.7.3 Wastewater and WWTP characteristics

More than 50 Danish full-scale WWTPs have been analyzed during the past 10 years, and of these, a comprehensive study of the filamentous microorganisms was carried out in 24 WWTPs. Information about plant configuration and type of influent wastewater is shown in Fig. 6.40. All plants have N and P removal, and most also perform enhanced biological phosphorus removal (EBPR). Half of the EBPR plants have implemented side-stream hydrolysis. Average wastewater characteristics are shown in Table 6.10.

6.7.4 Settling properties in Danish nutrient removal plants

Settling properties and overall abundance of filamentous microorganisms have been determined in all samples by the analyses of SVI, diluted SVI (DSVI), and filament index (FI). The distribution of the DSVI values for all the plants, analyzed 2–4 times per year, is shown in Fig. 6.41. Most samples were in the range of 80–120 mL g^{-1} SS, indicating good settling properties, while approximately 30% of samples analyzed had values exceeding 120 mL g^{-1} SS, indicating poor or very poor settling properties. This fraction is very similar to what has previously been observed in Danish WWTPs (Kristensen *et al.* 1994; Mielczarek *et al.* 2012), with a tendency for higher DSVI in winter (Mielczarek *et al.* 2012). The time series of the DSVI values in each plant during the whole sampling period is shown in Fig. 6.42. It can be observed that most plants had a typical DSVI level while some plants had a variable DSVI pattern throughout the years. Thus, the settling properties were consistently good in some plants, consistently average or poor in others, while in a few plants it varied more over time. Interestingly, the general settling properties have not improved measurably since the last large survey in 1994 (Kristensen *et al.* 1994). The reason is probably that most plants in general were running satisfactorily at that time, and that Microthrix, as the major problematic filamentous species (see later), had been efficiently controlled already at that time.

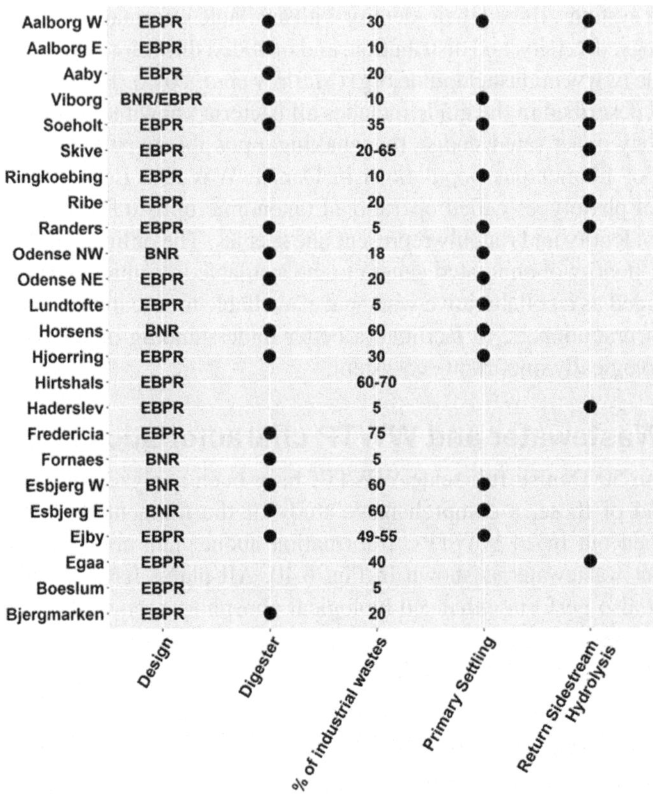

Figure 6.40 Plant configuration and type of influent wastewater.

Table 6.10 General characteristics of the wastewater (measured before process tank inlet), based on the data from 24 WWTPs in the years 2006–2014.

Parameter	Value
COD, mg L^{-1}	570 ± 310
Total N, mg N L^{-1}	45 ± 16
Total P, mg P L^{-1}	8 ± 3
pH	7.6 ± 0.4
Temp. °C	13 ± 4

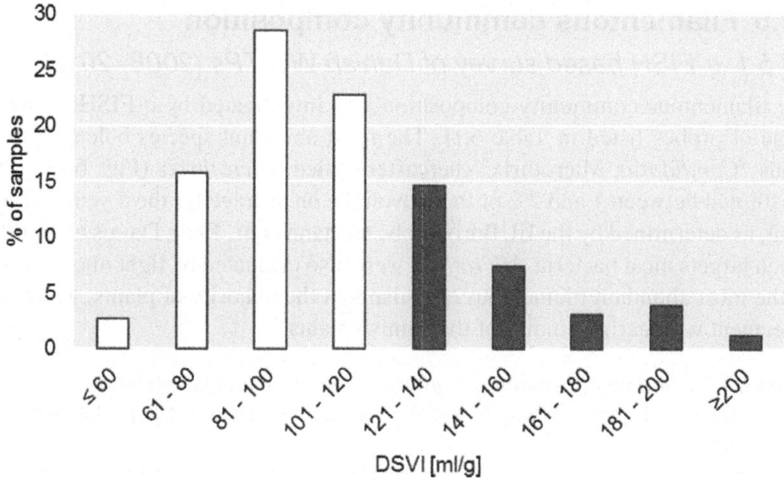

Figure 6.41 Distribution of DSVI values from 24 Danish WWTPs in the years 2006–2014. The dark columns indicate deteriorated settling properties.

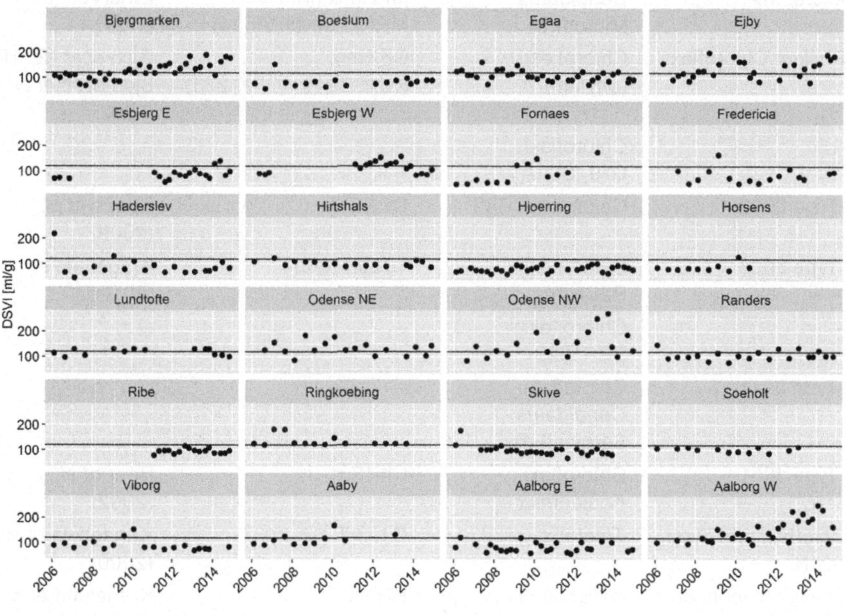

Figure 6.42 DSVI in 24 Danish WWTPs in the years 2006–2014 (except Fornaes WWTP, for which SVI values are shown). Horizontal line denotes DSVI = 120 mL g^{-1} SS.

6.7.5 Filamentous community composition

6.7.5.1 q-FISH based survey of Danish WWTPs (2008–2012)

The filamentous community composition was investigated by q-FISH, using the range of probes listed in Table 6.11. The most abundant species belonged to the genus '*Candidatus* Microthrix' (hereafter called *Microthrix*) (Fig. 6.43). They constituted between 1 and 7% of the biovolume on average for the 5 years for each plant, as determined by the EUBmix probe (Amann *et al.* 1990; Daims *et al.* 1999), which targets most bacteria. *Microthrix* were also evaluated by light microscopy to be the most abundant filamentous organisms in the majority of plants, which is in agreement with earlier studies of the Danish plants.

Table 6.11 Overview of selected filamentous bacteria in Danish WWTPs, morphotypes with corresponding MiDAS classification and FISH probe used in the study.

FISH Probe Coverage/ Morphotype	Best Corresponding MiDAS Taxonomic Classification	FISH Probe Used in this Study	Reference
–	*Trichococcus*	–	–
Microthrix	*Candidatus* Microthrix	MPAmix (MPA60 + MPA223 + MPA654)	Erhart *et al.* (1997)
Phylum Chloroflexi	Chloroflexi (phylum)	CFXmix (GNSB941 + CFX1223)	Gich *et al.* (2001) Bjornsson *et al.* (2002)
–	Chloroflexi, C10_SB1A	–	–
Type 0092	Chloroflexi, B45	CFX197	Speirs *et al.* (2009)
Type 0803	Chloroflexi, P2CN44	T0803–0654	Kragelund *et al.* (2011)
–	Chloroflexi, SBR1029	–	–
Type 0914	Chloroflexi, *Candidatus* Sarcinathrix	–	–
Type 1851	Chloroflexi, *Kouleothrix*	Chl1851	Beer *et al.* (2002)
Thiothrix and Type 021N	*Thiothrix*	G123 T	Kanagawa *et al.* (2000)
Haliscomenobacter	*Haliscomenobacter*	HHY-654	Kragelund *et al.* (2008)
N. limicola II and others	–	ALF968	Neef (1997)

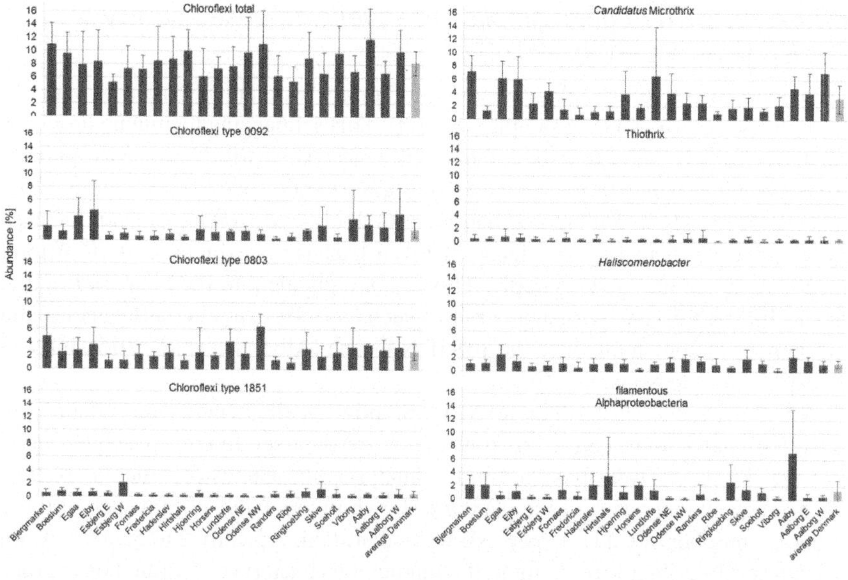

Figure 6.43 Filament abundance in 24 Danish WWTPs in the years 2008–2012, assessed by FISH. Each bar represents the average of all samples from the given WWTP (adapted from Mielczarek *et al.* 2012).

The phylum Chloroflexi contains many filamentous genera, three of which were analyzed by qFISH. They correspond roughly to Types 0092, 0803, and 1851 (Eikelboom, 2000a, b). Among them, Type 0803 was the most abundant in the majority of the plants, in quantities very similar to *Microthrix*, followed by Type 0092. Type 1851 was found in all plants in very low abundance (<1%). Furthermore, undefined Chloroflexi filaments (not targeted by the probes used) were present in all plants. Taken together as a group, Chloroflexi was most abundant in nearly all plants (>80% of the samples analyzed). Few other filamentous species were quantified, and in some plants, *Haliscomenobacter* and unknown filamentous Alphaproteobacteria were present in an abundance of a few per cent. *Thiothrix* populations were very low in nearly all plants (<0.5%).

The q-FISH analysis showed that the filamentous bacteria made up a significant fraction of the biomass in most plants. The average values in each plant ranged from 7.7% to 26.3%, however, it should be kept in mind that the actual abundance of filamentous bacteria was higher since not all filaments were targeted by the set of probes used in this study. The high fraction of filaments can explain the poor settling properties in the affected plants, as many of them produce open flocs that hinder good sludge settling.

6.7.5.2 16S rRNA gene amplicon sequencing-based survey of Danish WWTPs (2006–2014)

A comprehensive 16S rRNA gene amplicon survey was carried out on all samples from the period shown (514 samples in total). This sequencing technique produces short DNA reads with a length of approx. 500 base pairs, which can be assigned to the genus name, using the MiDAS taxonomy described earlier. In each sample, more than 10,000 reads were obtained and identified, allowing a relative quantification to take place, where the abundance of given bacteria is expressed as % of total read abundance in a given sample. This is not equivalent to the biomass fraction due to 16S rRNA gene sequencing biases such as DNA extraction efficiencies and differences in copy number of the 16S rRNA gene (Albertsen *et al.* 2015), but can be applied as a reasonable estimate.

All genera that, according to MiDAS, possess filamentous morphology in activated sludge were extracted from the entire data set and are presented here. In total, 18 genera were included in the analyses, the names of which as well as the average abundances in each plant are shown in Fig. 6.44. Within each of these genera, a number of OTUs ('species') were identified, typically 2 to 5, but they are not further described here. In total, the filamentous bacteria constituted on average 13.0 ± 5.0% of the total microbial community. The two most abundant genera were *Trichococcus* and *Microthrix*, each in an average abundance of, typically, 2–6% of the total population. The genus *Trichococcus* contains both filamentous and non-filamentous species, however, it is presently not known which of those are present in the Danish plants. *Nostocoida limicola* type I filaments belongs to this genus (Liu & Seviour, 2001), but it is still unknown whether it is abundant in the Danish plants. Future studies are needed to resolve this issue by linking different phylotypes in the genus to specific morphologies.

The top 4, 5, 7, and 9th most abundant genera belonged to the phylum Chloroflexi, and some of them roughly correspond to Types 0092, 0803, 0914, and 1851 (Fig. 6.44, Table 6.12). The genus with the name C10_SB1A has a morphotype like a Type 0092 (unpublished data), and it was very abundant in some plants, where it was identified as the filament causing severe bulking problems. We still do not know anything about its physiology and ecology.

A few other genera were present in low abundance (*Defluviicoccus* (only filamentous cluster III was included in the analysis), *Thiothrix*, *Haliscomenobacter*, *Meganema*, and *Sphaerotilus*) and they never were linked to bulking episodes. The foam forming *Gordonia* was present in some plants at relatively low abundance, rarely causing foaming incidents. The bacterial genera in the bottom of the graph ('*Candidatus* Nostocoida', *Acinetobacter* and *Skermania*) occured very sporadically in the Danish WWTP, thus their average abundance across the plants is 0.

The three dominant groups of filamentous bacteria, *Trichococcus*, *Microthrix*, and Chloroflexi, were dominant in 7–16 out of the 24 plants analyzed, each constituting a major fraction of the filamentous population (35–50% on average)

(Table 6.12). The maximum abundances observed in any plants during the 9 years for *Trichococcus, Microthrix*, and Chloroflexi were 16%, 18%, and 40% of the entire populations, respectively. This shows that one type of filament can be extremely dominating in full-scale WWTPs under certain conditions with a substantial effect on floc properties. However, such a high filamentous fraction may also have a strong effect on other performance parameters, such as removal of nutrients.

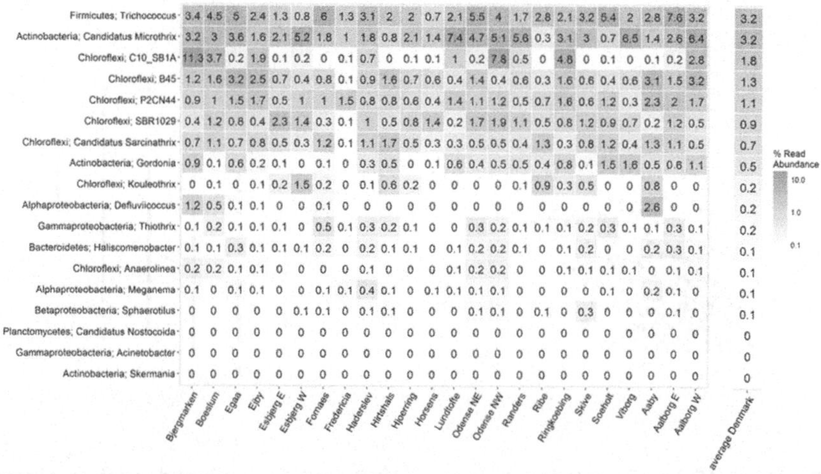

Figure 6.44 Filament abundance in 24 Danish WWTPs in the years 2006–2014, assessed by 16S rRNA gene amplicon sequencing. Each number represents the average of all samples from the given WWTP. Phylum and genus names are shown. The column to the right shows the average value from all samples.

Table 6.12 Occurrence of the dominant filamentous bacteria in 24 Danish WWTPs.

Filament(s)	Dominant in: No. of Plants (% of all plants)	Co-dominant in: No. of Plants (% of all Plants)	Abundance in Dominant Plants (% of Total Community ± SD)	Abundance in Dominant Plants (% of Filamentous Community ± SD)
Trichococcus	10 (42%)	8 (33%)	4.5 ± 1.7	34.5 ± 8.8
Microthrix	7 (29%)	9 (38%)	4.9 ± 2.3	41.9 ± 11.5
Chloroflexi (phylum)	16 (67%)	8 (33%)	6.5 ± 3.6	49.7 ± 8.6

Numerous studies have shown that the quantity of filamentous microorganisms affects sludge settling properties. We have attempted to make a correlation between the summed relative abundances of all filamentous bacteria present in each sample and the corresponding DSVI values for all 514 samples analyzed (Fig. 6.45a). A high level of scatter and no clear correlation between the two parameters ($R^2 = 0.24$) can be observed. This is probably due to the differences in morphology of the filaments present (bend, curled, or stiff), to whether the filament primarily grows inside or outside the flocs (e.g. Wágner *et al.* 2015), and to other factors affecting settling, such as presence of inorganic particles.

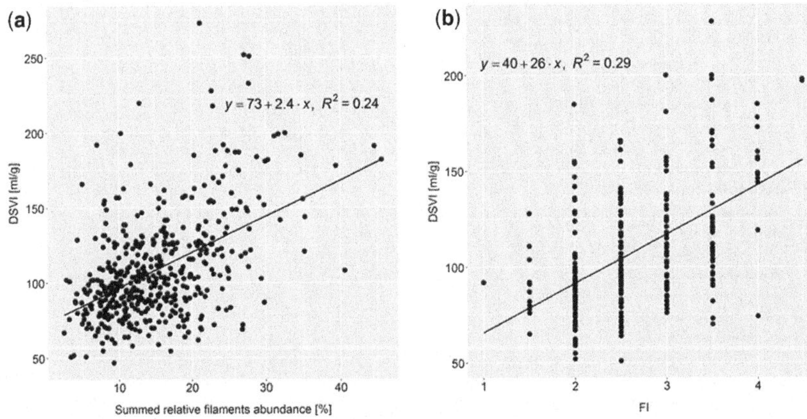

Figure 6.45 (**a**) Correlation between DSVI and (a) summed relative filament abundances assessed in a total of 514 samples from 24 WWTPs (years, 2006–2014); (**b**) FI assessed in a total of 330 samples from 37 WWTPs (years, 2008–2012).

A different way to analyze the effect of filaments on settling properties is the use of the filaments index (FI), in which the number of filaments is assessed by light microscopy on a scale from 0 (no filaments) to 5 (excessive amounts of filaments) (Eikelboom, 2000a, b). This has been performed on the subgroup of 330 samples, and the correlation between FI and DSVI is shown in Fig. 6.45b. These results also show a high level of scatter and the absence of strong correlation ($R^2 = 0.29$).

Even though the correlations of DSVI with both total filament abundance and FI were weak, as indicated by R^2 values, in both cases, we can observe a clear positive trend, indicating that filamentous overgrowth, whether assessed by sequencing or microscopic examination, has a strong tendency to cause sludge settling problems, reflected by high DSVI values.

The existence of a correlation between the relative filaments abundance and DSVI is more explicit when WWTPs are analyzed individually. We investigated the EBPR plant Bjergmarken in greater detail as it has had good settling properties followed by deterioration of sludge settling since 2009 with DSVI up to 200 mL g^{-1} SS. The variation in filament abundances through the 9 years is shown in Fig. 6.46. The most abundant genera were *Trichococcus*, *Microthrix*, and several genera belonging to Chloroflexi (C10_SB1A, B45, P2CN44, '*Candidatus* Sarcinathrix'). Both *Microthrix* and some Chloroflexi genera showed large variations over time. The DSVI values observed correlated well with the summed filament abundance in the plant ($R^2 = 0.6$, Fig. 6.47), suggesting that the relationship between filamentous microorganisms and sludge settling properties may be an intrinsic characteristic of a given WWTP.

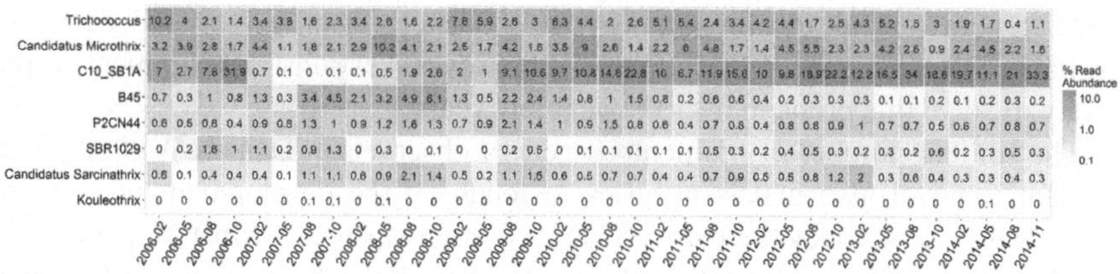

Figure 6.46 Time series of the predominant filamentous bacteria over the years 2006–2014 (year and months shown) in Bjergmarken WWTP experiencing settling problems. The numbers show the relative abundance of the total bacterial community.

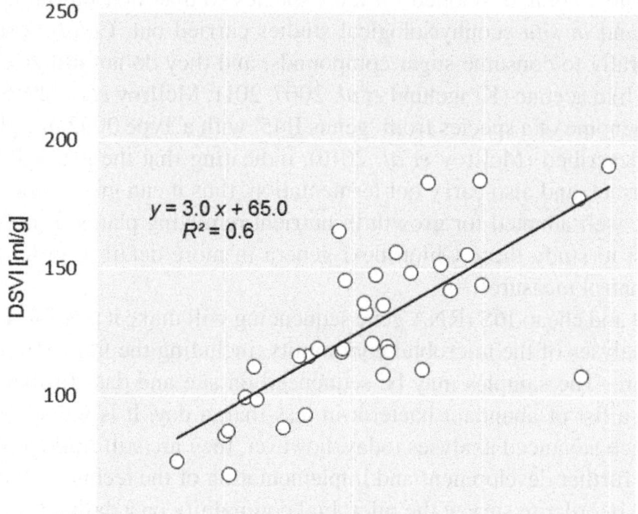

Figure 6.47 Correlation between the summed abundance of *Microthrix* and C10_SB1A genus (belonging to the phylum Chloroflexi) and DSVI in Bjergmarken WWTP.

6.7.6 The future: surveillance and control by DNA analyses

The use of molecular tools has been a revolution in the analysis and description of the activated sludge microbial community. All known filamentous microorganisms can now be identified by fast, cheap, and reliable 16S rRNA gene sequencing techniques, and most of them can also be visualized by the FISH probes available. There are still some problems in the resolution of specific species (OTUs) within some genera due to the relatively short sequences obtained by the currently available DNA sequencing technologies. However, very fast development in the area will solve this problem over the next few years.

The limited number of filamentous genera observed in Danish WWTPs with nutrient removal also makes it easier to focus on the development of efficient control strategies. Today there are efficient control measures against M. parvicella, *Gordonia,* and *Skermania* (Rossetti *et al.* 2005; Kragelund *et al.* 2010). They are applied in Denmark when bulking and foaming problems are observed. The big challenge is the Chloroflexi filaments, especially the ones that have been identified recently by sequencing and are abundant in many WWTPs. These species provide a general loose floc structure, which causes poor, but not very bad settling properties, and only occasionally real bulking troubles. For many of these there is no ecological information available (e.g. C10_SB1A), but FISH probes are now being developed for most species so that their morphology can be described and *in situ* ecophysiological studies carried out. *Chloroflexi* filaments seem generally to consume sugar compounds, and they do not utilize short chain fatty acids like acetate (Kragelund *et al.* 2007, 2011; McIlroy *et al.* 2016). The first complete genome of a species from 'genus B45' with a Type 0092 morphology has just been described (McIlroy *et al.* 2016), indicating that the genus B45 can use oxygen, nitrate, and also carry out fermentation, thus it can grow under anaerobic conditions, well adapted for growth in nutrient removing plants. It is high on the priority list to study these Chloroflexi genera in more detail in order to develop efficient control measures.

The fast and cheap 16S rRNA gene sequencing will make it possible to carry out 'online' analyses of the microbial community, including the filamentous bacteria, in the future. The samples may be sequenced on site and data further processed to provide a list of abundant bacteria in less than a day. It is already possible to perform such advanced analyses today, however, they are still limited for research use. Thus, further development and implementation of the technology in WWTPs are needed in order to survey the microbial community on a daily or weekly basis, use the knowledge to recognize early warning signs, and potentially also implement control measures, if these are known. Does that mean that we do not need to perform microscopic sludge examinations anymore? It can still be very useful as it tells us a lot about the sludge characteristics, such as change in floc structure, the relative amount of filaments inside and outside the floc, etc., which advantage the relative abundance number does not offer. Furthermore, if sequencing data suggests that a particular plant has only 2–3 predominant filamentous bacteria, these may easily be recognized by microscopy. However, one great advantage of using sequencing is that not only are the filaments identified and quantified, but all other important players in N and P removal are also included in the analyses for further info and use.

We do not know much about the stability of the microbial communities in the WWTPs, including the filaments, and what controls them. Therefore, there is a need to carry out surveys and time-series analyses with at least weekly resolution and correlate this information with 'metadata', that is, all the data available that describe the wastewater composition, plant operation, and performance. We have recently started a 'Biobank' project, in which many participating Danish WWTPs are asked to freeze activated sludge samples weekly and store them for potential later DNA analyses. If problems occur or interesting changes in microbial population or plant operation take place, the samples can be analyzed, and correlations may be established for future knowledge generation and development of suitable control measures.

6.8 FRANCE

N.Durban, L. Juzan, Y. Fayolle and S. Gillot

6.8.1 General situation

According to the national database provided by the Ministry of Ecology, Sustainable Development and Energy (BDERU, 2012), in 2012 metropolitan France accounted for 19,500 wastewater treatment plants (WWTPs), designed to treat the wastewater of 98 million of population equivalent (P.E.). The actual load reached a value corresponding to 76 million of p.e. in 2012. Activated sludge processes represent 36% of the French WWTPs and treat 73% of the total load generated as urban wastewater. In addition, the French National Agency for Water and Aquatic Environments (Onema) pointed out that 91% of the WWTPs with a capacity higher than 2000 P.E. used the activated sludge process (Golla *et al.* 2010). This process is well controlled and appropriate to reach the level of treatment specified by the European Urban Waste Water Treatment Directive, especially concerning nutrient removal in sensitive areas.

Metropolitan France accounted for around 6500 activated sludge plants operated under low or very low food to microorganisms (F/M) ratio in 2012 (BDERU, 2012). They provide high treatment performances in terms of nitrogen removal thanks to the high sludge retention times (SRT > 15 days at 12°C).

As shown in Fig. 6.48, 11% of the activated sludge systems with low and very low F/M ratio treated 76% of the French urban wastewater in 2012.

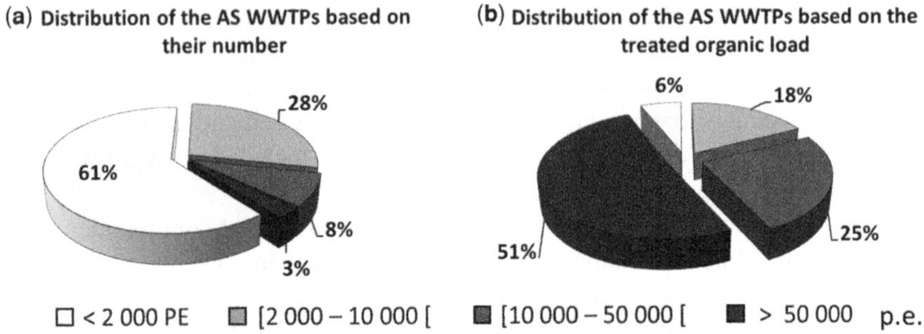

Figure 6.48 Distribution of the activated sludge WWTPs operated under low and very low F/M ratio based on their number (**a**) and on the treated organic pollution (**b**).

6.8.1.1 Bulking and foaming situation before 2005

Activated sludge processes operated under low F/M ratios are known to encounter bulking and foaming issues due to the growth of specific filamentous bacteria, such as M. parvicella or Type 0092 (Jenkins *et al.* 2004). Several surveys related to settling and foaming problems have been performed in France during the last thirty years. In terms of bulking, the first survey, carried out in 1984 on 3860 activated sludge plants (regardless of the F/M ratio), showed that 25% of these plants were subjected to bulking and, in 91% of the cases, the WWTPs were operated at a low or very low F/M ratio. Around 15 years later, a second survey performed on 1000 activated sludge plants showed that an equivalent proportion of WWTPs (25%) had sludge volume index values highlighting settling problems (above 200 mL g^{-1}) (Graveleau *et al.* 2005). As far as foaming was considered, 20% of the 6000 activated sludge plants (considering all F/M ratio) were subjected to foaming (Pujol *et al.* 1991) and these events occurred in majority in WWTPs operated at low or very low F/M ratio (97%) (Pujol *et al.* 1991; Pujol & Boutin, 1989). The dominant filamentous bacteria identified in the different surveys performed in France before 2005 are reported in Table 6.13.

M. parvicella, Type 0041/0675 and Type 0092 were the dominant filamentous organisms in activated sludge bulking and/or foaming in France. This result is coherent with recent publications on filamentous bacteria diversity among Danish activated sludge plants (Mielczarek *et al.* 2012). Indeed, low F/M ratio (<0.2 kg BOD_5 kg^{-1} MLVSS d^{-1}) and slowly biodegradable substrate promote the growth of these filamentous bacteria (Eikelboom, 2000a, b; Jenkins *et al.* 2004; Richard, 1989).

The frequency of bulking or foaming problems did not evolve significantly between 1984 and 2000. A new survey was carried out in 2012, in order to update the previous results as well as to inventory the control strategies employed. Particular attention was paid to the process configuration implemented in the surveyed WWTPs, especially in relation to phosphorus removal.

Table 6.13 Filamentous bacteria identified during bulking and foaming events according to previous surveys with their rank of occurrence and their frequency (%).

	Bulking		**Foaming**
	Pujol and Boutin (1989)	**Graveleau *et al.* (2005)**	**Pujol *et al.* (1991)**
Type 0041/0675	1	4 (6%)	2 (26%)
Type 0092	2	1 (47%)	4 (5%)
M. parvicella	3	2 (26%)	1 (55%)
Thiothrix sp. and Type 021N	4–5	3 (11%)	–
Sphaerotilus natans	6	0% as major 2% as secondary	–
Mycolata (*Nocardia* sp.)	–	0% as major 5% as secondary	3 (14%)

6.8.2 Current settling and foaming problems and control measures (2012 survey)

The national survey carried out in 2012–2013 aimed at estimating the occurrence of bulking and/or foaming problems due to filamentous bacteria in activated sludge plants operated at low or very low F/M ratios. A questionnaire was sent to 1000 plants taking into account the heterogeneity of French WWTPs, that is, the proportion of treated load (Fig. 6.48), and geographical locations (Durban, 2015). To this aim, five samples were defined according to the WWTP's size; 370 responses were obtained, as shown in Table 6.14. Sample 5 corresponds to the 19 WWTPs treating more than 250,000 P.E., this group was studied separately.

Table 6.14 Response rate according to the WWTP's size.

Maximum Organic Load Treated (P.E.)	**Sample 1**	**Sample 2**	**Sample 3**	**Sample 4**	**Total**	**Sample 5**
	<2000	**[2000; 10,000]**	**[10,000; 50,000]**	**[50,000; 250,000]**	**<250,000**	**≥250,000**
N. WWTPs	3955	1829	552	150	6486	19
N. questionnaires sent	609	282	86	23	1000	19
Response rate (%)	29	41	67	48	36	63

6.8.2.1 Characteristics of the surveyed WWTPs

The majority of the WWTPs include a nitrification/denitrification step (87% for WWTPs below 250,000 P.E. and 100% for the others) (Durban, 2015). Concerning phosphorus, 73% of WWTPs with a capacity lower than 2000 P.E. did not implement any specific treatment whereas 64% of the WWTPs between 2000 and 250,000 P.E. use metallic salt addition to remove phosphorus (Fig. 6.49a). Excluding Sample 5, WWTPs mainly use iron salts (Fig. 6.49b). The proportion of WWTPs treating phosphorus only through enhanced biological removal (EBPR) is very low (around 3%).

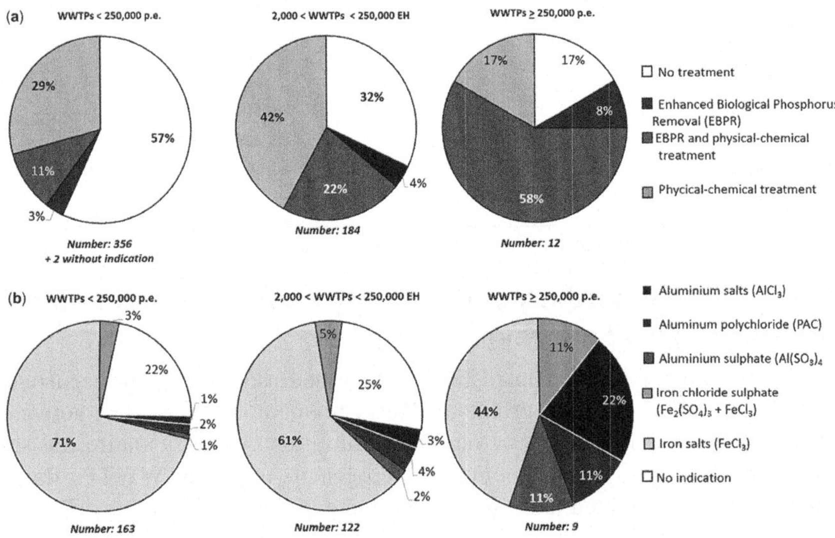

Figure 6.49 Distribution of the WWTPs according to (**a**) phosphorus removal and (**b**) type of chemical product added.

Biological treatment coupled with metallic salt addition is the main process for the WWTPs treating more than 250,000 P.E. (58%) (Fig. 6.49a). The combination of these two processes guarantees a good phosphorus treatment while reducing cost due to chemical addition. These WWTPs generally used metallic salt composed of aluminium or iron in the same proportion for physical–chemical phosphorus treatment (Fig. 6.49b).

An important point to notice is that French wastewater treatment plants are operated at relatively low effective load (in comparison to their design load): 71% and 53% of the design hydraulic and organic capacities respectively, which is coherent with recent data provided by Pujol and Arnaud (2015). A few values of F/M ratios and SRTs were provided in the collected data: the average F/M ratio was 0.07 kg BOD_5 kg^{-1} MLVSS d^{-1}. The average SRT was 29 days for WWTPs with a capacity below 250,000 P.E. and 19 days for Sample 5.

6.8.2.2 Occurrence of settling and/or foaming problems

The settling and/or foaming problems were divided in three categories:

Non-biological origin: a settling and/or foaming problem was observed but was not induced by an excessive growth of filamentous bacteria;
Biological origin: a settling and/or foaming problem was observed as well as an excessive growth of filamentous bacteria;
Unidentified origin: a settling and/or foaming problem was observed but the origin could not be identified.

Results reported in Fig. 6.50 show that settling and/or foaming problems concerned 37% of the WWTPs below 250,000 P.E., and that settling problems were dominating in 30% of the cases. This result is coherent with the surveys of Graveleau *et al.* (2005) and Pujol and Boutin (1989). Settling problems of unidentified origin are relatively important, especially for small WWTPs (<2000 P.E.). Microscopic observations are rarely carried out at this size of WWTPs, making it difficult to see if the occurrence of settling and/or foaming problems was due to excessive growth of the filamentous bacteria.

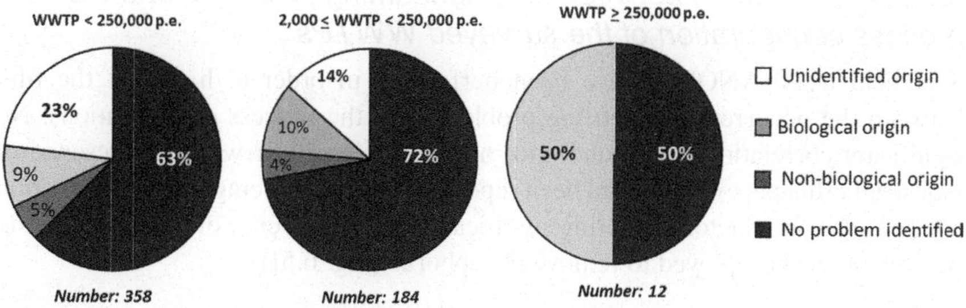

Figure 6.50 Proportion of WWTPs encountering a settling and/or foaming problem in 2012.

Fifty per cent of the WWTPs treating more than 250,000 P.E. encountered settling and/or foaming problems. In three cases out of six, an excessive growth of filamentous bacteria (with two due to M. parvicella) was observed and led to bulking and foaming. One case was due to recurrent bulking induced by M. parvicella and mycolata without foaming. In the two remaining cases, foaming was observed but no filament identified.

According to the survey results, bulking and/or foaming show a seasonal variation and are observed from November to April (winter time). In these large plants, sludge dewatering problems induced by bulking require the addition of larger amounts of polymers. To limit the costs associated with sludge treatment, a curative treatment using chlorination is set up when the diluted sludge volume index (DSVI) reaches a value of around 110–130 mL g^{-1}.

6.8.2.3 Factors influencing settling and/or foaming problems

Similarly to the results reported by Graveleau *et al.* (2005), the main factors inducing settling and/or foaming problems listed by the respondents to the questionnaire are the following (the average frequencies are indicated in brackets):

- Dilution of the influent by clean water infiltration (24%)
- Aeration shortage (22%)
- An important proportion of industrial influent in the wastewater (15%)
- A high concentration of suspended solids in the bioreactors due to sludge extraction/treatment problems (13%)
- Hydraulic overloads (10%)

Contrarily to the previous surveys, the plant's obsolescence does not seem to be a factor inducing settling problems anymore, as upgrades and/or new constructions have been performed to comply with the legislation.

6.8.2.4 Correlation between settling/foaming problems and the process configuration of the surveyed WWTPs

Statistical tests (ANOVA) have been performed in order to highlight the link between the occurrence of settling problems and the process configurations. No significant correlation was found with nitrogen removal or with the presence of industrial effluents even if it had been reported as an influencing factor (see section 6.8.2.3). The only factor impacting significantly the occurrence of settling/foaming was the process employed to remove phosphorus (Fig. 6.51).

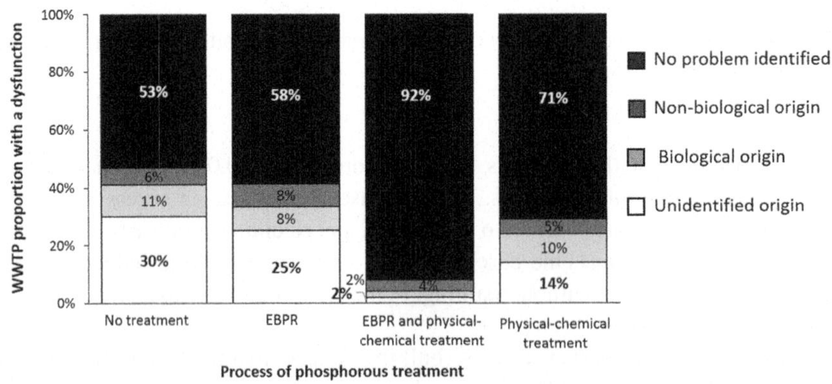

Figure 6.51 Occurrence of bulking/foaming according to the process used to remove phosphorus (WWTPs below 250,000 P.E.).

The influence of enhanced biological P removal was not statistically significant. As is well-known, the addition of metallic salts significantly enhanced settling and therefore reduces settling problems, especially when coupled with EBPR.

6.8.2.5 Control measures

According to the collected data during the last survey, corrective measures to face settling and/or foaming problems mainly consist in modifying operating parameters (64% of the responses) rather than setting up a curative treatment (17%) or both (19%).

Among modified operating parameters, reducing the mixed liquor suspended solids concentration is the most frequent measure (increasing the flow of extracted sludge or the sludge recirculation rate). Secondly, the aeration is upgraded in the cases where oxygen shortage has been identified. In terms of curative treatment, chlorination and metallic salt addition are dominant (27 and 33% of the answers, respectively).

The employed control measures are in agreement with the recommendations reported in the conclusions of the previous French survey (Graveleau *et al.* 2005; Tandoi *et al.* 2006): control the aeration/mixing conditions, sludge return septicity and sludge concentration.

6.8.3 A case study: metallic salt addition in an industrial size pilot-plant subject to M. parvicella bulking and foaming

Durban *et al.* (2016) report the impact of aluminium and iron chloride addition on the performance and the activated sludge microbiology of a BNR industrial-scale pilot plant. The pilot plant was located in the largest wastewater treatment plant of the Parisian area (Seine Aval) and consisted of an activated sludge system designed to treat the wastewater of 1500 P.E. It was continuously fed with pre-treated influent from the WWTP Seine Aval, with an average flow rate of 313 $m^3\ d^{-1}$.

In order to control recurrent M. parvicella bulking and foaming, metallic salts (Me) were added and two commercial solutions were used, containing (i) only $AlCl_3$ [30.5% (w/v) – series 1] and (ii) $AlCl_3$ and a small fraction of $FeCl_3$ [24% (w/v) of $AlCl_3$ and 8% (w/v) of $FeCl_3$, – series 2 called Al-Fe-Cl_3]. The applied dosages and the main results of these tests are reported in Table 6.15. It was shown by the authors that metallic salt addition induced no negative impact on the plant removal efficiencies, except a decrease of P-removal when additions were stopped (potentially due to a wash-out of PAOs during the dosage). The improvement of the settling conditions was certainly due to flocculation mechanisms.

A concentration between 0.2 and 0.5 mmol Me g^{-1} MLSS promotes a stabilization of settling conditions and M. parvicella growth was not affected. This dosage could be used as a preventive treatment. At a salt concentration higher than 0.7 mmol Me g^{-1} MLSS, the DSVI value decreases quickly and filamentous

Table 6.15 Metallic salt addition – applied dosage and main results.

Chemical Solution	Phase	Dosage (mmol Me kg^{-1} MLSS d^{-1})	Duration (*d*)	Observations
$AlCl_3$ *(30.5%)*	Initial state	0	18	DSVI ≈ 157 mL g^{-1} – filamentous index (FI) ≈ 5 with M. parvicella as dominant filamentous bacteria.
	Trial 1	41	70	Stabilization of the DSVI value below 150 mL g^{-1} and foam disappearance without significant impact on filamentous bacteria or floc morphology.
	Intermediate state	0	24	Increase in DSVI and FI – Progressive reappearance of foam.
	Trial 2	137	14	Significant decrease in DSVI (<110 mL g^{-1}) and FI (<3) with no longer free filamentous bacteria in the outside of the flocs, which became larger and more compact. The filamentous bacteria are embedded into the biologival aggregates. Foam disappeared in less than 1 day.
	Final state	0	54	Settling characteristics progressively deteriorated.
$AlCl_3$ *(24%)* **$FeCl_3$** *(8%)*	Initial state	0	44	DSVI ≈ 173 mL g^{-1} – FI ≈ 4 with M. parvicella as dominant filamentous bacteria.
	Phase I	16	21	Stabilization of DSVI and foam disappearance without significant impact on filamentous bacteria or floc morphology.
	Phase II	94	35	Significant decrease in DSVI (<100 mL g^{-1}) with no free filamentous bacteria in the outside of the flocs, which became larger and more compact. The filamentous bacteria were embedded into the biological aggregates.
	Phase III	16	24	Progressive re-opening of flocs with a DSVI increase during Phase III
	Final state	0	54	

bacteria were embedded in the aggregates. Microbiological analyses pointed out the decrease in M. parvicella quantity and suggested a bacterial growth limitation in addition to coagulation effects.

According to these results, aluminum salts should be added progressively and over a few weeks in order to efficiently control M. parvicella bulking without impacting the remaining biomass.

In addition, an interesting correlation between the sludge volume index and the copy number of M. parvicella 16S rRNA gene was highlighted in the work of Durban *et al.* (2016), as shown in Fig. 6.52.

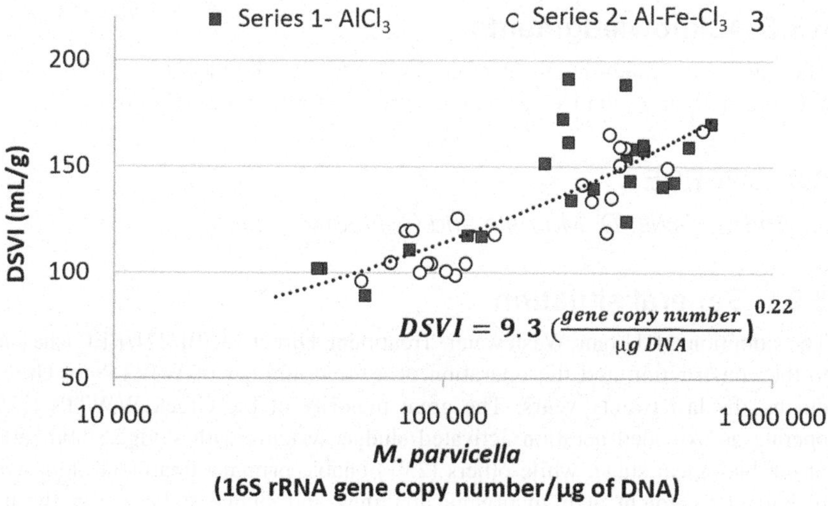

Figure 6.52 DSVI as a function of the number of gene copies of M. parvicella (Durban *et al.* 2016).

Given this correlation, measuring M. parvicella abundance by qPCR would be interesting in order to develop an early warning method to detect bulking and foaming and to follow the impact of control strategies.

This approach deserves to be deepened and extended. Other works are ongoing, developing for example the usage of phages that specifically target filamentous bacteria (Choi *et al.* 2011). Other teams are investigating the possible mechanisms involved in filament development, in order to define the intrinsic parameters for the control of the filamentous growth and to induce a hypothetical reversibility of this phenomenon (Seder-Colomina *et al.* 2015).

6.8.4 Conclusions

Several surveys carried out over the last 30 years showed that 25 to 30% of the French activated sludge wastewater treatment plants still encounter foaming and/or bulking problems. The dominant bulking filamentous bacteria were shown to be M. parvicella, Type 0041/0675 and Type 0092, in agreement with the operating conditions of the facilities (low to very low F/M ratios). The last survey performed in 2012–2013 pointed out that metallic salt addition in order to remove phosphorus significantly reduces the occurrence of bulking problems. Enhanced biological phosphorus removal coupled with metallic salt addition was also shown to be even more beneficial to settling. Control measures are still empirical, although new methods based on 16S rRNA data are being developed to early detect bulking and/or foaming issues and to follow the performance of implemented control strategies.

6.8.5 Acknowledgments

The authors are grateful to the persons involved in the completion of the last survey and especially to David Delage and Adrien Petit for their contribution.

6.9 GREECE

A. Andreadakis, D. Mamais and C. Noutsopoulos

6.9.1 General situation

The adoption of Urban Wastewater Treatment Directive (91/271/EEC) has led to the construction and the operation of a great number of WWTPs in Greece during the last twenty years. The great majority of the Greek WWTPs (77%) operate as extended aeration activated sludge systems with sludge stabilization in the biological stage, while others (23%) enable primary treatment along with biological treatment at high organic loadings and sludge stabilization through anaerobic digestion. Secondary treatment for C and N removal is applied in most of the WWTPs (either as pre-denitrification or as simultaneous nitrification/denitrification systems, e.g. Carrousel, oxidation ditches, etc.) and serve more than 75% of total P.E. (Table 6.16), while phosphorus removal (mainly by EBPR and/or by chemicals addition) is being implemented in WWTPs serving 16.3% of total equivalent P.E. Advanced treatment (sand filtration, micro-filtration, etc.) is applied in 15% of the WWTPs (serving 7.6% of the total P.E.). Chlorination is the main disinfection method (80% of the WWTPs) while UV (10%) and ozonation (3%) are alternative methods also applied.

6.9.2 Reason for dysfunctions and filamentous bacteria identified

The major settling problems encountered in Greek WWTPs are filamentous bulking and foaming. A national survey aiming to record bulking problems and dominant filamentous bacteria in 20 of the largest WWTPs of Greece was conducted by the Sanitary Engineering Laboratory of NTUA. Based on the findings of this study, 80% of the activated sludge systems exhibited occasionally filamentous bulking and/or foaming problems, while one out of three of the surveyed plants experienced settling problems on a rather permanent basis (>80% of the time). Furthermore, 35% of the WWTPs encountered intense filamentous bulking problems (DSVI > 250 mL g^{-1} SS) for almost 25% of the year and one out of ten plants exhibited intense bulking problems on a permanent basis. Most than half of the WWTPs exhibited a filament index (FI) greater than 3 (in a scale from 0–5 with a maximum FI equal to 5) during the winter period, while FI_{50} and FI_{80} values of all WWTPS surveyed (values corresponding to the 50% and the 80% of the time) were equal to 2 and 3.5 respectively.

Table 6.16 Types of wastewater treatment in Greek WWTPs and population served (Greek Ministry of Environment, 2015).

Treatment	Thousand Inhabitants Equivalent	% Treated Wastewater
Primary	–	–
Secondary (C,N removal)	8287	76.1
Secondary (C,N, P removal)	1771	16.3
Tertiary treatment	825	7.6
TOTAL	10,883	100

The filamentous microorganism that dominates most of the activated sludge units in Greece is by far M. parvicella, followed by Type 0092, *Gordonia* spp., Type 0041 and *N. limicola* (Fig. 6.53). Microthrix parvicella occurrence exhibits a seasonal pattern with the microorganism being more often dominant during the winter and spring periods, while its dominance is being reduced during summer and autumn periods (the opposite is observed for Type 0092). This seasonal pattern of biomass composition is associated with a similar seasonal variation of the settling characteristics in terms of FI's and SVI values (Noutsopoulos *et al.* 2007). Dominance of M. parvicella is more pronounced in the case of high DSVI (>150 mL g^{-1} SS) and Filament Index values (Fig. 6.54). Operation of 85% of the surveyed WWTPs as extended aeration activated sludge units resulted in microbial communities which were dominated by low F/M filamentous microorganisms. Filamentous microorganisms mostly found in conventional activated sludge systems (Type 021N, *S. natans, Thiothrix*) were absent from almost all BNP systems as a result of the metabolic selection imposed. Microthrix parvicella dominance is positively related with the presence of primary settling tanks (dominant in 73% of WWTPs with primary settling tanks and only in 33% of the WWTPs without primary treatment), while its proliferation is also favored in carousel and oxidation ditches and intermittent aeration systems. According to our experience (Noutsopoulos *et al.* 2002, 2006, 2012), effective control of M. parvicella growth can be achieved by imposing conditions of high organic loading throughout the bioreactor. This can be achieved by introducing plug flow (by proper compartmentalization) throughout the bioreactor or by operating in an SBR mode or alternatively by operating at low SRTs (unfortunately due to nitrification limitations this is almost impossible when nutrient removal is required).

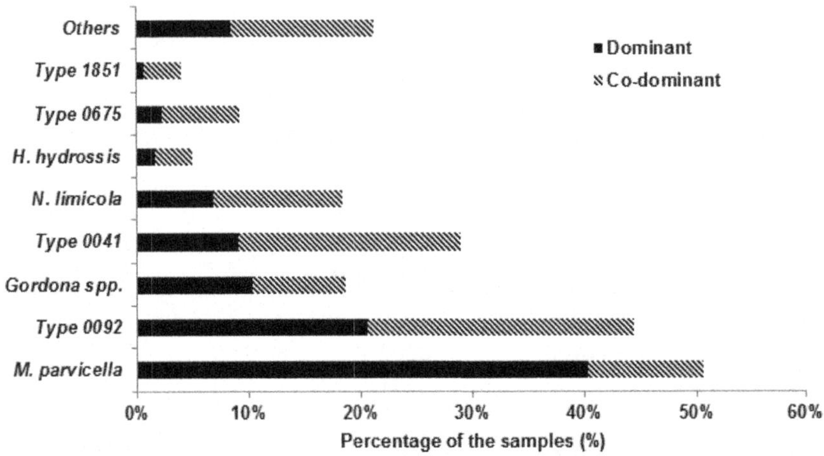

Figure 6.53 Dominant filamentous microorganisms in Greek WWTPs surveyed (Noutsopoulos *et al.* 2007).

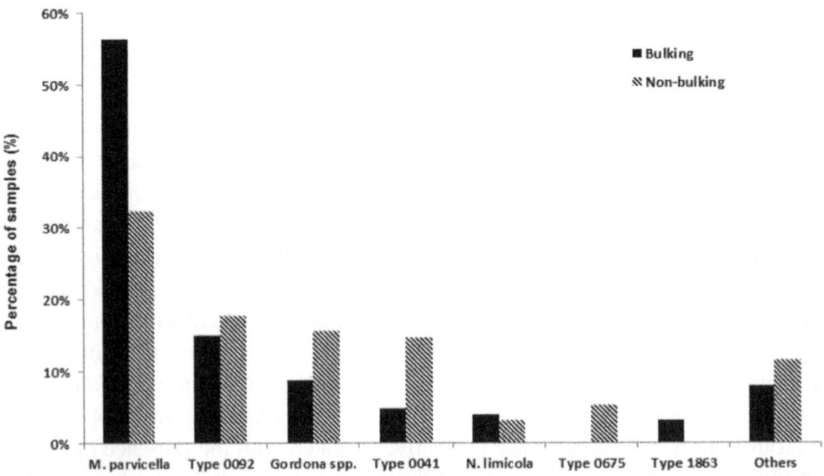

Figure 6.54 Occurrence of dominant filamentous microorganisms in bulking and non-bulking samples.

FISH has also been applied for the assessment of the dominant filamentous bacteria in foam samples of three WWTPs. In all cases, foam samples were dominated by M. parvicella and *Gordona* spp. with filament counts ranging between 2.3 and 8.6×10^7 intersection g^{-1} VSS (Mamais *et al.* 2012). According to the findings of this study, it is concluded that significant amounts of both filamentous microorganisms are being recycled to the activated sludge units through the sludge liquors.

6.9.3 Solution adopted

A series of measures have been applied to combat filamentous bulking and foaming problems in Greece. Some of these, such as changing RAS flow-rate, sludge age, are rather not-targeted measures not directly facing the problems. Selectors have been incorporated in many of the most recently constructed WWTPs; however their efficiency is unknown to the authors. Non-specific measures have also been implemented, such as chlorination, addition of aluminium salts (in the form of polyaluminum chloride), foam spraying, etc. However there isn't any survey undertaken in Greece to record in a systematic way the efficiency of measures followed to control filamentous bulking and/or foaming problems.

The authors have participated in the assessment and implementation of several filamentous bulking and foaming control strategies in several activated sludge systems. In most cases, non-specific methods were proposed and their effectiveness was evaluated through excessive monitoring of the microbial consortia and the settling and foaming characteristics of biomass (Mamais *et al.* 2012). An example of a successful application of polyaluminum chloride to control a serious filamentous foaming incident in Preveza WWTP is presented in Fig. 6.55. After gradually increasing the applied dose from 1.5 g Al^{3+} kg^{-1} SS d^{-1} to 2.5 g Al^{3+} kg^{-1} SS d^{-1} the proliferation of M. parvicella and *Gordonia* spp. was successfully controlled, the quality of final effluent was satisfactorily improved and that accumulated on the surface of biological reactor filamentous foam was significantly decreased (Fig. 6.55).

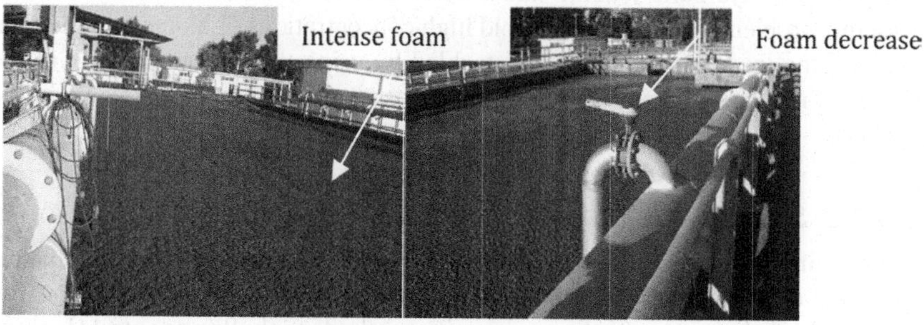

Figure 6.55 Decrease of filamentous foam as a result of polyaluminum chloride addition (applied dose 2.5 g Al^{3+} kg^{-1} SS d^{-1}).

6.9.4 A Greek case study

The case study refers to Ioannina WWTP which serves a population equivalent of approximately 135,000. It treats mainly domestic wastewater, with a small proportion of industrial waste. The plant has been constructed in two phases and now provides for primary and secondary treatment including nutrient removal followed by tertiary treatment via filtration and chlorine disinfection. It is a low F/M activated sludge plant which aims to perform biological removal of organic carbon, nitrogen and phosphorus. Primary sludge, after gravity thickening, is being digested anaerobically, while the mixture of digested primary sludge and surplus activated sludge undergoes mechanical dewatering through centrifuges.

6.9.4.1 Settling and foaming problems

The plant was subjected periodically to moderate–high DSVI values which exhibited a specific seasonal pattern. The $DSVI_{50}$ and $DSVI_{95}$ (values corresponding to 50% and 95% of the time) of the system were equal to 162 mL g^{-1} SS and 240 mL g^{-1} SS respectively during a period of 3 years. The settling ability of the biomass deteriorated mostly during the winter and spring periods. The dominant filamentous bacteria during these periods were M. parvicella, *N. limicola,* Type 0041, Type 0092, Type 0803, Type 1851 and Type 0675.

During January 2009, the plant exhibited very intense settling problems with DVSI values between 260 and 350 mL g^{-1} SS with an average value equal to 300 mL g^{-1} SS. In the same period, filamentous microorganisms dominated the microbial community of the activated sludge system and the filament index ranged between 4 and 4.5 (in the 0–5 Eikelboom scale). Bioflocs exhibited significant bridging and open structure mainly due to the presence of filamentous microorganisms outside the flocs. The dominant filamentous microorganisms, during this period, were in descending order: Type 0803, Type 0041, Type 1851, Type 0092 and M. parvicella. As a result of the unfavorable settling characteristics of biomass, the sludge blanket was elevated to 0.5 m from the weir level, thus limiting significantly the clarifiers' hydraulic loading and overall WWTP treatment capacity. To deal with this problem and in order to avoid high concentrations of biomass in the final effluent, the incoming flow rate was tackled by using the storage capacity of the sewerage network, a practice which imposed several operational problems in the network.

6.9.4.2 Setting up and applying a control strategy

Considering that the microbial consortium and the filamentous microorganisms prevailed in the biocenosis of the activated sludge system, chlorination appeared to be the best choice among the non-specific methods for bulking control (Jenkins *et al.* 1993). Therefore chlorine addition in the form of NaOCl solution (density 1.2 kg L^{-1}, 13% w/w) was started at day 30 (day 1 is the first day of monitoring of the settling properties of the activated sludge system on a daily basis). Throughout the application period, sludge quality through microscopic examination, DSVI measurements and effluent quality were monitored on a daily basis. Chlorine was fed to the RAS pumping station at a dose of 3.3 g Cl_2 kg^{-1} MLSS d^{-1}. The dosing point (sludge recirculation pumping station) was selected in order to achieve satisfactory mixing. More specifically, chlorine concentration in RAS was equal to 5.4 mg L^{-1} and the frequency of exposure of the solids inventory to chlorine was equal to almost 3 times per day. Chlorination at the aforementioned conditions was practiced for 12 days. During this period DSVI values decreased from 300 mL g^{-1} SS to approximately 225 mL g^{-1} SS and the filament index decreased from 4.5 to 3.5 (Fig. 6.56a). However the microscopic characterization of the biomass was still not satisfactory. Biomass was characterized by irregular and diffuse flocs with extensive bridging. The unsatisfactory results of chlorination were also evident by the high presence of free-living ciliates and filamentous bacteria in the bulk of biomass. Dominant filamentous microorganisms in the biocenosis of the activated sludge system remained, in descending order: Type 0803, Type 1851, Type 0041, Type 0092, *N. limicola* and M. parvicella (Fig. 6.56b). Specific filament indices of Type 0803 and Type 1851 were practically the same as those of the period before chlorination, besides the fact that both microorganisms are often found free in the bulk solution and were expected to be affected by chlorination. Due to the limited chlorination effects, chlorine dose was increased to 4.1 g Cl_2 kg^{-1} MLSS d^{-1} for 15 days (day 42–day 57). Following the increase of chlorine dose, the settling characteristics of biomass were significantly improved. More specifically, DSVI values abruptly decreased from 225 mL g^{-1} SS to 120 mL g^{-1} SS, while a filament index value of 2 was achieved. The chlorination effect was also evidenced through microscopic examinations with the progressive deformation and detachment of cells from the sheath and eventually the empty and broken sheaths (Fig. 6.57). As a result of the decrease of the filamentous microorganisms predominantly grown in the bulk solution, interfloc bridging was moderate and the flocs were more compact with good filament backbone. Along with the decrease in filamentous microorganisms, an equivalent decrease (but not complete loss) of the free-living ciliates was also recorded. Fifteen days after ending chlorine addition (period days 57–72) few free living filamentous microorganisms were recorded in the bulk solution (most with obvious signs of the chlorination effect), while protozoa started again to be present in the system. The increase of the chlorine dose from 3.1 to 4.1 g Cl_2 kg^{-1} MLSS d^{-1} resulted in the washout of Type 0803 after a period of 20 d and the significant control of the presence of Type 1851 after a period of 45 d. After the end of this period, M. parvicella (practically unaffected by chlorination) was the dominant filamentous bacterium but in limited numbers. As a result of the satisfactory improvement of the settling properties, the sludge blanket in the secondary clarifiers was efficiently decreased and the solids flux increased (along with influent wastewater flow) by 50–60%.

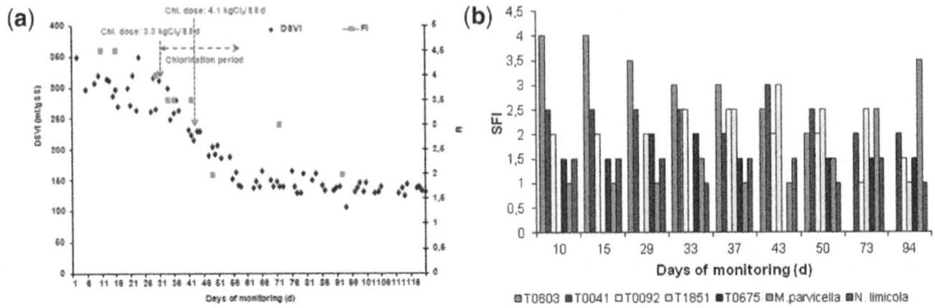

Figure 6.56 Evolution of settling properties through the monitoring period; (**a**) DSVI, (**b**) dominant filamentous microorganisms.

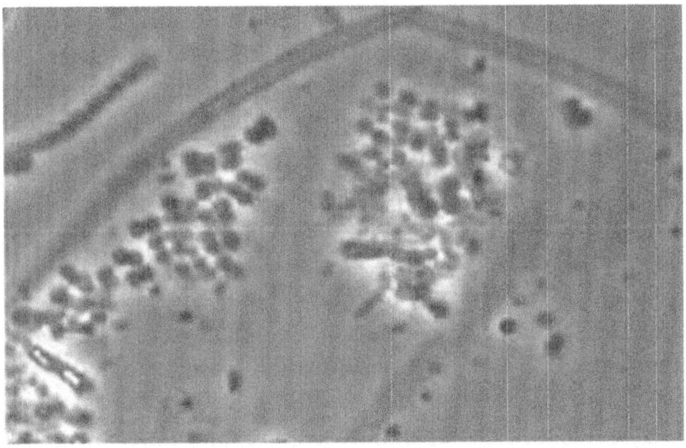

Figure 6.57 Severe chlorination effect on filamentous microorganisms present in the bulk solution (Sanitary Engineering Laboratory, 1000×, phase contrast).

Chlorination effects in final effluent quality were closely monitored throughout the application period to prevent any possible chlorine overdose. Throughout the evaluation period, turbidity and organic carbon concentrations in the treated wastewater were practically unaffected. Oxygen utilization rates were also unaffected by chlorination, along with nitrogen and phosphorus removal. More specifically, average ammonium–nitrogen final effluent concentrations before, during and after chlorine additions were equal to 0.6 mg L^{-1}, 0.9 mg L^{-1} and 0.9 mg L^{-1} respectively, whereas nitrification rates ranged between 1.8 and 2 mg N g^{-1} VSS h^{-1}. Finally, the possibility of THM formation as a result of chlorine addition for a period of 27 d, was also evaluated. THM concentration in the final effluent before (day 10) and after chlorination (day 75) were very low, of the order of 0.2–0.3 μg L^{-1}, with bromoform, dibromochloromethane and bromodichloromethane being below the detection limit (0.01 μg L^{-1}).

6.10 ITALY

S. Badoer, C. Davoli and V. Tandoi

6.10.1 General situation

There are 18,786 municipal wastewater activated sludge plants in Italy (96.7% in operation, Istat, 2014); the large majority of domestic wastewater (70.9% domestic, 19.1% industrial) is treated (Table 6.17) by secondary (35.0%) or advanced treatment (60.1%). Despite the wastewater advanced treatment plants being only 10% of the total existing plants, they treat the largest part of the wastewater (more than 60%): a precise inventory of industrial wastewater treatment plants is not available.

Table 6.17 Waste influent in urban wastewater treatment plants, for plant typology (Census of Water for Civil Use, ISTAT, 30 June 2014).

Treatment	Thousand Inhabitants Equivalent	% Treated Water
Imhoff	1190	158
Primary	2522	335
Secondary	26,304	3498
Advanced	45,177	6008
TOTAL	75,192	100

The plants have a mainly aerobic or anoxic/aerobic configuration and less than 5% of the influent wastewater is treated only partially. At present very few EBPR plants are present in Italy, and this is because, apart from sensitive areas, the limit for $P\text{-}PO_4$ discharge is still quite high (10 mg L^{-1}) and also because this technology does not appear to have convinced the market. A considerable number of activated sludge plants in Italy still suffer periodic dysfunctions (especially bulking and foaming problems), even though the situation has improved in the last 10 years, when the first edition of this book was published (Tandoi *et al.* 2006) and 30–40% of plants presented bulking and foaming phenomena.

Microscopic observation of activated sludge is now common practice, especially in larger plants, and some data are available on the main filamentous bacteria present. These are identified using the manuals of Eikelboom and van Buijsen, 1983; Jenkins *et al.* 2003 or Italian versions of these manuals (Cingolani & Ciccarelli, 1996; Madoni *et al.* 1996; Rossetti & Tandoi, 1999).

6.10.2 Characterizing the activated sludge and the qualification circuit

While biomolecular studies are in progress in the country, the large majority of plant operators and management, regularly apply the traditional microscopic characterization of activated sludge for its properties and description of the main filamentous bacteria. To diffuse this approach through plant operators, CNR-IRSA, University of Palermo and Gruppo IREN, have been organizing yearly training courses since 1992. Three training courses have also been organized by CNR-IRSA on the application of the FISH technique for the identification of the main activated sludge microbial components. Considering that traditional microscopic observation is still the widest applied technique, over recent years, in order to guarantee homogenization of the results, three Italian Institutions (CNR-IRSA, Iren (ex-ENIA) and Università di Parma) have organized a qualification circuit (www.irsa.cnr.it), sending two selected activated sludge samples to participants, and comparing their results with those of eight national experts, according to the directions of UNI CEI EN ISO/IEC 17025, 2005. A score is attributed according to the agreement of the results of participants with those of the expert team; the properties that have to be evaluated are the filament abundance, filament effect of flocs, floc diameter class, inter flocs supernatant aspect, floc morphology, dominant filament, secondary filaments (two have to be indicated). In the four circuits organized (2008, 2010, 2012, 2014) 57 laboratories were qualified for activated sludge characterization on 70 participants (Davoli *et al.* 2016).

6.10.3 Filament surveys

Several surveys have been made on the filaments responsible for bulking and foaming problems, in domestic plants in Southern Italy (Di Marino *et al.* 1992) and Central Italy (Rossetti *et al.* 1994), and in industrial plants (Molinari & Canziani, 1992). Wider surveys were made all around the country (Madoni *et al.* 2000) and in Northern Italy WWTPs (Pasinetti & Emondi, 2005). All these surveys had been made several years ago, by traditional microscopic description, establishing the basis for proper comparisons; FISH was introduced later and it is now improving its utilization.

A very comprehensive survey was made in the Puglia Region, when filamentous bacteria and chemical physical parameters (COD; BOD, TKN, Redox potential, etc.) were accurately reported, in 172 plants in the Region (Di Marino *et al.* 1992). Approximately 30% of the 172 plants examined had problems, and severe filamentous bulking (SVI > 300 mL g^{-1}, filament abundance 5 or 6) was the most common problem encountered (24 cases).

The main microorganisms causing problems were *Thiothrix* (dominant in 9 and secondary in 10 cases), Type 021N (dominant in 7 and secondary in 8 cases), M. parvicella (dominant in 4 and secondary in 4 cases), and *Nocardia* (dominant in 2 and secondary in 8 cases). *Sphaerotilus natans* and Type 1701 were dominant only in one case and others (*Haliscomenobacter hydrossis*, Type 0041; *Nostocoida limicola* and Type 1863) were present only as secondary filaments. The sewage very often had a low redox potential which may favour the abundance of *Thiothrix* and Type 021N. Foaming was a problem in 5 cases (caused by M. parvicella and *Nocardia*) and mainly resulted from long sludge age and recirculation of scum to the head the plant.

In the Lazio Region (Central Italy), 39 activated sludge plants were monitored (Rossetti *et al.* 1994). Problems with the activated sludge process were extensive: 46% of surveyed plants had filaments at abundance levels *very common*, 30% *abundant* and 5% *excessive.* The main filamentous bacteria present in the plants were M. parvicella and nocardioforms (dominant in 32 and 15% of the plants respectively) and Type 0092, Type 0041 and Type 0675. The wastewater characteristic was quite different from Southern Italy, probably because the wastes were not as septic as in Southern Italy.

In Northern Italy, five plants with more severe problems were selected from amongst 15 N/DN plants treating wastes from various industries (fine chemicals, tannery, petrochemical, pharmaceutical, food processing). The main filaments were Type 0041 (dominant twice, and secondary three times), *Nostocoida limicola* (dominant once, secondary three times), Type 0092 (dominant once, secondary twice), *Haliscomenobacter hydrossis* (dominant once), and a group of other organisms found only as secondary filaments (Type 0675, *Nocardia* spp., M. parvicella, Type 1701, Type 1851, Type 1863). These data again confirm that the organisms that causes the main problems of bulking and foaming, M. parvicella and *Nocardia*, play a minor role in activated sludge fed with industrial wastewater (Molinari & Canziani, 1992).

Five hundred plants had been contacted in Italy, which elicited 167 answers, complete with data and sludge samples (Madoni *et al.* 2000); 66% of plants were affected by bulking or foaming, and 33% suffered both problems. Microthrix parvicella, Type 0041 and Type 021N were the most common filaments in bulking sludges, while foams were mostly dominated by M. parvicella and GALOs. Microthrix parvicella and Type 0092 dominated conventional and BNR plants, while *Nostocoida limicola* and Type 1701 dominated only in conventional plants. The main problems remarked on by plant operators was the deterioration of effluent quality, due to the effluent suspended solids increase.

In Northern Italy, a survey has been done on 86 WWTPs (Pasinetti & Emondi, 2005), mainly industrial (70), in a wide F/M range (0.2–0.8 kg COD kg^{-1} SS d^{-1}). The situation has improved from the past years: foaming is a problem in only 36% of domestic WWTPs while no bulking problems have been recorded in the others. Bulking is a problem in only 21% of industrial WWTPs (no foaming problem and never GALO or M. parvicella are detected in such WWTPs). In domestic plants the more abundant organisms were GALO, Type 0041/0675, M. parvicella, *Nostocoida limicola, Thiothrix,* Type 0092. A wide variety of filamentous bacteria were found in industrial plants: the more abundant were Type 0041/0675, *Nostocoida limicola, Thiothrix,* plus others such as *S. natans,* Type 1701, Type 0803, Type 1851, Type 1863, Type 0914, *H. hydrossis.*

6.10.4 Control methods

The introduction of systematic activated sludge monitoring, either from a microbiological or operating point of view, changed the approach of curing dysfunctions, starting with a whole picture of the plant operating and design parameters, to individuate the dysfunction causes. Both specific and non-specific methods for controlling bulking and foaming phenomena have been utilized in Italy.

6.10.4.1 Non-specific methods

Coagulants: Coagulants such as $FeCl_3$ (50 mg L^{-1}), Ca $(OH)_2$ (50 mg L^{-1}), had been used in the past, especially in small treatment plants, and often in conjunction with cationic polyelectrolytes (1–2 mg L^{-1}, Vismara, 1998).

Oxidant compounds: Sodium hypochlorite is used for bulking control at a dosage of 5–15 mg L^{-1} in the RAS. Usually the sludge is exposed at least three times per day to the effect of the chlorine (Cingolani & Ramadori, 1992; Beccari *et al.* 1995). The use of H_2O_2 at 10–15 mg L^{-1} doses (Vismara, 1998), is very limited because its cost is higher and it is not as readily available as sodium hypochlorite at the plants. Nowadays, the approach mostly utilized, is to figure out the reason for the dysfunctions (not always due to filamentous bacteria presence) and to find suitable operational conditions to avoid them.

6.10.4.2 Specific methods

Specific methods of bulking control are usually based on the correlation between filaments and cause (Jenkins *et al.* 2006). Success was obtained by using 'contact zone' (2 solved cases on 2 tested), nutrient addition (7 solved cases on 8 tentatives), antifoaming addition (6 solved cases on 7 tentatives), coagulant addition (24 solved cases on 33 tentatives), sludge age decrease (40 solved cases on 75 tentatives), dissolved oxygen decrease (19 solved cases on 30 tentatives), mechanical removal of foam (17 solved cases on 27 tentatives) (Madoni *et al.* 2000).

6.10.4.3 Case studies with specific solutions

Several specific solutions had been successfully adopted at full-scale plants: these can be summarized as follows:

- *Very low sludge age utilization*

  A substantial decrease of sludge age was necessary at Varese-Olona treatment plant, which experienced significant foaming problems on the aeration basin and final sedimentation tank: SVI values were typically 250 mL g^{-1} and microscopic observations showed very small flocs dominated by *Nocardia* and M. parvicella.

  The strategy adopted (Bargna *et al.* 1993) was to reduce the sludge age from 15 to 2 days, over a two month period. A progressive improvement in both the quality of the effluent and the settling characteristics of the sludge occurred. The SVI stabilized under 200 ml g^{-1} and COD removal efficiency increased. The flocs gradually became more compact and *Nocardia* and

M. parvicella eventually disappeared. Nitrification capacity was lost but the nitrogen removal by biomass synthesis was enough to produce an effluent that complied with the Italian limit for N (15 mg L^{-1}).

- *Type 021N and Thiothrix reduction by converting an anoxic selector in an existing basin*

  The Area Nolana Treatment Plant is designed for 450,000 P.E. and currently receives one third of its design flow. Several rectangular aeration basins constitute the biological section. Sludge age is normally 4 days and the organic load is 0.3 kg BOD_5 kg^{-1} MLSS d^{-1}. Like many plants in Southern Italy, the RAS is introduced into the primary settling tanks, and the waste sludge is taken from the bottom of the primary settler (a mixture of primary and secondary sludge). This procedure causes two problems:

  - it is very difficult to control activated sludge; the sludge age is normally higher than expected
  - fermentation occurs in the primary tank

  The effluent of the primary tank normally has a very low ORP (–125 mV) and contains high concentrations of short chain fatty acids and sulfides, which can promote the growth of the filaments Type 021N and *Thiothrix*.

  Significant problems occur in the summer when the seasonal industries (tomato canning) produce a large increase of VFAs in the sewage. In this period the settling properties of the activated sludge deteriorate dramatically (DSVI up to 900 mL g^{-1}) and the content of filaments Type 021N and *Thiothrix*, increases significantly.

  Chemical addition ($FeCl_3$ 40 g m^{-3} for 10 d) and sodium hypochlorite addition (3 g Cl_2 kg^{-1} SS d^{-1} to the RAS for 15 d^{-1}) did not produce any settling improvement.

  An anoxic selector (250 m^3, 1/24 of the volume of oxidation basins) was installed (Cesaro, 1997) utilizing a basin to transfer the mixed liquor, by three screw pumps to the aeration basin. The recirculation flow and the influent mixed together in this selector. This simple solution was implemented by using existing facilities.

  The selector hydraulic retention time (influent plus recirculation flow) was 7.4 min, and the NO_3-N in the recirculation flow 5–7 mg L^{-1}. At the same time, the sludge age was reduced (to 2.5 days) and the aeration basin DO increased (3.5 mg L^{-1} and 5.8 mg L at the head and the effluent end respectively).

  The modifications introduced were kept for 80 days (industrial waste overload). The DSVI (initial value 800 mL g^{-1}) decreased after 15 d and after 35 d reached 190 mL g^{-1}. The ORP in the selector was approximately –110 mV, indicating anoxic conditions.

  Seventy-five per cent of the influent RBCOD (161 mg L^{-1}) was taken up in the anoxic selector. The selector mainly affected Type 021N populations, which decreased from abundance level 6 to 1–2. The effect was less pronounced on the *Thiothrix* population, which decreased from an abundance level 6 to 3–4. It is well known that Type 021N affects more the settling properties, because of the bridge effect on the flocs.

- *Selector investigations on four large WWTP in Central Italy*

  Full-scale application have been tested in four large WWTP in Central Italy, suffering seasonal bulking and foaming problems: the effect of anoxic selectors has been investigated (Davoli *et al.* 1997). Plant A, C and D operated in plug flow conditions, while plant B was CSTR. The selectors worked with a soluble COD concentration in the range 14–69 mg L^{-1}, and were effective in two cases, Plants A and C, where there was a lower amount of soluble COD measured in the selector effluent.

- *M. parvicella elimination by selective removal of oil and greases*

  A specific case of bulking due to M. parvicella and Type 021N in a piggery slaughterhouse was solved taking into account the capacity of the first bacterium to utilize oils and greases present in large amount in the wastewater, deriving from the pig carcass processing. Removing these compounds and bypassing them downstream of the production cycle to a proper disposal, was enough to solve the problem. After the action, M. parvicella and Type 021N never exhibited an excessive growth anymore, the waste being well balanced between organic substances, nitrogen and phosphorus: other filamentous bacteria, such as *Thiothrix*, Type 021N and *S. natans*, which are favored by nutrient imbalance, never proliferated significantly and the sludge was characterized by good settleability and Sludge Biotic Index (Pedrazzani *et al.* 2016).

- *M. parvicella elimination by poly aluminium chloride at Fusina treatment plant*

  The Fusina (VN) biological treatment plant treats wastewater from the industrial and civil area of Venice, for approx. 400,000 equivalent inhabitants. During past winters the plant suffered intense biological scum formation, due to the presence of the bacterium M. parvicella (Miana *et al.* 2002).

  The plant was carefully monitored in the years 2001–2002, in order to individuate the reasons for the M. parvicella proliferation, and to find solutions for containing its proliferation. A positive correlation was found, once again, between the temperature and the presence of the organism in activated sludge (Fig. 6.58), and consequently the percentage of aeration basins covered by scum, in the winter period. No correlation was found with either surfactant or oil content in the influent; their concentrations in the influent did not change substantially during the course of the year.

  The solutions adopted. In February 2002, a specific compound was tested, polyaluminium chloride, whose efficacy has been reported (Váňa & Wanner, 1996; Eikelboom, 1997).

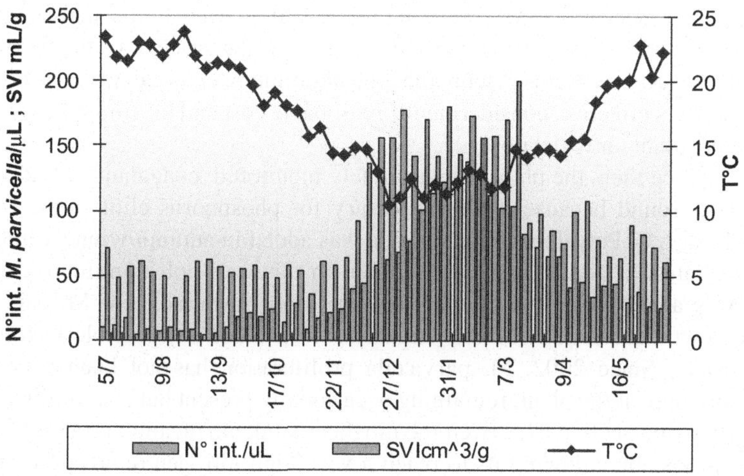

Figure 6.58 M. parvicella, SVI and temperature trends during the period of experimentation 2000–2001 (Badoer *et al.* 2003).

The dosage of 1.5 g Al kg^{-1} MLSS d in return sludge, for a period of 42 days, caused, already after 2 weeks dosage, a dramatic reduction of the number of free filaments either present in the mixed liquor or the foam covered surface (Fig. 6.59). The SVI decreased from 188 to 88 mL g^{-1} and analogously the suspended solids in the final effluent (from 60 to 24 mg L^{-1}) (Badoer *et al.* 2003). The M. parvicella filaments during the treatment appeared damaged and eventually were incorporated into the biological flocs.

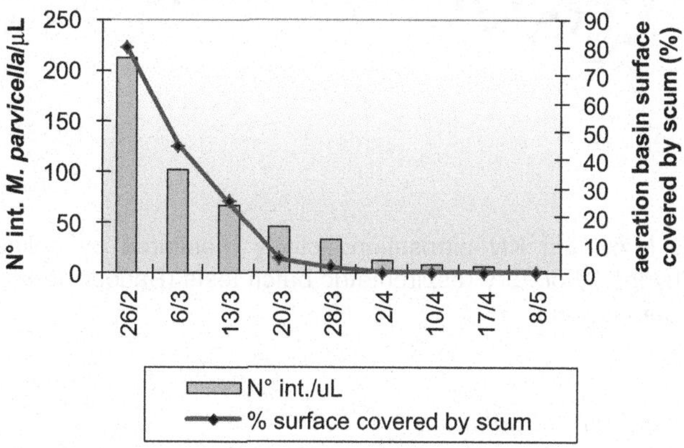

Figure 6.59 M. parvicella presence and aeration basin percentage covered by foam, during the polyaluminium chloride addition, February/May 2002 (Badoer *et al.* 2003).

Considering that M. parvicella reaches high levels in the sludge only for a limited period of time, normally in winter, the strategy of monitoring its presence, intervening with the polyaluminium chloride when necessary, appears effective, considering the reasonable cost and the limited days of the coagulant (or toxic) dosage.

Since then, the plant was accurately monitored: coagulant was added all year round because it was necessary for phosphorus elimination (below 1 mg L^{-1}). Polyaluminium chloride was added in autumn/winter while iron sulfate was added in spring/summer. In this way, polyaluminium chloride was added when the conditions were more favourable for M. parvicella growth, and the alternation with iron salts avoided any possible acclimation effect. Since 2002, M. parvicella proliferation has not been a problem anymore at the plant, remaining occasionally present but at a minimal level (few intersection µL). Even the nitrification/denitrification process became more stable allowing us to reach a very high nitrogen removal efficiency. The nitrification capacity of the plant is continuously monitored by laboratory respirometric measurements and by an automatic online titrator: the nitrosation kinetic activity is in the range 0.019–0.072 kg N-NH4 kg^{-1} VSS d^{-1}; FISH analysis showed AOB and NOB presence (estimated by FISH) in the range 5–14 and 2–6% respectively (AOB/DAPI, Fig. 6.60, Badoer *et al.* 2015).

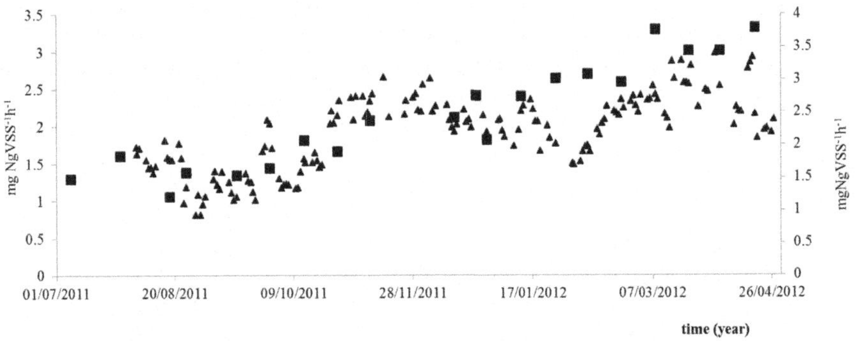

Figure 6.60 Maximum KN nitrosation activity measured by online automatic titrator and by the laboratory respirometric batch tests (Badoer *et al.* 2015). (With the Editor's permission.)

- *Additional specific solutions*

Given that activated sludge process dysfunctions can be due to a wide range of causes, Table 6.18 reports some particular cases solved in the period 2011–2016 (anonymous plants).

Table 6.18 Particular solutions adopted for solving bulking and foaming in full-scale Italian WWTP in Italy in the period 2011–2016 (authors' personal data)

Plant Typology	Problem	Filamentous Bacteria Present	Solution Adopted	Note
Urban (8000 m^3)	Large amount of foam on aeration basins and settling tank	Nocardioforms In limited amount and in activated sludge and foam at the same level of abundance	Decreased the air immission	In the SVI test the sludge does not settle but goes in the upper part of the cylinder. Typical overareation
Pharmaceutical Denitrification/ nitrification plant (5000 m^3)	Severe bulking problem	Type 0675 and Type 0092	Toxic stream in the plant individuated and treated separately by chemical / physical process	This situation suggests that these two filamentous bacteria are more robust and resists to the toxic effect better then floc-formers
Pharmaceutical aerobic plant (4000 m^3)	Severe bulking problem	*N. limicola* II (Alfaproteobacteria), Type 0041	Anaerobic selector, periodic chlorination	The solution adopted, even though not definitive, allows one to operate the plant
Food processing aerobic plant. (6000 m^3)	Severe bulking problem	*N. limicola* II (Alfaproteobacteria), at abundance level: 6	Oxidizing agents unsuccessful; Settling operation optimized	The low ratio VSS/ SS (50–60%), allows one to operate the process even with an excessive amount of filaments

6.11 MALAYSIA

N. Syamimi Zaidi, K. Muda, J. Sohaili and M. Sillanpää

6.11.1 General situation

In Malaysia, extensive use has been made of primary treatment systems such as communal septic tanks, Imhoff tanks and unreliable low cost secondary systems such as oxidation ponds (as in Table 6.19). As for large urban areas, Individual Septic Tanks (IST) are being implemented. It is estimated that there are over one million individual septic tanks in Malaysia. These tanks only partially treat sewage, discharging effluents that are still rich with the content of organic materials into the river. This phenomenon has the potential to create public health and environmental problems, particularly in urban areas.

Table 6.19 Public sewage treatment plants in Malaysia (www.iwk.com.my).

Types of Sewage Treatment Plant	Number of Plants (as at October 2014)	Population Equivalent
Imhoff tank	679	507,648
Oxidation ponds	403	1,681,176
Mechanical plants	4902	18,665,408
Network pump stations	982	4,852,844
Communal septic tank	3625	405,432

Indah Water Konsortium (IWK) is a wholly government-owned company which has been entrusted with the task of developing and maintaining a modern and efficient sewerage system for all Malaysians. This company is responsible for planning and rationalizing the public sewerage facilities to reduce the number of treatment plants using the 'multipoint concept' or regionalization. Sewerage pipeline networks will then be laid in urban areas currently serviced by IST to convey the domestic sewage to modern secondary treatment facilities. In Malaysia, 38% of public sewage treatment plants in the country are mechanical plants. These plants operate using mechanical equipment that accelerates sewage break down. It is hoped that in the long term, Malaysia's sewerage system will be made more efficient through the standardization of the types of plants used. These extensive programs are nothing short of a revolution in the management of domestic sewage in Malaysia. The entire sewerage infrastructure can expect to undergo changes. Estimates have been made of the number and type of public treatment plants currently in Malaysia. The trend will be moving towards 'mechanical plants' (as in Fig. 6.61) such as Extended Aeration (EA), Oxidation Ditch (OD), Rotating Biological Contactors (RBC), Sequenced Batch Reactors (SBR) and Trickling Filters. Careful management of this change will ensure the future of Malaysia's public sewerage systems.

Figure 6.61 Mechanical sewage treatment plants authorized by Indah Water Konsortium (IWK).

The most common biological process used in the mechanical plants is the activated sludge process (ASP). In general, the activated sludge process is comprised of two stages, namely, biological degradation, which occurs in an aeration tank, and the physical stage, which takes place in a secondary clarifier. The key feature of the biological degradation process is the responsible bacteria, which convert the incoming organic pollutants into the purified water, which later has to be separated from the activated sludge by gravity. The performance of ASP therefore, relies on bacterial activity and efficient solid–liquid separation as to guarantee good effluent quality being discharged from the system (Clauss *et al.* 1998). The diversity of the biological community in the activated sludge is very large, containing many species of bacteria, protozoa, fungi, metazoa and algae. Nevertheless, among those microbial populations, filamentous microorganisms have been reported to be able to hinder the proper function of the treatment processes (Pujol & Boutin, 1989; Ng *et al.* 2000). The existance of these filamentous microorganisms will normally cause solid–liquid separation problems that are commonly associated with the sludge bulking occurrences.

Therefore, the overall scope of this survey was to assess the extent and severity of sludge bulking problems, which may cause dysfunction of sewage treatment plants (particularly activated sludge) in Malaysia. The applied control strategies that are currently being used in Malaysia have also been presented in this chapter.

6.11.2 Causes of plant dysfunction

The application of activated sludge treatment systems is very common in Malaysia. However, its stable operation always has the potential to be plagued by the occurrence of sludge bulking. Like other countries, treatment plants in Malaysia are also faced with this bulking issue as one of the main causes of the plants' dysfunction. Nevertheless in Malaysia, the scenario of sludge bulking is not too critical and can be easily predicted and pre-controlled. The most common types of treatment system that are used in sludge bulking scenarios are Extended Aeration (EA) and Sequenced Batch Reactors (SBR), as shown in Figs 6.62 and 6.63, respectively.

Theoretically, the primary factor that induces bulking in treatment plants is the excessive proliferation of filamentous microorganisms that interfere with the settling process. This is because the filamentous microorganisms physically prevent the close packing of activated sludge flocs during the settling process. Kappeler and Gujer (1994) mentioned that a volume fraction of only 1–20% was more than sufficient to cause bulking. Kaewpipat and Grady (2002) reported that the number of filamentous microorganisms in sludge bulking can be too low to be detected but proved that with the minor existence of filamentous microorganisms, bulking can eventually occur. The excessive proliferation of filamentous microorganisms can lead to the continuous deterioration of effluent quality. Much of the sludge will be washed out with the final treated effluent. Sludge bulking can also collapse the overall performance of a treatment system (Martins *et al.* 2004a;

Lou & de Los Reyes, 2008). The presence and further proliferation of filamentous microorganisms are known to be enhanced by specific operational factors such as low dissolved oxygen (DO) concentration (Gaval & Pernelle, 2003; Rossetti *et al.* 2005; Guo *et al.* 2012), low F/M ratio, low organic loading rate (Martins *et al.* 2004b; Comas *et al.* 2008), low pH, increased concentration of sulfides (Guo *et al.* 2012), low nutrient contents (Thompson & Forster, 2003; Martins *et al.* 2011), and low temperature (Madoni *et al.* 2000; Juang, 2005; Van den Akker *et al.* 2010).

Figure 6.62 Common extended aeration (EA) in Malaysia.

Figure 6.63 One of the sequenced batch reactors (SBR) located at Kuala Lumpur, Malaysia.

In Malaysia, the filamentous sludge bulking always occurred at the treatment plants where there was rapid discharge of industrial/illegal wastewater. Such sources of wastewater contributed to the high concentration of oil and grease in the treatment plant. High concentration of oil and grease may contribute to the high food-to-microorganisms (F/M) ratio. When the F/M ratio is high, the microorganisms experience logarithmic growth (excess food and/or organic compounds), and the maximum rate of metabolism and a large removal of biochemical oxygen demand (BOD) will occur (Gray, 2010). In such a situation, even if dissolved oxygen (DO) levels in the activated sludge treatment plants are controlled, the conditions favorable for sludge bulking still occur. This may occur when DO concentration is not sufficiently high during a high F/M ratio. The supplied oxygen during this condition will be used for the biodegradation of high organic loading, leaving only a small oxygen concentration for the aerobic flocs. As a result, the formed weak flocs may disperse, making them difficult to settle. Eventually, the limited oxygen supply will favor the proliferation of filamentous microorganisms, thus inducing a sludge bulking occurrence.

6.11.3 Implemented control strategies of filamentous sludge bulking in Malaysia

Since the occurrence of sludge bulking in Malaysia was classified as not critical, there are no fix available control strategies listed with regard to such an occurrence. Generally, when the responsible authority for the treatment plants (i.e. operator of IWK) is alarmed with any sign of the sludge bulking occurrence, they only mitigated by decanting the sludge. This method is quite similar to one of the existing methods of controlling the sludge bulking under a specific approach which is minimizing the sludge retention time (SRT).

Sludge retention time (SRT) is an important factor affecting the performance of treatment systems as SRT gives control over sludge activity via biomass properties. SRT can be controlled by altering the sludge decanting rate. Theoretically, a low SRT (<0.5 day) produces sludge with a high growth rate and a high volume of waste sludge and vice-versa. A conventional activated sludge system has an SRT of 3 to 4 days, which gives good settling properties to the activated sludge. Nevertheless, longer SRT (> 6 days) causes the accumulation of dead or inactive microorganisms to occur in the system, thus affecting the sludge activity. This negative condition is prone to the growth of filamentous microorganisms hence inducing a high risk of sludge bulking occurrences. Prior to this, the approach of sufficiently minimizing the SRT is taken (Martins *et al.* 2004a; Rossetti *et al.* 2005; Xie *et al.* 2007).

SRT is a design criterion for WWTPs including activated sludge systems. This parameter is strongly related to microorganism growth. By reducing the SRT, there will be a subsequent increase of the F/M fed ratio. This condition is therefore able to minimize the risk of sludge bulking occurrences. The growth of filamentous microorganisms, particularly M. parvicella species are being inhibited under reduced SRT (Xie *et al.* 2007; Comas *et al.* 2008). Researchers also reported that strategies of reducing SRT from 15 to 7 days have resulted in significant minimization of sludge bulking with little effect on the effluent water quality (Xie *et al.* 2007). Due to minimization of SRT, the SVI value decreased from 200 mL g^{-1} to approximately 100 ml g^{-1}. This shows an improvement in the solid–liquid separation and settling of the activated sludge (Xie *et al.* 2007).

Instead of implementing the method of sludge decanting, the occurrence of sludge bulking in treatment plants in Malaysia is also being mitigated by the method of purging and by the addition of hydrogen peroxide. Albeit the hydrogen peroxide method has a history of about 7–8 years. Conceptually, hydrogen peroxide can be used to correct a serious filamentous bulking situation or, preferably, to prevent one from occurring until adjustments can be made to remove the cause. When applied to the return activated sludge, hydrogen peroxide supplies dissolved oxygen which helps restore the microbial activity necessary for effective operation, while selectively oxidizing the filaments which retard settling. Effective dosage of hydrogen peroxide is a function of time and concentration, and varies from plant to plant. To correct a serious bulking problem, immediate results may be obtained by adding 100–200 mg L^{-1} H_2O_2 to the biosolids recycle line. Once control of bulking is obtained, the dose may be reduced to 25–50 mg L^{-1} H_2O_2 to prevent re-occurrence.

6.11.4 Future scenario of sludge bulking occurrences in Malaysia

Although currently the occurrence of sludge bulking in Malaysia is not critical, it cannot be promised that the situation will be the same in the future. Throughout the years, the sources and constituents of the wastewater may vary greatly and become very complex in alignment with the modern structure of human living styles. These unpredictable sources and constituents of the wastewater that are being discharged into the treatment plants may cause the sludge bulking and trigger its severity. The activated sludge itself was intended to be very easily induced by the filamentous microorganisms, which have eventually initiated the sludge bulking occurrences. To this extent, the existing control strategy (i.e. sludge decanting, purging) may not be sufficient to control the severity of sludge bulking. Hence, application of magnetic fields is seen as a potential solution to control the filamentous sludge bulking. To our knowledge, research on the effects of magnetic fields on filamentous microorganism proliferation and sludge bulking in activated sludge systems is lacking and has not yet been fully explored.

From past research, the application of magnetic fields was evidenced in exhibiting good improvement to the system performances by enhancing the biomass concentration (Yavuz & Çelebi, 2000; Ji *et al.* 2010; Łebkowska *et al.* 2011; Zaidi *et al.* 2014) and sludge settleability (Liu *et al.* 2008; Xu *et al.* 2009; Zaidi *et al.* 2014). The magnetic field was also reported to positively influence the bacterial activity in the wastewater, resulting in the improvement of organic compound biodegradation. In recent research, wide ranges of magnetic field intensities from weak to strong intensity showed various effects to the microorganisms (Yavuz & Çelebi, 2000; Zhang *et al.* 2002; Łebkowska *et al.* 2013). Some of these microorganisms were positively influenced while others were adversely affected. This suggested that different microorganisms have their unique levels of magnetic field susceptibility. They may respond differently, either being inhibited or enhanced, similar to the filamentous microorganisms. Currently, there are no confirmed findings on the suitable range of magnetic field intensity that can be applied in order to inhibit the filamentous microorganisms' activity. Therefore, a thorough study on the effect of a magnetic field on the susceptibility level of filamentous microorganisms is essential.

During sludge bulking problems, poor effluent quality is normally due to high withdrawal of suspended solids and turbidity in the discharged effluent, which can be due to the interference of filamentous microorganisms with floc settlement and the subsequent compaction of the sludge. This results in a much thinner sludge being returned to the reactor, making the maintenance of the desired biomass concentration within the reactor system more difficult. Implementing a magnetic field in the wastewater treatment process is expected to reduce the existence of filamentous microorganisms in sludge, thus encouraging the close packing of activated sludge flocs during the settling process and further increasing the settling velocity. The significant effect of a magnetic field in activated sludge settling through improvement of suspended solids' removal and turbidity reduction have also been reported (Hattori *et al.* 2001; Chin *et al.* 2006; Zaidi *et al.* 2014). Zaidi *et al.* (2014) discovered that by implementing a magnetic feld of intensity 88 mT, the settling velocity of the activated sludge and the subsequent turbidity reduction were improved.

To sum-up, the application of a magnetic field may stand a great chance to control the sludge bulking in a more effective way. Magnetic field application can also maintain the concept of green technology, which is the main focus for sustainable development in wastewater treatment systems. Nevertheless, further investigation regarding the relationship between the magnetic field and the filamentous microorganisms as well as other treatment operational factors need to be critically conducted.

6.12 SOUTH AFRICA

N. Deepnarain, S. Kumari and F. Bux

6.12.1 General situation

The initial development and application of conventional activated sludge processes (ASP) began in South Africa, during the early 1920's. Excess filamentous bacteria were first reported in bulking sludge, at the Johannesburg Bruma works in the early 1930's, whereby low DO was considered a contributing cause (Osborn, 1988). Early systems were built based on diffused air, simple mechanical aeration and spiroflow concepts; however, the former aeration design gave the best results and was further progressed into plug flow AS types (Hamlin & Wilson, 1951). Although biological nutrient removal (BNR) systems have been the preferred technology for the last two generations, bulking and foaming problems have persisted, primarily due to proliferation of low F/M filamentous types (Ekama, 2015; Barnard, 1976). First generation BNR systems included the 4-stage Bardenpho or 5-stage Phoredox/Bardenpho configuration, however, bulking continued at these works. An increase in scum formation was also noted, together with abundant M. parvicella populations particularly during winter and autumn seasons. Causes for this were; a paradigm shift from plug flow systems to completely mixed reactors, changes in mixing mode, poor reactor design, reduced DO levels and increased industrial waste (Pitman, 1994; Casey, 1994a). Second generation BNR works were configured to conform to the wastewater effluent NH_4-N standards of 10 mg L^{-1} by using extended aeration systems with longer sludge ages, up to 30 days (Pitman, 1998). Bulking and foaming was still very common in second generation BNR systems; however with limited success, chlorination was investigated as a common non-specific control measure in many full-scale WWTWs (Lakay *et al.* 1988; van Leeuwen *et al.* 1992; Osborn, 1988; Thirion *et al.* 1983; Wiechers, 1983).

At present, South Africa has 16 types of wastewater treatment technologies that are categorized into three generic types; (i) trickling filters, (ii) activated sludge and BNR processes, and (iii) pond and lagoon systems (Department of Water Affairs, 2013). However, the implementation of new technologies had exposed many operational problems in full-scale systems, such as improper sludge settling and the requirement of longer sludge ages to facilitate nitrification in winter seasons (Ekama & Wentzel, 1999; Ekama, 2010; Hu *et al.* 2003). These operational problems often lead to failure in achieving the required wastewater effluent discharge standards.

Currently there are 1286 municipal and privately owned WWTWs in South Africa (Roux *et al.* 2014). In 2011, a 'Green Drop' based regulatory assessment was performed throughout the country and only 40 audited WWTWs received a prestigious award based on their exceptional performance. Of these, nineteen were from the Western Cape, eleven from KwaZulu-Natal and five from the Gauteng province (Department of Water Affairs, 2012). Some of the critical performance indicators included the design and operating capacity of the WWTWs, as well as compliance with effluent standards. Moreover, 90% of all systems had shown non-compliance to more than three effluent determinants and 143 had a high risk of failure (Department of Water Affairs, 2013). The Green Drop audit is an incentive based regulatory programme set to improve the level of wastewater treatment and management in South Africa.

6.12.2 Overview of filamentous bulking and foaming in South African wastewater treatment works

Despite extensive research on filamentous bacteria, these organisms still pose serious bulking and foaming problems in WWTWs. A survey conducted by Blackbeard *et al.* (1986), reported that out of the 111 AS processes surveyed, 33 were BNR systems, from which 27 WWTWs experienced bulking episodes. In their study, filamentous bacteria that were known to proliferate under polar F/M ratios, were commonly identified. Among these, Eikelboom Type 0092, Type 1851 and Type 0041 were dominant in approximately 75% of all WWTWs. A recent survey in the Western Cape Province indicated that these filaments are still dominant in current BNR systems. The dominant filamentous bacteria in descending order of frequency were: Type 0092, Type 1851, *Gordonia* spp., M. parvicella and Type 021N (Welz *et al.* 2014).

Investigations directly related to foaming problems indicated that filamentous Type 0092, M. parvicella and *Gordonia amarae* were most dominant in SA-WWTWs (Blackbeard *et al.* 1986; Marrengane *et al.* 2011; Welz *et al.* 2014). In KwaZulu-Natal, Bux and Kasan (1994), had reported a high incidence of the latter filaments, particularly in winter when foaming was often severe. The major precursors for sludge bulking and foaming in SA-WWTWs are: low DO and low F/M ratios (longer sludge age), which are inter-related to an aging process infrastructure. Nevertheless, through proper reactor design/upgrade and operation, the growth of filamentous bacteria can be suppressed, as described in the following case study.

6.12.3 Case study

6.12.3.1 Filamentous bacteria and sludge bulking in seven South African municipal works

An ongoing project at the Institute for Water and Wastewater Technology (IWWT-Durban University of Technology) currently investigates seven BNR-WWTWs. All works selected (i.e. WWTWs *I-VII*) treated primarily domestic wastewater. The common operational features and filamentous populations of the selected WWTWs are listed in Table 6.20. Filamentous bacteria are subjectively scored using the filamentous abundance (FA) scale, from 0 to 6, as per Jenkins *et al.* (2004). This survey had further reflected a geographical representation of various filaments and their effect on sludge bulking, situated in three South African provinces (viz. Gauteng, KwaZulu-Natal and Western Cape). This study had also observed the impact of aeration type and operational design on the selection and dominance of filamentous bacteria in SA-WWTWs (Table 6.20).

6.12.3.2 Filamentous bacteria abundance and floc structures

Filamentous bacteria were identified using conventional staining techniques and identification keys (Eikelboom, 2000a, b; Jenkins *et al.* 2004). Fluorescence *in situ*

Table 6.20 Process configuration, operational parameters and filamentous abundance from seven WWTWs in South Africa (2015).

WWTW	WWTW *I*	WWTW *II*	WWTW *III*	WWTW *IV*	WWTW *V*	WWTW *VI*	WWTW *VII*
Province	Western Cape	Western Cape	Western Cape	Gauteng	Gauteng	KZN	KZN
Configuration	5-stage Phoredox	Modified UCT	UCT	3-stage Phoredox	3-stage Phoredox	MLE	Modified UCT
Wastewater type	Domestic	Domestic	Domestic industrial (5%)	Domestic Industrial (5%)	Domestic Industrial (5%)	Domestic	Domestic Industrial (5%)
Aeration type	Surface aeration	Surface aeration	Surface aeration	fine and course bubble diffused aeration	fine and course bubble diffusedaeration	Surface aeration	Surface aeration
Primary settling tank	No	No	No	Yes	Yes	Yes	No
Total design capacity (ML d^{-1})	12.5	7.5	14	45	40	25	7.2
Current capacity	75%		50%			83%	65%
Balancing tank	No	No	Yes	Yes	Yes	No	No
Anaerobic digester	Yes	Yes	Yes	No	No	Yes	No
Chemical dosing	None	None	None	Ferric chloride	Ferric chloride	None	Aluminum chloride
SVI (mL g^{-1})	155.85 ± 7.20	157.05 ± 2.27	167.62 ± 12.15	50.13 ± 13.52	98.09 ± 3.27	167.95 ± 7.18	178.33 ± 6.19
MLSS (mg L^{-1})	4837 ± 637.55	5346 ± 752.26	5465 ± 657.11	4384 ± 481.66	4514 ± 526.05	5105 ± 750	5071 ± 597

Sludge age (days)	30	25	15	20	20	25	30
F/M (kg COD kg^{-1} MLSS d^{-1})	0.13	0.11	0.14	0.11	0.11	0.14	0.18
Microbial effluent Compliance (DWA, 2012)	94%	94%	91%	74%	74%	67%	42%
Filamentous abundance (FA: 1–6)							
Total filamentous population	**5**	**5**	**5**	3	**4**	**6**	**6**
Type 0092	**4**	**4**	**5**	3	**4**	3	3
Thiothrix spp.		1	3		3	**4**	3
Type 021N				2	1	2	**5**
Type 0041	3	3	3	2	3	3	3
Type 1851	**4**	3	3			3	3
Type 021N				2		2	
Type 1701		1				2	2
N .limicola	1	1				1	2
Gordonia spp.	2	2	1			2	2
M. parvicella	2	2	2	1		3	2

NB: Filamentous bacteria were ranked very few to excessive (FA: 1–6 respectively) as per filamentous abundance scale (Jenkins *et al.* 2004).

hybridization (FISH) had confirmed the identity of the dominant filamentous bacteria in all WWTWs. Dominant filaments were highly abundant in all works (FA: ≥4; Table 6.20), which resulted in poor sludge settling characteristics (DSVI > 250 mL g^{-1}), with the exception of WWTW *IV* and *V*, which often experienced a lower filament abundance (FA: ≤3) and a better sludge settling quality (DSVI < 150 mL g^{-1}). During severe bulking conditions, filaments had extended out of the flocs, causing interfloc-bridging (Fig. 6.64a, b).

Works *VI* and *VII* had shown a diversity of known bulking filamentous bacteria, frequently ranked FA: ≥3, with *Thiothrix* spp. and Type 021N being most abundant. These filaments significantly impacted on floc structures by protruding from the floc-core and extending into the bulk solution. Weak irregular shaped flocs were formed, thus impeding the sludge settling characteristics (Fig. 6.64a, b). All WWTWs showed similar weak floc structural characteristics (open flocs), except for WWTWs *IV* and *V* situated in the Gauteng Province, in which Type 0092 prevailed at moderate levels (FA: 3–4), within sturdy compact flocs (Fig. 6.64c, d).

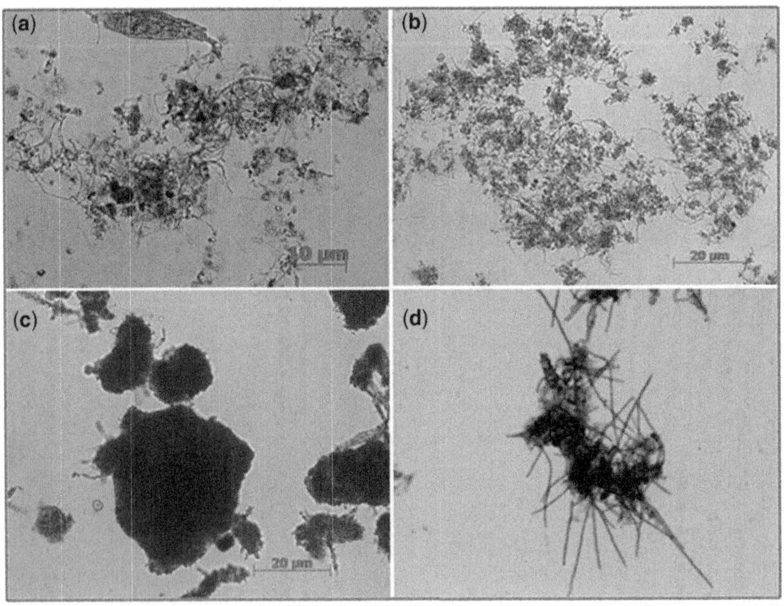

Figure 6.64 Floc structural characteristics in activated sludge works: (**a**) weak flocs depicted from WWTW *III*, (**b**) open, irregular and diffuse flocs, also showing interfloc bridging from WWTW *VII*, (**c**) round, firm and compact flocs from WWTW *IV* (all 400×, phase contrast); (**d**) Neisser staining of Type 0092 from WWTW *V* (1000× direct illumination).

6.12.3.3 Gauteng Province

Wastewater treatment works selected from the Gauteng province (WWTW *V*-old module) was designed to treat 45 ML d^{-1}, average dry weather flow (ADWF), which is also affiliated with an upgraded, duplicate system (WWTW *IV*-new module), which treats a further 40 ML d^{-1} ADWF. Both modules are flexible in operation and can be easily switched between five different operational modes (i.e. 3-stage Phoredox, Johannesburg, modified Johannesburg, UCT and modified UCT configurations). During this study, both modules operated as a 3-stage Phoredox configuration. For the secondary treatment, mixed liquor is channelled into two aerobic bioreactors, which are sectioned with coarse and fine bubble aeration, whereby DO concentration was maintained at ~2–3 mg O_2 L^{-1}. Increased aeration efficiency and diffusion possibly allowed floc-formers to out-compete filamentous bacteria types, particularly the dominant filamentous Type 0092. Whereas, the remaining WWTWs selected in this case study, were configured with submerged surface aeration, which produced DO levels <1.5 mg O_2 L^{-1}.

Both modules exhibited excellent sludge settling characteristics (DSVI < 150 mL g^{-1}), however the newer module, WWTW *IV* occasionally settled more quickly (DSVI < 90 mL g^{-1}). It appeared that with over-aeration, pin-point flocs had also formed and a high effluent suspended solids was often observed. In the older module (WWTW *V*), the total filamentous length of dominant Type 0092 was observed further extended out of some flocs, into the bulk solution (Fig. 6.64d). Less compact sludge flocs were formed and a slight impediment in sludge settling velocity was also observed. Whereas, in the newer module, Type 0092 filaments was found completely retracted within the flocs (Fig. 6.64c). However, no bulking was observed in either module. The difference in floc structure could be attributed to the difference in aeration blowers as the older module was equipped with older air compressors, whilst the newer module was equipped with a turbine diffuser, which offers greater diffusion and dispersion of fine air bubbles.

6.12.3.4 KwaZulu-Natal Province

Wastewater treatment work *VI* operates as an MLE process, treating 25 ML d^{-1} ADWF of domestic wastewater. High bulking and foaming was frequently observed in the anoxic and secondary clarifiers. *Thiothrix* spp., was the most dominant group (FA: 4), whereas, *M. parvicella*, Type 0041, Type 1851 and Type 0092 were also commonly identified (FA: ≤ 3). Proliferation of these organisms was possible, due to their ability to survive and adapt to the wide parameter ranges usual in activated sludge systems (Blackbeard *et al.* 1988; Bux & Kasan, 1994; Lacko *et al.* 1999). Poor mixing, low DO concentration, low F/M ratios (0.02–0.14 kg COD kg^{-1} MLSS d^{-1}) and long sludge ages were possible causes for their prolific growth.

Wastewater treatment works *VII* treated 3.4 ML d^{-1} ADWF and the system operated as a UCT process configuration, with an average influent flow rate and organic loading rate of 3100 m^3 d^{-1} and 3 kg COD m^{-3} d^{-1}, respectively. The recycle pumps, mixers and surface mechanical aerators often malfunctioned, moreover, F/M ratios and DO levels were often very low, ranging between 0.08 and 0.18 kg COD kg^{-1} MLSS d^{-1} and <1.5 mg O_2 L^{-1}, respectively. To this end, the process had shown a diversity of low F/M filamentous bacteria and dominant Type 021N (FA: 5) and was often observed extending out of the AS flocs (Table 6.20; Fig. 6.64b). Waste treatment works *VII* had shown higher bulking and foaming events, particularly during winter. Deepnarain *et al.* (2015) applied a cumulative logistic model (CLM) on WWTW *VII*, to predict the potential abundance of the dominant filamentous types in relation to physical-chemical and operational parameters. Further analysis included polybar probability plots. This was done to gain a more comprehensible prediction of the collective effect of these parameters on each of the dominant filamentous bacteria. The probabilities investigated clearly indicated the role of DO and F/M ratios over Type 1851 and a few others. Moreover, the CLM illustrated that increased Type 021N population was attributed to low COD (< 750 mg COD L^{-1}) and high NH_4-N levels (≥ 32 mg L^{-1}) (Deepnarain *et al.* 2015). Output analysis from this model coincided with literature findings; however with further validation, the model could be a useful tool for regulating specific filamentous types.

Thiothrix spp. and Type 021N were excessively prevalent in WWTWs *VI* and *VII* respectively (Table 6.20). These filaments are able to oxidize sulfur and temporarily store them as an intermediary product within the cells (Vaiopoulou *et al.* 2007). Septic sludge was directly added to the preliminary treated wastewater, at both WWTWs and thus, could further explain the prevalence of *Thiothrix* spp. and Type 021N at the respective WWTWs.

6.12.3.5 Western Cape Province

Wastewater treatment works *I* operated as a 5-stage Bardenpho configuration and WWTW *II* is capable of switching between different operational modes (i.e. UCT, MUCT, JHB and 3 stage Phoredox). During this study, WWTW *II* and *III* operated as a modified UCT operational mode. All three systems experienced bulking (DSVI > 250 mL g^{-1}), with peak values occurring in winter and moderate foams also occurred occasionally. All works selected from the Western Cape operated at low F/M ratios (< 0.14 kg COD kg^{-1} MLSS d^{-1}), long sludge ages (15–35 days) and reduced DO levels (< 1.5 mg O_2 L^{-1}). Type 0092 was found dominant and a few other low F/M filamentous populations were also observed as secondary filaments (viz. Type 0041; Type 1851; M. parvicella) (Table 6.20). The presence of these filaments is commonly grouped together as aerobic- anoxic- anaerobic- zone growers and are usually present in conditions of high sludge age, possibly utilizing particulate substrates for growth (Martins *et al.* 2004c; Welz *et al.* 2014).

6.12.3.6 *Prevalence of Type 0092 in South African biological nutrient removal works*

The prevalence of Type 0092 was not surprising, as it was found to be a frequent morphotype in nutrient removal works, in many worldwide surveys on filamentous bacteria that had been associated with low F/M and long sludge ages (Martins *et al.* 2004c; Tandoi *et al.* 2006). According to Eikelboom (2000), Type 0092 was observed most often inside the flocs with limited effect on sludge settling velocity; however some reports showed that they were also dominant in bulking and foaming activated sludge (Blackbeard *et al.* 1986; Welz *et al.* 2014).

In this case study, Type 0092, was highly prevalent and ranked FA: ≥3 in all seven WWTWs. Madoni *et al.* (2000) reported that Type 0092 are common substrate competitors which dominated in WWTWs designed for organic carbon and nitrogen removal (Madoni *et al.* 2000; Welz *et al.* 2014). A recent study, conducted by Welz and co-workers, focused on the dominance of Eikelboom Type 0092 from eleven WWTWs in the Western Cape province of South Africa. From their evaluation, three WWTWs were configured as a Modified Ludzack-Ettinger (MLE) process, designed for organic carbon and nitrogen removal. Welz *et al.* (2014), further reported that Type 0092 were not so common in the KZN region (Welz *et al.* 2014). These findings were also congruent with the results in this study (Table 6.20), as Type 0092 were common (FA: ≥4) in the Western Cape WWTWs, however, lower levels (FA: ≤3) were observed in WWTW *VI*, WWTW *VII* and a few other AS works situated in KZN region (data not shown).

A few research studies had shown that Type 0092, was frequently reported in BNR and enhanced biological nutrient removal (EBPR) works, which were classified as 'all-zone' growers (Speirs *et al.* 2009; Wanner & Grau, 1986), as they can take up both oxygen and nitrates as electron acceptors.

6.12.4 Remedial methods to control filamentous bulking and foaming in South Africa

6.12.4.1 *Physical and chemical treatment methods*

Foams/scums are often water-sprayed or manually removed and scraped off the surface of clarifiers, these methods are proven to be very tedious and dangerous to works operators. To control sludge bulking and foaming in activated sludge works, chemical methods such as chlorination and hydrogen peroxide are still implemented, however, this is only temporary and less cost effective. Chlorine was one of the first technologies proposed to control filamentous bulking and it is still commonly used as a disinfectant in effluent wastewater in the form of chlorine gas. Lakay *et al.* (1988) and van Leeuwen *et al.* (1992) had reported a laboratory scale system showing a chlorination dose of 6–8 g Cl_2 kg^{-1} MLSS d^{-1} that was successful to control filamentous bulking. However, this method affected nutrient removal efficiency and was resistant to specific filamentous types such as Type 021N and M. parvicella.

The use of ozone treatment to reduce filamentous bulking has proven to be partially successful in full-scale WWTWs (Leeuwen, 1988). The treatment did not have any adverse affect on the BNR process viz. nitrification and phosphorus removal. Van Leeuwen (1988) showed that an ozone dose of 2 mg O_3 L^{-1} reduced DSVI from 180 to less than 100 mL g^{-1}. Even doses up to 30 mg O_3 L^{-1} did not affect the nutrient removal process.

Saayman *et al.* (1996, 1999) investigated chlorination, hydrogen peroxide (H_2O_2) as well as ozonation to control the growth of filamentous bacteria that had caused bulking in the BNR works. From their investigation, it was found that a chlorine dose 1.42 g Cl_2 kg^{-1} MLSS d^{-1} reduced the DSVI from 180 to 105 mL g^{-1} within 10 days. In their study, chlorine was the most effective method in controlling filamentous bulking, however, it was reported to damage the biomass and caused difficulties in P removal. While H_2O_2 and ozone was found less effective, but it had also controlled bulking temporarily.

6.12.4.2 The effect of anoxic–aerobic selectors on filamentous bacteria

The University of Cape Town, South Africa, had put forward some significant research on possible reasons for the proliferation of filamentous bacteria in SA works (Casey *et al.* 1994a, b, 1995; Ekama *et al.* 1996a, b; Gabb *et al.* 1996a, b; Still *et al.* 1996). Their research had shown that proliferation of low F/M filamentous bacteria was generally improved by the continuous cycle of anoxic–aerobic conditions; however these microbes proliferated during reduced intermittent aeration conditions and alternating anoxic–aerobic conditions (Casey *et al.* 1994a, 1999). The significance of the anoxic selector is to reduce the readily biodegradable COD (RBCOD) following the aeration tank, thus also selecting for floc forming bacteria. However, with high RBCOD/NO_x-N ratios in the anoxic tanks, excess COD following aerobic conditions had significantly promulgated filamentous bacteria (Peng & He, 2015). Furthermore, NO accumulation left by incomplete denitrification can be fatal, as this intermediate interferes with electron transfer to oxygen, thus inhibiting floc formers during substrate utilization under aerobic conditions (Casey *et al.* 1999). It was suggested that controlling the nitrate and nitrite levels at the anoxic to aerobic transition, may control low F/M filamentous bulking (Lakay *et al.* 1999; Musvoto *et al.* 1999; Tsai *et al.* 2004).

6.13 SPAIN

F. Estevez, E. Rodríguez and E. Reina

6.13.1 General situation

Full water cycle management, intended to guarantee the necessary quality and quantity of water supply, is mandatory for the correct development of a society. The Spanish water governance system, formed by the different River Basin Districts, is a management system based on planning, public participation and technological development and innovation.

All this, added to the development of regulations which are more and more restrictive for the protection of this medium, have favoured water treatment in Spain throughout the years. This makes up nearly 100% of the consumption at present, as shown in Fig. 6.65.

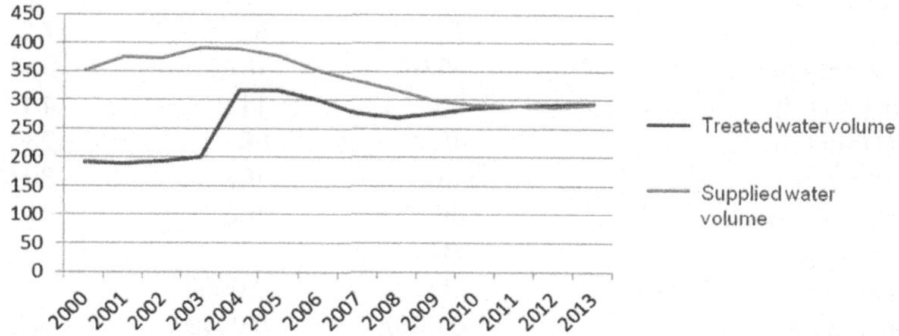

Figure 6.65 Evolution of water consumption and treatment in Spain. (Units L/inhabitant · day). Instituto Nacional de Estadística (INE).

6.13.1.1 Wastewater composition

It is a challenging task to define the composition of wastewater in Spain as there are so many factors that influence such composition, depending as it does on the geographical demarcation, the weather, the climate, etc.

For reference, the data shown in Table 6.21 below correspond to the profiling of wastewater generated in Seville and its metropolitan area with the maximum, average and minimum values in 2014. It is necessary to remark that this city's sanitation system is formed by unitary networks, as is most common in Spain, and the reason why wastewater is collected together with rainwater.

6.13.1.2 Treatment level

Wastewater treatment depends on the pollutant load of the incoming water, but even more on the conditions of discharge into the receiving medium. In Spain there are a high number of sensitive areas, as defined by the 91/271/EEC (Fig. 6.66); hence most wastewaters are managed through treatments that produce some nutrient elimination.

Industrial processes in Spain generate 7% of the total production of wastewaters (Fig. 6.67). Composition of these wastewaters and the treatment that we apply to them vary depending on the industrial process in which the waters were involved. For example, taking into account that the metallurgic activity is the most important one in the Spanish industrial economy, we can observe that its wastewaters are mainly made up of inorganic substances and so the proper treatment of these waters will be based on physical processes.

Table 6.21 Characteristics of wastewater generated in Seville during 2014. Information supplied by EMASESA (Metropolitan Water Company of Seville).

Seville's Wastewater Profiling	Values Registered in 2014		
Pollutant	**Average**	**Maximum**	**Minimum**
SS (mg L^{-1})	245	526	134
COD (mg L^{-1})	555	927	248
BOD (mg L^{-1})	279	504	107
pH (Ud)	7.7	8.1	7.5
Conductivity (μS cm^{-1})	1536	1875	953
P total (mg L^{-1})	9.1	14.2	4.1
N total (mg L^{-1})	48	66	23
TOC (mg L^{-1})	135	187	89
Oils and fats (mg L^{-1})	66	106	41
Mercury (mg L^{-1})	<LC	–	–
Cadmium (mg L^{-1})	<LC	0.00032	–
Chromium (mg L^{-1})	0.0062	0.0148	<LC
Lead (mg L^{-1})	0.014	0.066	<LC
Zinc (mg L^{-1})	0.163	0.332	0.088
Aluminium (mg L^{-1})	1.4	4.5	0.4
Cobalt (mg L^{-1})	<LC	–	–
Copper (mg L^{-1})	0.04	0.10	0.02
Iron (mg L^{-1})	1.2	4.0	0.4
Potassium (mg L^{-1})	24	31	18
Manganese (mg L^{-1})	0.07	0.12	0.05
Nickel (mg L^{-1})	0.0034	0.0117	0.0012

Limits of quantification: mercury and cadmium 0.001 mg L^{-1}; chromium and lead 0.005 mg L^{-1} and cobalt 0.01 mg L^{-1}.

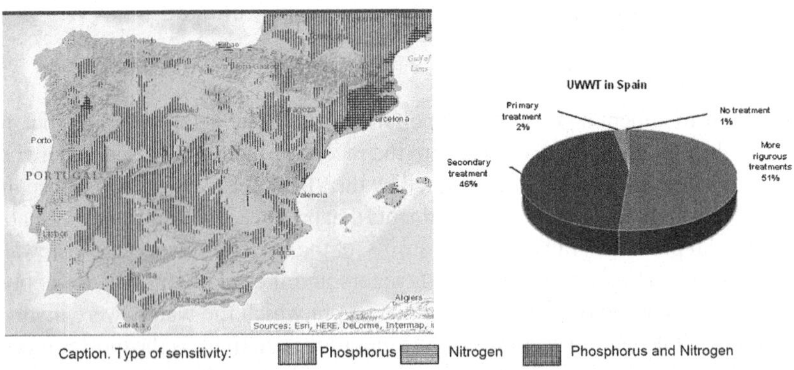

Figure 6.66 Map of sensititve areas and percentage distribution of type of treatment in Spain (European Environment Agency, 2012).

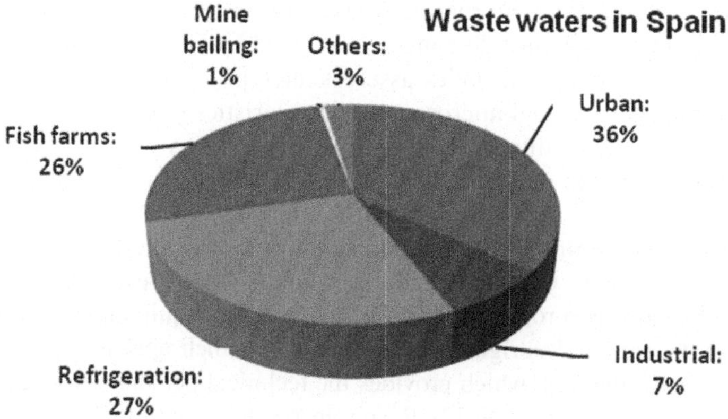

Figure 6.67 Water uses in 2014. Water uses in 2014 (Ministerio de Agricultura, Alimentación y Medio Ambiente, 2014).

The agro-food industry is the second most important in Spain and, unlike the first, its effluents are mainly organic, thus the water treatment and regeneration are based on biological processes in order to remove dissolved substances.

6.13.1.3 Level of application of microbiological control: interlaboratory exercises

As has been pointed out regarding urban wastewaters, 97% of them are submitted to biological treatments, either by means of activated sludge or any other means. The same applies to industry originated wastewaters. So, one of the most important tools for these treatments is the microbiological analysis of the system.

Profiling and microbiological follow-up of activated sludge have been developed in Spain in the last 35 years and we deem it necessary to refer to the fact that progress has been outstanding during this period and at present this is the routine analysis in most WWTPs (Serrano & Arregui, 2014). This has been made possible partly because of the University–Industry symbiotic efforts of recent times which have had GBS (Bioindication Group Seville) as the connecting link (Serrano *et al.* 2006; Serrano & Arregui, 2014), as well as certain other institutions, which especially stand out in spreading these connections, such as the Complutense University of Madrid for its numerous works on the suitability of protists as bioindicators; the University of Barcelona, which has carried out a high number of studies about ecology of activated sludge; the Polytechnic University of Valencia, working on the use of FISH techniques to identify filamentous bacteria present in the sludge; and several companies such as EMASESA, FACSA or ESAMUR, which have fostered the organization of technical meetings and the publication of different works on the topic.

The most frequently used routine indexes are SBI (Madoni, 1994, 2011), and the sludge index (IF-Índice de Fango), developed by GBS (Jiménez *et al.* 2001; Rodríguez *et al.* 2008). This index assesses the type of floc assigning numerical values to macroscopic and microscopic characteristics, obtaining a value which indicates a specific quality of the floc. IF is defined as a very reliable index for its application by non-expert technicians in the biology of the system (Arregui *et al.* 2012).

As a system for training on and assessing the quality of microbiological analysis carried out in Spain, GBS has organized interlaboratory exercises since 2003, developed following a routine that complies with the requirements issued by the G-ENAC-Procedure (Rodríguez & Reina, 2014), which appears as an excellent training, qualifying tool, which provides the technical personnel with knowledge and attitudes (Arregui *et al.* 2012). Respirometric tests and FISH techniques have also become more frequent at the WWTP.

6.13.2 Separation problems and control methods applied

The most frequently used identifying keys at Spanish WWTPs are those of Eikelboom (1975), as the most usual facility in the laboratories is optical microscopy although there has been progress toward the implementation of Eikelboom's (2006) industrial filamentous keys and the use of FISH techniques (Alonso *et al.* 2011; Infante & Rodríguez, 2010, 2011; Infante, 2012; Infante *et al.* 2012a). A significant improvement in this sense, during the last few years, has been the access to easy-to-use kits in the market, which will make identification easier through FISH at the WWTP premises. This identification is at present only made in specialized institutions such as universities and certain companies.

The various studies carried out by GBS in a number of both urban and industrial WWTPs show the difference exisiting between the two sectors. In urban WWTPs the prevailing floc structure (IF) has mostly been correct, whereas it has been deficient in industrial plants.

Throughout the period between 2005 and 2015, GBS has organized numerous interlaboratory tests and different follow-ups at WWTPs, which have made it possible to get data about the presence of the most common filaments in Spain, both in urban and industrial WWTPs.

6.13.2.1 Major filaments responsible for problems in Spanish Activated Sludge Plants

Filamentous bulking is the most usual problem for 46% of urban WWTPs and for 48% of industrial WWTPs assessed by GBS.

From the 108 urban WWTPs spread all over Spain analysed by GBS, we obtain the filamentous microorganism distribution detailed in Table 6.22, which shows that the prevailing filaments most usually found are nocardioforms, Type

021N-*Thiothrix*, *H. hydrossis*, *N. limicola* and M. parvicella. Regarding the distribution of secondary filaments, the most usual were Type 021N-*Thiothrix*, *N. limicola* and *H. hydrossis*.

Table 6.22 Prevailing and secondary filaments found in urban WWTP samples analysed by GBS.

% of Presence	**Prevailing Filament**
23	*Nocardioforms*
14	T021N-*Thiotrix*
11	*H. hydrossis*
10	*N. limicola*
9–5	M. parvicella, Type 1701, Type 1863
<4	Type 0581, Type 0041/0675, *S. natans*, Type 1702, Type 0092, *Meganema*, *Monilibacter batavus*, Type 1851, *Bacillus*

% of Presence	**Secondary Filament**
22	Type 021N-*Thiothrix*
14	*N. limicola*
12	*H. hydrossis*
9–5	Type 0041/0675, Type 0092, Nocardioformes, Type 1701, M. parvicella
<4	*S.natans*, Type 1851, *Streptococcus*, Type 1863, Type 0803, Type 1702, *Bacillus*, Type 0581

The floc quality of the analysed samples was assessed with Average IF for 52% of the samples, followed by 35% of Good IF, 10% Bad and 4% Optimal.

As for the 114 samples of industrial WWTPs assessed by GBS, among which several sectors are to be found, such as agro-food, metallurgic, petrochemical, etc., the assigned floc categories were 48% for Average IF, 30% Bad, 13% Very Bad and 4% Good. The filaments found as prevailing in these samples are shown in Table 6.23.

The breakdown of organisms by types of industrial WWTP is shown in Table 6.24.

6.13.2.2 Other solid separation problems

Although filamentous bulking is the most usual problem in both urban and industrial WWTPs, a lower level of viscous bulking and pin-point processes as well as in turbidity clarification has been detected.

Viscous bulking has been detected in 2% of the urban samples and 24% of the industrial samples. As already known, floc forming bacteria aggregation depends to a large extent on extracellular polymers, which mediate in the cell cohesion and adhesion and play an essential role in keeping the immobilized cell complete (Liu *et al.* 2004). However, the stress generated by factors such as toxic substances, nutritional unbalance, inadequate physical conditions, cellular low viability, toxic metabolite production, etc. alters this EPS level and, therefore, floc structure. Viscous bulking usually occurs in wastewater rich in easily degradable organic compounds, characteristic of the agro-food sector's industrial sewage, and under nutritional deficiency conditions (Jenkins *et al.* 2003). It has also been related to toxicity states (Tandoi *et al.* 2006). This is why it is especially interesting to routinely follow up the soluble COD from the influent to the reactor, as it very often appears as an important fraction within the total, in the case of viscosity.

Table 6.23 Prevailing filaments found in the analysed samples of industrial WWTPs.

% of Presence	Prevailing Filament
30	Type 021N-Thiothrix
11	Type 0041/0675
11	*H. hydrossis*
Organisms appearing in percentage between 9% and 5%	Nocardioforms, Type1701, Type 0092
Organisms appearing in percentage below 4%	*N. limicola*, Type1851, *Thiothrix-morphotype4, Monilibacter batavus*, Type 0803, Type 0211, *Meganema perideroedes*, M. parvicella, Type 0961, Type 1863, *Fungi*, Type 0581, *S. natans*, Type 0581, *Alysiosphaera europeae, Isosphaera*

Another viscous bulking enhancing factor in WTTP has been the high temperature of the influent water.

The pin-point floc cases in urban samples have not been significant, although they have always been connected to either prolonged aeration WTTPs or overdimensioned ones or those with importantly different flows throughout the year. Abundance of rainfalls during prolonged periods is another factor enhancing this problem in urban WWTPs.

Besides, the effect of heavy metals present in the metallurgic sector industry or the high variability of potentially toxic elements found in the industrial sector facilitate the development of very small flocs with scarce purification capacity, affecting 25% of the analysed industrial samples.

Finally, urban WTTPs reported processes of turbidity provoked by filamentous growth in 15% of the studied samples, associated with *Bacillus, Streptococcus* and Type 1863 (the latter linked to previous chlorination processes).

Table 6.24 Prevailing filaments in the analysed samples of industrial WWTPs by sector.

Sector	**Prevailing Filament Found**
Cheese dairy	*Meganema perideroedes*
Slaughterhouse	Nocardioforms, Type 0092, M. parvicella,
Textile–tinted	Type1701, Type 0041/0675
paper	Type 021N-*Thiothrix,* Type 0041/0675, *H. hydrossis,* Type 0581, *S.natans,* Type 1851
Olive oil	*N. limicola,* Nocardioforms
Undetermined agrofood	*Monilibacter batavus,* Type 0803
Milk dairy	Type 021N-*Thiothrix,* Type 0961, Type *0041/0675,* Type 1701, *H. hydrossis,* Type 1863, Type 0803, Type 0211
Eggs-manufacture	*H. hydrossis*
Tannery	Nocardioforms, Type 021N-*Thiothrix, H. hydrossis*
Bioethanol	*H. hydrossis*
Petrochemical	Type 1701, Type 0041/0675, Type 0581, Type 1851, *Alysiosphaera europeae,*Type 0092, *Isosphaera, Thiotrix-morphotype4, N. limicola, Monilibacter batavus*
Oil	*Nocardioforms*
Metallurgic	Type 0092
Terpen	Nocardioforms
Citric	Type 021N-*Thiothrix*
Wine	Type 021N-*Thiothrix*
Vegetable Preserves	Type 021N-*Thiothrix*
Milling	*H. hydrossis,* Type 021N-*Thiothrix,* Type 0041/0675
Beer brewing	Type 021N-*Thiothrix,* Type 0041/0675, Type 0211, *H. hydrossis*

6.13.2.3 Reasons for dysfunctions in Spanish activated sludge plants

The causes of the different problems detected in urban WTTPs can be classified as follows:

- *Environmental conditions*: important differences in environmental temperature and seasonal rainfall.
- *Design conditions*: over- or under-design, wrong hydraulic distribution, limited or not adjusted to aeration need. In industries, it is extended to install MBR and SBR affected by contamination or by strongly seasonal flow problems.

- *Influent conditions*: high biodegradability, nutritional deficiency, presence of sulfides or presence of uncontrolled industrial discharges.
- Operational conditions inadequate to the needs of the cultivation or to the influent characteristics, overload in the sludge line.

These limitations are shared by the industrial WTTPs, together with toxicity problems, high temperature and high variability, both in the flow and in loads, in influents and low secondary decantation capacity.

6.13.2.4 Control measures applied

The first control measure the operator must implement is the comprehensive control of the WTTP's operational parameters in order to eliminate the causes generating the growth of the filament in question. For instance, in the case of nocardioforms, adjustment of oxygenation excess during the times of influent lower charge, lamination of the flow in order to maintain a more homogeneous charge during the day, draining maintenance in the line of sludge lower than 0.3% of dry matter, concentration in primary decantation of less than 0.8% and oxygenation decrease combined with an increase of excessive sludge to reduce foaming processes characteristic of these organisms (Rodríguez, 2009).

Nowadays, the use of control probes of type Redox, ammonium, nitrates and oxygen is more extended as well as the application of programmes to manage and supervise those data and mathematical simulators to control the process. Its final aim is to have integrated computing systems available that allow for telecontrolled biological processes and optimizing of the solutions thanks to the bioindicator parameters.

Other tools which are not so extended, yet they are used, are CFD (Computational Fluid Dynamics), CTR (Resistence time distribution) and physical–chemical parameter mappers to control hydraulic short circuits, dead zones and real vs theoretical retention times (Castell, 2013).

After obtaining some adjusted operational parameters, we can go on to define other control strategies that can be classified as specific and non-specific methods.

Specific methods combine both kinetic and physiological selection and they become a method of great development, as they fight the root of the problem by eliminating the factors generating phase separation.

For example, anaerobic/anoxic selectors eliminate Type 021N, Type 0961, *S. natans* and *Thiothrix* spp., whereas aerobic/anoxic processes act on M. parvicella, Type 0092, Type 0041–0675, Type 0914 and Type 1851 (Martins *et al.* 2004c). As an instance, in the case of the Community of Madrid (Suárez, 2014), anaerobic selectors were used to control the excessive proliferation of nocardioforms, Type 1863 and M. parvicella in three WWTPs.

For very biodegradable influents, strategies such as the application of anaerobic biological systems prior to aerobic biological systems of activated sludge can be effective as these systems basically eliminate the easily biodegradable water so that they reduce the risk of bulking in an important part. This is quite an extended option in the paper industrial WWTP sector (Infante *et al.* 2012b).

Finally, several systems are under study to reduce nitrogen and phosphorus of supernatants and so avoid nutritional imbalance (Claros *et al.* 2012).

Likewise, non-specific methods take advantage of the fact that filamentous bacteria are more susceptible to toxic attack because they have a bigger surface and grow toward the floc's outer part. However, its effect is temporary.

The most frequently used non-specific methods to reduce filamentous bacteria in Spain are:

- *Adding polymers (aluminium salts)*: application of PAX salts to treat M. parvicella because it affects this filament's substrate uptake (Nielsen *et al.* 2005). PAX dosing continued for 109 days with the following daily doses: 4.6 g Al kg^{-1} MLSS for 9 days (Shock Dose); 2.0 g Al kg^{-1} MLSS for 16 days; 4.0 g Al kg^{-1} MLSS for 24 days; 3.0 g Al kg^{-1} MLSS for 50 days; 1.0 g Al kg^{-1} MLSS for 10 days (Albert, 2009).
- *Adding coagulants*: ferric chloride (7–50 g Fe m^{-3})
- Ultrasound treatment
- *PH adjustment*: adding lime (10–30 g m^{-3})
- *Treatment with toxics*: chlorine (10–20 mg L^{-1}: 8.5 kg Cl_2 t^{-1} MLSS), after 4 h they provoke damage by eliminating the cells, nocardioform fragmentation, or slight oxidation of sulfur granules, in the filaments present in the sample, together with protists' mobility reduction (Rodríguez, 2009): adding biocide for 9 d to control *T. eikelbomii* sp., distributed into three periods. Four days: 6.30 and 8.69 g Cl_2 kg^{-1} MLVSS d^{-1}; three days: 8.82 g Cl_2 kg^{-1} MLVSS d^{-1}; two days: 6.33 and 6.76 g Cl_2 kg^{-1} MLVSS d^{-1}. Deflocculation and important damage in filaments. Control by means of bacterial viability to finish chlorination (Montoya *et al.* 2009) or hydrogen peroxide (100–200 mg L^{-1}).
- *Adding cationic polyelectrolyte*: 1 mg L^{-1}.
- Adding liquid oxygen.
- *Mechanical removal in case of foaming*: suction by means of turbine (Ferrer, 2005), water spray on the surface to redissolve foam and ozone application (Ramírez *et al.* 2009).
- *Reduction of sulfides*: dosing at the beginning of calcium nitrate collector network (*Nutriox*), reactant which inhibits the action of sulfate-reducing bacteria (Albert, 2009).
- Adding nutrients to compensate the influent; dosing phosphoric acid into the channel distributing the biological reactors.

6.13.3 Spain case study

Two cases are described, one referring to a viscous bulking process combined with excessive growth of nocardioforms and Type 021N. The second is a case of filamentous bulking caused by *S. natans*, both taking place in two WTTPs of conventional activated sludge of complete mix and plug flow respectively, both without nutrient elimination.

6.13.3.1 Viscous bulking with Nocardioforms and Type 021N growth

The effluents collected in the aforementioned WTTP present very high soluble loads due to rice industry discharges, with an average of 45–50% of the COD/BOD_5 entering the biological reactor.

The influent and operational characteristics are shown in Table 6.25.

Table 6.25 Influent and operational characteristics.

Parameter (average value)	Manner/value
Type of process	Complete mix
Nitrification/denitrification	No
% of anoxic zone devoted to denitrification	–
COD average of water entering the biological reactor	276 mg L^{-1}
BOD_5 average of water entering the biological reactor	148 mg L^{-1}
Ammonium average input/output	35–29 mg N L^{-1}
% Sludge flow Recirculation vs. influent flow	50%
MLSS and MLVSS average of the present sludge in biological reactor	1260 mg L^{-1}, 84% MLVSS
MLSS and MLVSS average of recirculation sludge	2400 mg L^{-1}, 85% MLVSS
Aerobic area total hydraulic retention average time (h)	3–4 h
Average Θ: Sludge Age	3 d
F/M: mass load	0.5
SVI average	320 mg L^{-1}
C/N/P in biological process	BOD/N/P:148/35–43/4.6–5.1 CODS/N/P:103/35–43/4.6–5.1
Nitrogen total average in influent	47 mg N L^{-1}
Total phosphorus soluble in influent	5.8 mg P L^{-1}
Total nitrogen average in effluent	35 mg N L^{-1}
Total phosphorus soluble in effluent	3.1 mg P L^{-1}
Nitrite in effluent	0.11 mg N L^{-1}
If activated sludge present foaming: foaming colour	Light colour

The activated sludge characteristics are reported in Table 6.26.

Table 6.26 Microbiological characteristics of the activated sludge sample under study.

Sample Estimated Average Results	
Sludge index	Regular
Bacterial category	4 (5–20 filaments per floc)
Identification of filaments	Nocardioforms/*Type 021N*
Approximate density of protists	$>10^6$ Ind L^{-1}
Shannon index	2.5
Madoni index	8
Madoni class	I
N° species found	9
Prevailing group	Sessile bacterivores
Performance according to Madoni	Stable sludge, very well colonized, excellent biological activity and very good performance
Estimated V_{30}	584 mg L^{-1}
MLSS	1074 mg L^{-1}
SVI	546 mg L^{-1}

The main problem of this sludge is the bad decantation. Figure 6.68 shows several aspects key in this cultivation: wrong decantation (interfloccular bridges of Type 021N/viscous), nutritional deficiency and high level of polyhydroxyalkanoate PHA). Influent organic loads with high levels of soluble fraction foster PHA accumulation which acts as carbon reserve.

Combination of filaments, presence of viscosity and metabolic change brought about by the discharge clearly show a system in nutritional stress state where oxygen levels are moderate (1 mg L^{-1} O_2).

This situation affects considerably the process of aggregation of flocs. The protist community presents acceptable conditions. The floccular aggregation process is affected with macroscopic conditions which show a serious problem of bulking and slight turbidity of the clarifying process as a consequernce of an influent with a biodegradable fraction higher than normal and a nutritional imbalance of carbon/nitrogen. This problem has not worsened the effluent quality, thanks to the good capacity of the WTTP's clarifiers.

After finding this alteration, heavy metal levels in the sludge were checked, detecting abnormally high levels for copper, lead, chromium and mercury, whose toxicity is related to a process of viscosity in the activated sludge.

In the assessed case, two simultaneous situations took place: toxicity due to metals and a high soluble fraction in the influent. In order to try to solve the problems, the following measures were adopted:

Table 6.27 Main characteristics of the biological process.

	Raw Water			Treated Water			Biological Reactor					
	Q	S.S.	COD	S.S.	COD	T.	MLSS	V–30	SVI	RASS	Flow Fexc.	C.M.
Day	$m^3\ d^{-1}$	$mg\ L^{-1}$	$mg\ L^{-1}$	$mg\ L^{-1}$	$mg\ L^{-1}$	NTU	$mg\ L^{-1}$	–	$mL\ g^{-1}$	$mg\ L^{-1}$	$m^3\ d^{-1}$	d^{-1}
29–Jul	46.580	164	395	20	65	17	–	–	–	–	–	–
30–Jul	46.120	154	534	30	69	25	620	470	758	1.240	0	1.14
31–Jul	50.850	152	511	30	74	27	570	400	705	1.800	0	1.74
01–Aug	47.060	138	494	76	118	60	647	827	1.278	1.820	695	1.37
02–Aug	46.620	146	460	109	154	77	757	857	1.133	1.680	596	1.07
03–Aug	46.100	–	–	–	–	59	657	550	859	1.360	160	
04–Aug	46.360	–	–	–	–	53	–	–	–	–	–	–
05–Aug	45.620	98	376	28	59	22	–	–	–	–	–	–

- MLSS decrease to adjust mass load and reduce possible accumulative toxicity.
- Decrease retention times of primary decantation to limit fermentative processes.
- Increase recirculation to avoid anoxic processes.
- Increase oxygenation levels.
- Add sodium hypochlorite to control filamentous growth combined with adding cationic polyelectrolyte to improve compaction.

After one month of keeping these operational measures, a significant decrease in viscosity and filamentous bulking was observed.

Figure 6.68 Macroscopic characteristics: V_{30} and viscosity test (**a, b**). Microscopic characteristics: interfloc bridge formation. *In vivo* 100×. Phase contrast (**c**). Nutritional deficiency. Neisser staining. 1000×. Bright field (**d**). PHB high level. PHB staining. 1000×. Light field (**e**). Methylene blue staining 1000×. Bright field (**f**).

6.13.3.2 Filamentous bulking caused by Sphaerotilus natans

During the first days of a vacational period (August), with influent into the WTTP of very low load and flow, an excessive growth of S. natans was reported, which provoked a high increase in the Sludge Volumetric Index (SVI) and bad decantation capacity of the sludge. V30 evolved from 400 mL L^{-1} up to 850–900.

This episode occurred in a conventional activated sludge WWTP at half load without nutrient removal. The consequence of this process brought about a very badly clarified influent, with treated water values of 76–109 mg L^{-1} SS and COD of 120–150 mg L^{-1}. The usual values were approximately 10 mg L^{-1} SS and 50 mg L^{-1} COD. Before the episode, MLSS values were around 600 mg L^{-1}, decanted COD was approximately 315 mg L^{-1} and BOD_5 was over 160 mg L^{-1}, mass load was 1.1–2.2 d^{-1}. The V_{30} reached 850–900 from 400 mL L^{-1}. Table 6.27 describes in detail the main characteristics of the process.

The microscopic level observed floccular disaggregation. Regarding protists, sessile ciliates were not detected, although some specimens of *Aspidisca costata* and *Trachelophyllum pusillum* were observed.

The prevailing filament was *S.natans* with a level of more than 20 filaments/flocs (normal values were at levels < 5 filaments/flocs), as Fig. 6.69 shows.

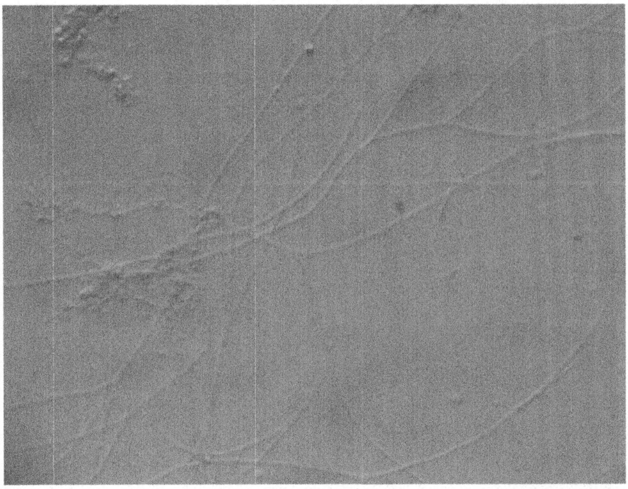

Figure 6.69 Aspect of biological cultivation. *In vivo*. Phase contrast, 200×.

It was decided to recover the cultivation, decrease the excessive sludge purging, even stopping it for 2 days, as there was no increase in MLSS. A slight increase started and 1000 mg L^{-1} of MLSS were reached in 5 days, remaining at some 800 mg L^{-1} of MLSS during the following 10 days.

During this process, the biomass started to change: there appeared a wider variety of protozoa, such as *Opercularia* sp., *Vorticella microstoma, Carchesium polypinum, Aspidiscacostata, Trachelophyllum pusillum.* Regarding the presence of filaments, *S. natans* remained, although at a lower density.

After 20 d, the biomass kept the protozoan diversity, although a population increase took place. Regarding the present filaments, it was observed that *S. natans* virtually disappeared, being replaced by Type 021N reaching up to the levels of 5–20 filaments/floc. MLSS were at 800 mg L^{-1} and V_{30} was 350 mL L^{-1}. The quality of the treated water was recovered around 10 d before, with values of 10 mg SS L^{-1} and COD 45 mg L^{-1}.

6.13.4 Acknowledgement

The authors would like to express their sincere thanks to the Metropolitan Water Company of Seville (EMASESA) and for all participants of the interlaboratory tests for providing samples and information about the general situation of bulking and foaming in Spain.

6.14 USA

T. G. Daigger and D. Jenkins

6.14.1 General situations

Each year, at the beginning of the US Congressional Session, the President of the United States delivers a 'State of the Union' message to a joint assembly of the Senate and the House of Representatives. In this address the President gives his (her) assessment of the condition of the United States and presents a summary of proposed programs. Since more weighty matters than activated sludge solids separation occupy the national agenda, we do not believe that this topic has ever been addressed in this forum. So it is left to us to report on the current state of this field in the United States.

Happily we can report that the state of the field is good! The advances in the understanding of the population dynamics of the microorganisms that influence activated sludge solids separation that have taken place through research over the 40 years since Eikelboom (1975) devised a filamentous organism 'identification' key have had a great influence on the design and operation of the activated sludge process in this country. Assessment and control of what is often known here as 'sludge quality' are firmly established in professional wastewater treatment practice. Microscopic characterization of activated sludge is routinely practiced at many wastewater treatment plants, and it forms an integral part of troubleshooting activated sludge operating problems. Activated sludge design often incorporates features that discourage filamentous organism overgrowth (selectors, RAS chlorination) and encourage bioflocculation (secondary clarifier flocculating center wells). Many specialized courses are offered every year throughout the country on various aspects of activated sludge population assessment and control. Thus the 'culture' of population dynamics has been embraced by the profession. This is not to say that all activated sludge solids separation problems have been resolved. It does mean, however, that when they occur there are a series of rational investigative and remedial tools available to address them.

6.14.2 Control of filaments in activated sludge systems

Our assessment of the current situation is that design and operating methods are available to address most of the filamentous bulking problems in activated sludge. The filamentous organisms most frequently causing bulking in activated sludge in the USA in 1984 are summarized in Table 6.28 (Richard *et al.* 1981; Strom & Jenkins, 1984). Unfortunately no more recent comprehensive summaries have been conducted so that it is not possible to say whether Table 6.28 accurately represents the current situation. Control methods are available for all of these organisms, as indicated in Table 6.28. Note that Table 6.28 does not contain 'use of Cl_2' as a control method because it is universally successful.

Table 6.28 Summary of most frequently encountered filamentous microorganisms in the USA in 1984, and control methods.

Filamentous Organism	Control Method
Type 1701/*S. natans*	raise DO, lower F/M anoxic, anaerobic selector
Type 021N/*Thiothrix* spp.	anoxic selector nutrient (N/P) supplementation remove SH-
H. hydrossis	raise DO, lower F/M
Types 0041 and 0675	decrease MCRT anoxic/anaerobic selector (?)
M. parvicella	eliminate anoxic/anaerobic zones increase aerobic zone DO decrease MCRT
Type 0092	decrease MCRT eliminate anoxic/anaerobic zones

Reflecting the adoption of filamentous organism control methods by the practicing profession, most US articles on bulking have dealt with field experience and case histories of design and performance. Table 6.29 provides a summary of publications on filament controls presented over the past years in USA technical journals. These publications demonstrate that tools are available to control excessive filament growths in essentially any activated sludge system.

An interesting exchange on the 'success' of bulking control using selectors is presented in a discussion by Parker *et al.* (1998) of a paper by Martens and Daigger (1997), and the closure by Martens and Daigger (1998). The discussers argued that a good settling activated sludge should be defined as having an SVI of <100 mL g^{-1} for 90% of the time (rather than <150 mL g^{-1}), and that this value should be the target for plants with selectors if it is desired to achieve the greatest benefit from selector installation. They further suggested that a plant-scale research

Table 6.29 Summary of recent filament control experience in the USA.

Filaments Studied	Filament Control Methods Investigated	Results	Reference
S. natans	Completely mixed anoxic selector studied at bench-scale using actual wastewater. Existing plant experienced severe bulking due to periodic discharges of readily biodegradable substrate.	Completely mixed anoxic selector effective at anoxic HRTs ranging from 17 to 164 min in system with a total HRT (aeration basin plus clarifier) of 17 h and a total ST of 6 d.	Murakami and Babcock (1998)
Nocardioforms	Addition of 0.5 mg L^{-1} of a polyacrylamide cationic polymer to a full-scale WWTP with significant foam trapping characteristics.	Aeration basic foam coverage reduced significantly, and mixed liquor nocardioform levels reduced to a lesser extent.	Starr *et al.* (1996); Shao *et al.* (1997)
Nocardioform, Type 021N	Pilot-scale unit consisting of 4-stage anaerobic selector (25% of total bioreactor volume), followed by 3-stage high purity oxygen bioreactor. Full-scale plant consisting of single stage anaerobic selector (one-sixth of total bioreactor volume) followed by 5-stage high purity oxygen bioreactor.	Effective filament control achieved in pilot-scale unit at 50 min anaerobic HRT, 2.4 d total MCRT, pH 6.3, temperature 20°C. Effective filament control achieved in full-scale unit at 17 to 24 min anaerobic HRT, 3.5 d total MCRT, temperature = 20°C. Filament control correlated with EBPR.	Fainsod *et al.* (1996); Fainsod *et al.* (1999)
N.limicola II, Type 0041	The performance of anoxic selectors at 4 full-scale WWTPs were investigated.	Two factors were found necessary for effective filament control. One was an anoxic F/M of 0.8 to 1.2 kg BOD_5 kg^{-1} MLSS d^{-1}, or *an* anoxic SRT of 1 d, for temperatures above 18°C and 0.7 to 1.0 kg BOD_5 kg^{-1} MLSS d^{-1}, or an anoxic SRT of 1.4 d, for lower temperatures. The second was an aerobic SRT of no more than 12 d.	Martens and Daigger (1995); Martens and Daigger (1997)

(*Continued*)

Table 6.29 Summary of recent filament control experience in the USA (*Continued*).

Filaments Studied	**Filament Control Methods Investigated**	**Results**	**Reference**
Thiothrix II, M. parvicella, Nocardioforms, Types 0675 and 0092	Control of temperature of full-scale 2-stage carbon removal/nitrification activated sludge system treating pharmaceutical wastewater	Excessive filament growth observed at bioreactor temperatures of ≥40°C, while filaments controlled at temperatures of 38°C or below.	Bastain *et al.* (1999)
M. parvicella, Types 0092, 0041, and 0675	RAS chlorination	Dosing at 2 g Cl_2 kg^{-1} MLVSS, controlled SVI without adversely affecting N removal and with only modest effect on P removal. Higher Cl_2 doses controlled SVI but adversely affected N&P removal.	Ward *et al.* (1999)
Types 0041 and 0675	Nutrient addition, intermittent aeration, aerobic selectors, anoxic selectors to treat municipal and food processing wastewater.	Intermittent aeration was effective, but operating problems were encountered due to use of gravity sludge recycle from the clarifier. Nutrient addition was marginally effective. A completely anoxic selector, operating at an F/M of 1.0 to 1.3 kg BOD_5 kg^{-1} MLSS d^{-1} was effective. Complete denitrification led to better SVI control. Aerobic selectors were most effective. Aerobic selector performance improved with increasing selector F/M ratio over the range of 0.1 to 1.4 kg BOD_5 kg^{-1} MLSS d^{-1}.	Nakhla and Lugowski (1999)

Not identified	Aerobic and anaerobic selectors to treat cheese processing wastewater.	Single, completely mixed selector operating at F/M of 3 to 5 provided control, but severe foamingoCcurred due to low MLSS concentration. A single, completely mixed anaerobic selector operating at an F/M of 2.6 provided effective control of filaments.	Rein and Paulson (1997)
Not identified	Changed supplemental ammonia dosing strategy for N deficient wastewater to ratio strategy.	Effective in reducing SVI from 145 to 117 mL g^{-1} and reduced filament abundance.	Barber *et al.* (1997)
Type 021N (Predominant) and Types 0041 and 1701, *Thiothrix* II and *S. natans*	Anaerobic selectors and sludge chlorination of long HRT (3.7 d) activated sludge system treating ceiling tile wastewater.	Chlorine was not effective when applied to the RAS, but was effective at doses between 1 and 3 g Cl_2 kg^{-1} MLSS d^{-1} when applied to the aeration basin. An anaerobic selector operating at an F/M of 14.3 kg BOD_5 kg^{-1} MLSS d^{-1} was partially effective.	Adey and McClintock (1996)
*N. limicola*II (Predominant)	Staged aerobic selector (three equal sized compartments) followed by completely mixed aerobic bioreactor to treat readily biodegradable industrial wastewater. Periodic sludge chlorination.	Aerobic selectors operating at first stage F/M values of about 10 kg COD kg^{-1} MLSS d^{-1} and overall F/M values of about 3 kg COD kg^{-1} MLSS d^{-1} provided effective SVI control. An extended start-up was required for the sludge to acclimate and to acquire sufficient substrate storage capacity. Sludge chlorination was practiced periodically to maintain stirred SVI values below 100 mL g^{-1}.	Boe *et al.* (1996)
Nocardioforms	Classifying selector consisting of the addition of 0.5 mg L^{-1} of readily biodegradable nonionic surfactant and an aerated section with foam removal.	Effective at laboratory and full scale at controlling nocardioform foaming.	Pagilla *et al.* (1996)

program be mounted to determine the SVI that can be achieved by various types of selector (aerobic, anoxic, anaerobic) and control measures (such as controls of MCRT, recycle rate and OUR) that might improve their performance. In their response, the authors demonstrated that such performance is possible with a properly designed and operated selector. Further research would serve to better define the design and operating parameters that that will produce such excellent settling sludges.

A recent development is the emergence of computerized filament identification keys. One example of these in the USA is the 'Fil-identpro™', developed by The Stover Group (Stover & Associates, Inc., ANON). We may expect to see further developments like this in the future.

While additional research and full-scale experience will, undoubtedly, produce further refinements in filament control options, it is clear that currently an effective set of tools are available to control activated sludge settling properties. Experience indicates that, for the situation in the USA, an excellent settling sludge will be achieved when the following guidelines are followed:

- The design incorporates an appropriately sized selector. For completely mixed anoxic and anaerobic selectors, the selector must be sized to metabolize the biodegradable organic matter within it. This generally requires an anoxic or anaerobic SRT in the 1 to 1.5 day range. The size of the selector can be reduced somewhat if it is staged. Aerobic selectors are also effective. However, they must be staged and sized using appropriate F/M ratios.
- The aerobic zone following the selector is staged.
- Flexibility is provided to allow selector and aerobic zone SRTs to be adjusted independently. This can be accomplished, for example, by providing 'swing zones' at the end of the selector/beginning of the aerobic zone that can be either mixed or aerated.
- A robust and effective oxygen transfer system is provided in the aerobic zone to maintain adequate dissolved oxygen (DO) concentrations. The system should also include an effective DO control system, and sufficient mixing to prevent low DO concentration zones from developing in the aerobic zone.
- Excessive dissolved sulfide must be removed upstream of the biological process to prevent the growth of type 021N and *Thiothrix*.

The prudent designer will also provide activated sludge chlorination capabilities for those instances in which operational constraints allow unwanted filaments to grow. Also, it may be desirable to provide the capability to allow some filament growth in the activated sludge system. Some selector systems have been so effective in reducing filament growth that the activated sludge floc is essentially devoid of filaments and is fragile and easily sheared. Proven methods for growing filaments under controlled conditions include the flexibility or by-passing of a portion of the influent wastewater around the selector into the aerobic zone and the ability to aerate anoxic or anaerobic selectors.

The Tri-City WWTP in the State of Oregon, USA is representative of the use of selectors to control activated sludge bulking (Martens & Daigger, 1997). This is a 51,000 $m^3\ d^{-1}$ facility consisting of preliminary treatment, primary treatment, air activated sludge, effluent chlorination, waste activated sludge thickening, and anaerobic digestion. Its location in the Pacific Northwestern portion of the United States means that a definite wet and dry season is experienced. Basic secondary treatment (30 mg L^{-1} five-day biochemical oxygen demand, BOD_5, and 30 mg L^{-1} total suspended solids, TSS, on a monthly average basis) is required during the wet weather period, but the treatment goal is 10 mg L^{-1} BOD_5 and 10 mg L^{-1} TSS during the dry season. Consequently, several operating modes were provided in the activated sludge system, including conventional air activated sludge, an anoxic selector, complete mix activated sludge, and step feed activated sludge.

The bioreactor is configured as a conventional bioreactor with a length to width ratio of 5:1. Figure 6.70 provides a schematic of the bioreactor and illustrates that diffused aeration equipment is provided throughout but a mixer and baffle are provided to segment off the first 20% of the bioreactor as a selector. Coarse bubble diffusers are provided in the initial 20% of the bioreactor, and fine bubble diffusers are provided in the remainder of the bioreactor. Mixed liquor recirculation capabilities are also provided.

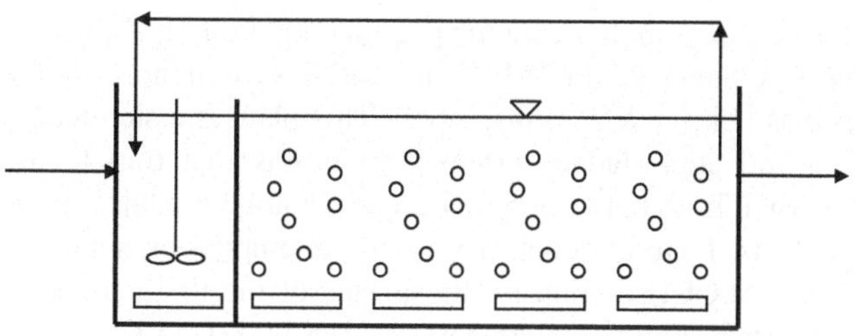

Figure 6.70 Schematic of Tri-City WWTP bioreactor illustrating anoxic selector operating mode.

The Tri-City facility replaced two existing WWTPs that both experienced severe bulking problems, which was one of the reasons that the anoxic selector features described above were incorporated into the facility design. In operation since 1986, facility operation has demonstrated the effectiveness of the anoxic selector. The facility has been operated in both the conventional and step feed activated sludge modes, resulting in sludge bulking as evidenced by SVI values well over 200 mL g^{-1}. In contrast, an SVI consistently below 120 mL g^{-1} and generally less than 100 mL g^{-1} can be maintained when operated in the anoxic selector mode. In fact, the selector is so effective that filaments can be eliminated to the point where a sludge that is too fragile is produced. When this occurs a small amount of air is allowed into the anoxic selector to grow low DO filaments. These features allow activated sludge settling characteristics to be controlled to essentially any desired value.

6.14.3 Other solid separation problems

We all know that filamentous bulking is only one of the microbially based activated sludge solids separation problems. It is for the others, especially foaming, non-filamentous (or 'viscous' or 'zoogloeal') bulking and dispersed growth that advances in our understanding are needed to produce significant reliable control measures.

6.14.4 Foaming

In the USA, bacterial foaming of activated sludge is largely due to nocardioforms and to a lesser extent to M. parvicella. With the increasing use of BNR the incidence of M. parvicella foaming may increase.

A relatively complete tool kit for solving nocardioform foaming problems has been developed, but the fundamental understanding of how some of these control measures work is not yet available (Table 6.30). We have found that control measures such as 'washout' and anoxic or anaerobic selectors are universally successful when coupled with methods to control foam trapping. In contrast, systems with effective selectors and foam trapping characteristics may continue to be plagued with excessive foaming. Thus, attention to the details of the physical configuration of the facilities is required during design to minimize foam trapping. The prudent designer will also provide remedial foam control measures, such as surface chlorine sprays and/or cationic polymer addition capabilities.

The San Jose/Santa Clara WPCP has used several nocardioform foaming control methods (Ekster & Jenkins, 1999). This plant is a step feed nitrification/dentrification activated sludge system with basins that trap foam (Fig. 6.71). Operating at MCRTs of between 5 to 7 days, it had been subject to nocardioform foaming incidents. Foaming control was first attempted by applying a fine mist spray of 3g L^{-1} NaOCl solution to the surface of the first aerobic zone using a shrouded spray bar (Fig. 6.72). At doses of 0.5 to 1.0 mg Cl_2 L^{-1} based on the wastewater flow rate this system effectively controlled nocardioform foam without altering bulk mixed liquid nocardioform organism levels, suggesting that it only kills nocardioforms at the surface of the basins. This method of nocardioform organism control did not compromise the BNR effluent in any way.

The second foam control method was to add 0.3 to 0.5 mg L^{-1} (based on combined wastewater and RAS flow) of a high charge catonic polymer to either the RAS or to the mixed liquor (Shao *et al.* 1997). Foam disappeared from the aeration basins in under 10 days. These two foam control methods were unnecessary when it became apparent that after converting the system to automatic MCRT control (based on automatic online MLSS and RAS SS measurements) there had been no nocardioform foaming incidents (even though the average MCRT was higher for automatic MCRT control than for manual control). Figure 6.73 illustrates this point and suggests that with manual MCRT control based on laboratory SS measurements, short periods of high MCRT exist that may allow nocardioforms to grow and become trapped in the aeration basins (Ekster & Jenkins, 1999).

Table 6.30 Nocardioform foaming control methods and status.

Control Measure	Result	Research Required
Remove foam traps	Allows foam to escape. Prevents foam accumulation.	Does a foam trapping system influence nocardioform growth habit? Does a foam layer reseed mixed liquor?
Selective foam wasting	Selectively removes foam and associated organisms from process, thereby minimizing foaming	How selectively can nocardioforms be removed from process? Can other organisms (such as nitrifers) be enriched in mixed liquor?
Eliminate foam recycle	Prevents reseeding of activated sludge	None
Surface Cl_2 spray	Kills surface accumulations rich in nocardioforms	None
Reduce MCRT	Washout of nocardioforms	None
Aerobic, anoxic, anaerobic selectors	Deselects nocardioforms	None
Cationic polymer addition	Flocculates nocardioforms into floc? reducing exposure to bubbles	Proof of concept

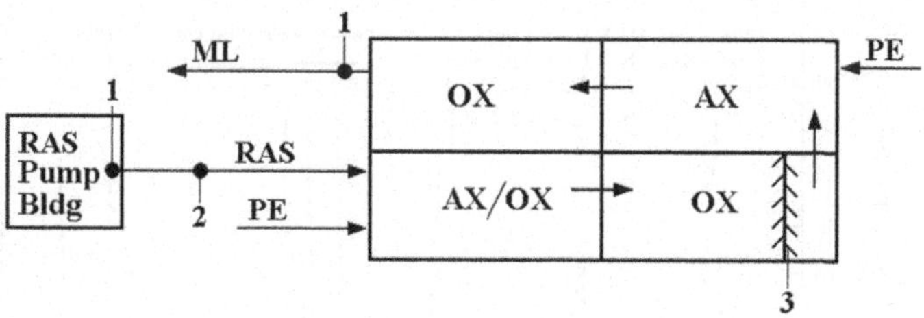

Figure 6.71 San Jose/Santa Clara step WPCP feed aeration basins. Some nocardioform control features shown. 1, TSS meter; 2, cationic polymer injection; 3, Cl_2 spray; AX, anoxic; OX, oxic; PE, primary effluent.

6.14.5 Viscous bulking and dispersed growth

It is in the area of viscous bulking and dispersed growth that the greatest research is needed. Viscous (non-filamentous zoogloeal) bulking refers to the production of slowly settling and poorly compacting activated sludge in the absence of large amounts of filamentous organisms. It is thought to be due to an overabundance of exocellular polymer that does not adhere well to the microbial surface.

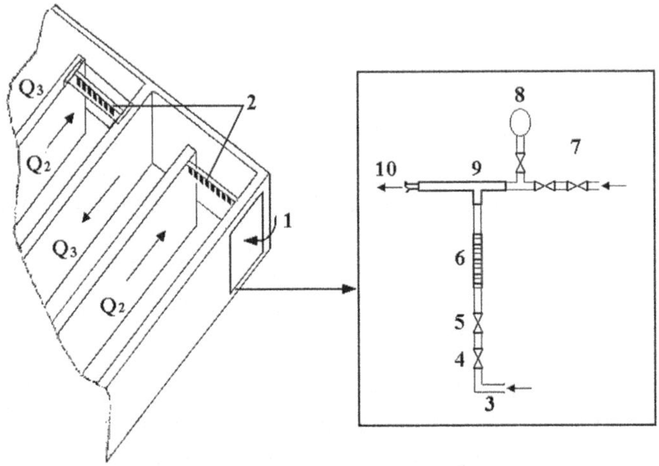

Figure 6.72 Suface mist spray chlorination system for nocardioform foam suppression; San Jose/Santa Clara WPCP; 1, dose panel; 2, hypalon should be closed at top and rear. At rear it extends a bit below liquid level to trap foam. Spray bar spaced spray nozzles; 3, from 12.5% NaOCl tank and pump; 4, check valve; 5, valve; 6, rotameter; 7, pressure regulator; 8, pressure gauge; 9, eductor; 10, to spray bars.

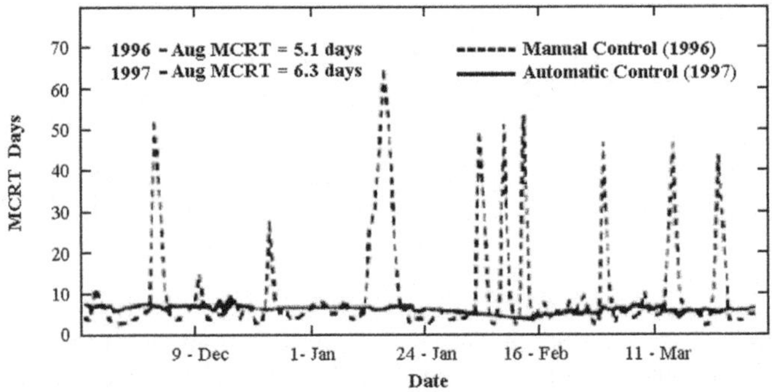

Figure 6.73 MCRT obtained with manual control (1996) and automatic control (1997) in the San JoseSanta Clara, CA, BNR Plant.

Microbial exocellular polymers consist of a wide variety of compounds including polysaccharides, proteins, DNA, and phospholipids. In activated sludge, proteins and polysaccharides appear to be major components of the exocellular material

(Higgins, 1995). Work by Novak's group (Higgins & Novak, 1995, 1997a, b, 1997c; Higgins *et al.* 1999; Murthy & Novak, 1998; Murthy *et al.* 1998; Novak, 1997; Novak *et al.* 1998) has shown that a range of solids separation and dewatering properties are influenced by both the absolute and the relative amounts of monovalent cations (e.g. Na^+, K^+, NH_4^+) and divalent cations (e.g. Ca^{2+}, Mg^{2+}). These properties include floc strength, settling rate, supernatant turbidity and soluble organic matter concentrations, filtration rates, capillary suction times, cake solids concentration, filtrate turbidity and soluble organic matter concentrations. A floc model (Fig. 6.74) is proposed in which proteins (lectins) attached to the bacterial surfaces bridge between polysaccharides molecules with divalent cations. These cations can also bridge between anionic sites on adjacent polysaccharide molecules. A network is formed that binds the exocellular polysaccharides and proteins into the floc matrix.

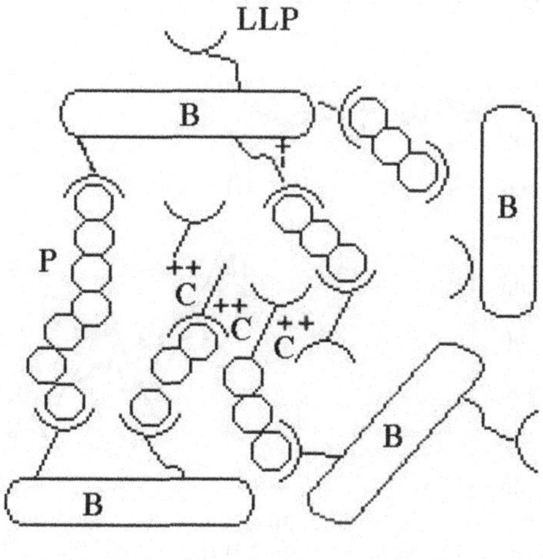

Figure 6.74 Proposed roles of polysaccharides and proteins (biopolymers) and divalent cations on bioflocculation (Higgins, 1995). B: bacteria; LLP: lectin-like protein; C^{++}: divalent cation; P: polysaccharide.

When insufficient divalent cations and/or an overabundance of monovalent cations are present the proteinpolysaccharide bridging is interfered with. The flocs lose their resistance to shear, resulting in dispersed particles that are difficult to settle and dewater, produce turbid supernatants (filtrates) and contain soluble polysaccharides that elevate the soluble organic matter content of the supernatant (filtrate). These types of effects seem to be most critical in solids dewatering and concentrating processes such as sludge filtration and centrifugation where particulate concentrations are the highest. However, many situations exist where these effects are important in the activated sludge process itself. These include the treatment of highly saline industrial wastewaters (especially where the salt derives from the use of NaOH to neutralize an acidic wastewater prior to biological treatment) and aerobic and possibly anaerobic digestion (where NH_4^+ concentrations can become substantial).

Several full-scale case studies (Higgins & Novak, 1997b; Murthy *et al.* 1998) have been presented where increased divalent/monovalent cation ratio in a wastewater has improved activated sludge and sludge dewatering solids separation process performances.

Murthy *et al.* (1998) investigated the effect of divalent cation (Mg^{2+} as $MgSO_4$) addition on activated sludge settling at an industrial wastewater treatment plant which received a high COD concentration (10,000 mg L^{-1}, mostly acetate). The wastewater contained an average Na^+ influent concentration of 2000 mg L^{-1} because it was neutralized (to pH 8.8) by NaOH addition. The effect of adding $MgSO_4$ to produce a feed concentration of 1 mmol L^{-1} (24 mg L^{-1} Mg^{2+}) is illustrated in Fig. 6.75. Table 6.31 also shows that polymer usage for WAS thickening by gravity belt was significantly lower when $Mg(OH)_2$ was added.

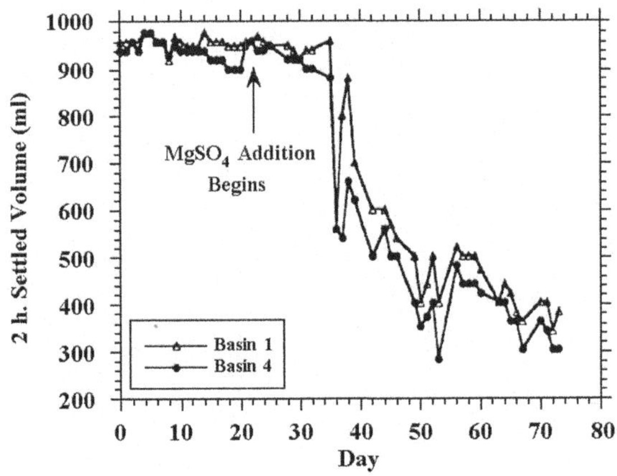

Figure 6.75 Effect of $MgSO_4$ addition on activated sludge settling (Murthy *et al.* 1998).

Table 6.31 Comparison of activated sludge operational parameters during month long trials with $Mg(OH)_2$ and NaOH base addition (Murthy *et al.* 1998).

Parameter	$Mg(OH)_2$ Trial[a]	NaOH Trial[a]
Weekly usage of base, lb wk^{-1}	6860	3360
SVI, mL g^{-1}	170	200
Aeration basin MLSS, mg L^{-1}	3180	3370
Digester MLSS, mg L^{-1}	17,600	16,800
Gravity belt polymer usage, lb wk^{-1}	150	580

[a] Values represent monthly averages, unless otherwise specified.

6.14.6 Conclusions

The 'culture' of activated sludge population dynamics and the concept of sludge quality are well established in the environmental engineering profession in the USA. Filament and floc characterization of activated sludge systems for process control and problem diagnosis and resolution are becoming common. Rational design and operating methods are available and are being used in practice to prevent and control filamentous bulking and bacterial foaming. The sludge quality problems of 'viscous' bulking and dispersed growth require further investigation.

REFERENCES

Adey A. H. and McClintock S. A. (1996). Control of filamentous bulking in a complete-mix activated sludge process treating high-strength industrial wastewater. Proceedings of the Water Environment Federation 69th Annual Conference & Exposition, Volume 3. Remediation of Soil and Groundwater Industrial Wastes, pp. 395–404.

Albert J. (2009). Medidas de control operacional durante episodios de bulking por M. parvicella y por el Tipo 021N. VI Jornadas de Transferencia de tecnología sobre Microbiología del Fango Activo. ISBN: 978-84-613-7691-9.

Albertsen M., Saunders A. M., Nielsen K. L. and Nielsen P. H. (2013). Metagenomes obtained by 'deep sequencing' – what do they tell about the enhanced biological phosphorus removal communities? *Water Science and Technology*, **68**(9), 1959–1968.

Albertsen M., Karst S. M., Ziegler A. S., Kirkegaard R. H. and Nielsen P. H. (2015). Back to basics – the influence of DNA extraction and primer choice on phylogenetic analysis of activated sludge communities. *PLoS ONE*, **10**, e0132783. doi:10.1371/journal.pone.0132783.

Alonso J. A., Ciriero E., Rodríguez E., Infante P., Serra M. C. and Huguet A. (2011). Biological scum by Candidatus Monilibacter spp. in a wastewater treatment plant in north-eastern Spain. The conference Microbes in Wastewater and Waste Treatment, Bioremediation and Energy Production Topic: Microbes in Wastewater Treatment, Goa, India, 24–26 Enero 2011.

Amann R. I., Binder B. J., Olson R. J., Chisholm S. W., Devereux R. and Stahl D. A. (1990). Combination of 16S rRNA-targeted oligonucleotide probes with flow cytometry for analyzing mixed microbial populations. *Applied and Environmental Microbiology*, **56**, 1919–1925.

Arregui L., Liébana R., Rodríguez E., Murciano A., Conejero F., Pérez-Uz B. and Serrano S. (2012). Analysis of the usefulness of biological parameters for the control of activated sludge wastewater treatment plants in an interlaboratory study context. *Journal of Environmental Monitoring*, **14**, 1444–1452.

Badoer S., Cuomo M., Miana P., Fabris M., Rossetti S. and Tandoi V. (2003). Biological scum by M. parvicella in a large wastewater treatment plant in Northern Italy: causes and solutions. Presented at IWA Regional Conference "Water key to sustainable development in Africa", Cape Town, 14–17 September 2003.

Badoer S., Miana P., Della Sala S., Marchiori G., Tandoi V. and Di Pippo F. (2015). Nitrifying bacterial biomass and nitrification activity evaluated by FISH and an automatic on-line instrument at full-scale Fusina (Venice, Italy) WWTP. *Environmental Science and Pollution Research*, **22**(24), 19786–19792.

Barber J. B., Bullard C. M., Meyers Jr. *et al.* (1997). Nitrogen Supplementation Strategy for Variable-Strength Industrial Wastewater. Proceedings of the Water Environment Federation 70th Annual Conference & Exposition. Volume 3, Remediation of Soil & Groundwater and Industrial Issues & Treatment Technology. Alexandria (VA), pp. 551–557.

Bargna R., Terzi C. and Ramadori R. (1993). Bulking – foaming control in the Varese – Olona plant. In: Prevention and Control of Bulking Activated Sludge, D. Jenkins, R. Ramadori and L. Cingolani (eds), Centro Studi Provincia di Perugia, Perugia, Italy, pp. 139–140.

Barnard J. L. (1976). A review of biological phosphorus removal in the activated sludge process. *Water SA*, **2**, 136–144.

Bastain O. J., Parks J. A., Bordacs K. and Philp J. (1999). Effect of high temperature on filament growth and nitrifier activity in an activated sludge process: a pilot study. Proceedings of the Water Environment Federation 72nd Annual Conference & Exposition. Water Environment Federation, Alexandria, VA, CD-ROM 1999.

BDERU (2012). La Base de Données Eaux Résiduaires Urbaines 2012.

Beccari M., Ramadori R., Rozzi A. and Tandoi V. (1995). Trattamenti biolotgici delle acque. In: Manuale di conduzione degli impianti di depurazione delle acque, R. Passino (ed.), Zanichelli/Esac, Bologna.

Beer M., Seviour E. M., Kong Y., Cunningham M., Blackall L. L. and Seviour R. J. (2002). Phylogeny of the filamentous bacterium Eikelboom Type 1851, and design and application of a 16S rRNA targeted oligonucleotide probe for its fluorescence in situ identification in activated sludge. *FEMS Microbiology Letters*, **207**, 179–183.

Bjornsson L., Hugenholtz P., Tyson G. W. and Blackall L. L. (2002). Filamentous Chloroflexi (green non-sulfur bacteria) are abundant in wastewater treatment processes with biological nutrient removal. *Microbiology*, **148**, 2309–2318.

Blackall L. L., Parlett J., Hayward A., Minnikin D., Greenfield P. and Harbers A. E. (1989). *Nocardia pinensis* sp. nov., an actinomycete found in activated sludge foams in Australia. *Microbiology*, **135**(6), 1547–1558.

Blackbeard J. R. and Ekama G. A. and Marais G. v. R. (1986). Survey of filamentous bulking and foaming problems in activated sludge plants in South Africa. *Water Pollution and Control*, **85**(1), 90–100.

Blackbeard J. R., Gabb D. M. D., Ekama G. A. and Marais G. v. R. (1988). Identification of filamentous organisms in nutrient removal activated sludge plants in South Africa. *Water S.A.*, **14**, 29–33.

Boe R. S., Givens D., Jenkins D. and Daigger G. T. (1996). Use of Selectors for Bulking Control in an Industrial Wastewater Activated Sludge Treatment Plant. Proceedings of the Water Environment Federation 69th Annual Conference & Exposition, Volume 3, Remediation of Soil and Groundwater Industrial Wastes, 1996, pp. 487–497.

Bux F. and Kasan H. C. (1994). A microbiological survey of ten activated sludge plants. *Water SA*, **20**(1), 61–71.

Callens W. (2007). Invloed van fysische parameters op de bezinkbaarheid van actief slib: analyse van praktijkgegevens. MSc thesis, Universiteit Antwerpen.

Caporaso J. G., Lauber C. L., Walters W. A., Berg-Lyons D., Huntley J., Fierer N., Owens S. M., Betley J., Fraser L., Bauer M., Gormley N., Gilbert J. A., Smith G. and Knight R. (2012). Ultra-high-throughput microbial community analysis on the Illumina HiSeq and MiSeq platforms. *ISME Journal*, **6**, 1621–1624.

Casey T. G., Ekama G. A., Wentzel M. C. and Marais G. v. R. (1993). Causes and control of low F/M filamentous bulking in nutrient removal activated sludge systems. Prevention and Control of Bulking Activated Sludge Workshop, Perugia, 69–76.

Casey T. G., Wentzel M. C., Ekama G. A. and Marais G. v. R. (1994a). Causes and Control of Low F/M Filamentous Bulking in Long Sludge Age Nutrient Removal Activated Sludge Systems. WRC Report No. 286/2/93 (UCT Research Report No. W83), Water Research Commission, PO Box 824, Pretoria 0001, RSA.

Casey T. G., Wentzel M. C., Ekama G. A. and Marais G. v. R. (1994b). Development and Evaluation of Specific Control Methods for Ameliorating Low F/M Filament Bulking. WRC Report No. 286/1/93 (UCTResearch Report No. W82), Water Research Commission, PO Box 824, Pretoria, 0001, South Africa.

Casey T. G., Ekama G. A., Wentzel M. C. and Marais G. v. R. (1995). Filamentous organism bulking in nutrient removal activated sludge systems. Paper 1: An historical overview of causes and control. *Water SA*, **21**(3), 321–338.

Casey T. G., Wentzel M. C. and Ekama G. A. (1999). Filamentous organism bulking in nutrient removal activated sludge systems. Paper 11: A biochemical/microbiological model for proliferation of anoxic-aerobic (AA) filamentous organisms. *Water SA*, **25**, 443–451.

Castell D. (2013). herramientas de control de proceso en edar. Casos practices. IX Jornadas de Transferencia de Tecnología sobre Microbiología del Fango Activo. ISBN: 978-84-616-7408-4.

Cesaro G. (1997). Utilizzo di un selettore anossico nella lotta al bulking da Type 021N e Thiothrix. In: La Gestione degli Impianti di Depurazione delle Acque di Scarico. Esperienze Nazionali a Confronto, 43–49, Centro Studi Provincia di Perugia, Perugia Italy.

Chin C.-J. M., Chen P.-W. and Wang L.-J. (2006). Removal of nanoparticles from CMP wastewater by magnetic seeding aggregation. *Chemosphere*, **63**(10), 1809–1813.

Choi J., Kotay S. M. and Goel R. (2011). Bacteriophage-based biocontrol of biological sludge bulking in wastewater. *Bioengineered Bugs*, **2**, 214–217.

Cingolani L. and Ciccarelli E., (1996). Depurazione Biologica a Fanghi Attivi; I microrganismi responsabili di bulking e schiume biologiche. Vol I, Centro Studi Provincia di Perugina, Perugia, Italy.

Cingolani L. and Ramadori R. (1992). Il controllo del bulking negli impianti a fanghi attivi. Centro Studi Provincia di Perugia, Perugia, Italy.

Claros J., Aguado D., Serralta J., Seco A. and Ferrer J. (2012). Posibilidades de implantación de un proceso Sharon en los sobrenadantes de una EDAR VIII Jornadas de Transferencia de Tecnología sobre Microbiología del Fango Activo.

Clauss F., Hélaine D., Balavoine C. and Bidault A. (1998). Improving activated sludge floc structure and aggregation for enhanced settling and thickening performances. *Water Science and Technology*, **38**, 35–44.

Comas J., Rodríguez-Roda I., Gernaey K. V., Rosen C., Jeppsson U. and Poch M. (2008). Risk assessment modeling of microbiology related solids separation problems in activated sludge systems. *Environmental and Modelling Software*, **23**, 1250–1261.

Cyrus Z. and Sladka A. (1970). Several Interesting Organisms Present in Activated Sludge. *Hydrobiologia*, **35**(3/4), 383–396.

Daims H., Bruhl A., Amann R., Schleifer K. H. and Wagner M. (1999). The domain-specific probe EUB338 is insufficient for the detection of all bacteria: development and evaluation of a more comprehensive probe set. *Systematic and Applied Microbiology*, **22**, 434–444.

Davoli D., Madoni P., Collini G. and Pergetti M. (1997). Approccio sperimentale al controllo del fenomeno di foaming biologico in un impianto di depurazione a fanghi attivi. *Acqua e Aria*, **4**, 77–83.

Davoli D., Madoni P., Guglielmi L., Pergetti M. and Barilli S. (2002). Testing the effect of selectors in the control of bulking and foaming in full scale activated-sludge plants. *Water Science and Technology*, **46**(1–2), 495–498.

Davoli C., Guglielmi L. and Stefanini C. (2016). Circuito di qualificazione per il riconoscimento microscopico su basemorfologica dei batteri filamentosi del fango attivo: risultati dei quattro circuiti effettuati. Notiziario dei Metodi Analitici & IRSA News, vol. 1, Maggio 2016, 21–23 (in Italian, English abstract).

Deepnarain D., Kumari S., Ramjith J., Swalaha F. M., Tandoi V., Pillay K. and Bux F. (2015). A logistic model for the remediation of filamentous bulking in biological nutrient removal wastewater treatment removal wastewater treatment plant. *Water Science and Technology*, **72**(3), 391–405.

de los Reyes F. (2010). Foaming. In: Microbial Ecology of Activated Sludge, R. J. Seviour and P. Neilsen (eds), IWA Publishing, London, pp. 215–258.

Department of Water Affairs (2012). Introduction to Green Drop Progress Report 2012/ National Overview. http://www.ewisa.co.za/misc/BLUE_GREENDROPREPORT/ GreenDrop2012/GreenDrop2012_Intro_Background.pdf

Department of Water Affairs (2013). A Strategic Overview of the Water Sector in South Africa. Pretoria. DWA Directorate: Water Services Planning & Information, version 1.5. http://nepadwatercoe.org/wp-content/uploads/Strategic-Overview-of-the-Water-Sector-in-South-Africa-2013.pdf

Di Marino R., Donadio F. and Verdiani B. (1992). Alcune esperienze sul controllo del bulking filamentoso e foaming. Indagini ed interventi proposti. Il bulking filamentoso: controllo e gestione. AGAC Ed., Reggio Emilia, Italy, 5 Maggio 1992.

dos Santos L. A., Ferreira V., Neto M. M., Pereira M. A., Mota M. and Nicolau A. (2015). Study of 16 Portuguese activated sludge systems based on filamentous bacteria populations and their relationships with environmental parameters. *Applied Microbiology and Biotechnology*, **99**(12), 5307–5316.

Durban N. (2015). Traitement du foisonnement filamenteux induit par M. parvicella par ajout de sels métalliques: impact sur les boues activées. L'Institut des Sciences et Industries du Vivant et de l'Environnement (AgroParisTech), Paris, France.

Durban N., Juzan L., Krier J. and Gillot S. (2016). *Control of* M. parvicella by aluminium salts addition. *Water Science and Technology*, **73**, 414–422.

Dyson Z., Tucci J., Seviour R. and Petrovski S. (2016). Isolation and characterization of bacteriophage SPI1, which infects the activated-sludge-foaming bacterium *Skermania piniformis*. *Archives of Virology*, **161**(1), 149–158.

Eales K., Nielsen J. L., Kragelund C., Seviour R. and Nielsen P. H. (2005). The in situ physiology of pine tree like organisms (PTLO) in activated sludge foams. *Acta Hydrochimica et Hydrobiologica*, **33**(3), 203–209.

Eales K. L., Nielsen J. L., Seviour E. M., Nielsen P. H. and Seviour R. J. (2006). The in situ physiology of Skermania piniformis in foams in Australian activated sludge plants. *Environmental Microbiology*, **8**(10), 1712–1720.

Eikelboom D. H. (1975). Filamentous organisms observed in activated sludge. *Water Research*, **9**, 365–388.

Eilkelboom D. H. (1997). Control of M. parvicella by addition of PAX 14. TNO Report R 97/305.

Eikelboom D. H. (1999). Procesbewaking door microscopisch slibonderzoek. TNO Milieu, Energie en Procesinnovatie, The Netherlands.

Eikelboom D. H. (2000a). Process Control of Activated Sludge by Microscopic Investigation. London. IWA Publishing, London, England.

Eikelboom D. H. (2000b). Prozessüberwachung von Belebungsanlagen durch mikroskopische Schlammuntersuchung, ATV-DVWK, Hennef 2000; ISBN 3-933707-76-5.

Eikelboom D. H. (2002). Process Control of Activated Sludge Plants by Microscopic Investigation, IWA, 2002.

Eikelboom D. H. (2006). Identification and control of filamentous microorganisms in industrial waste water. *Pollution Control*, **85**(1), 101–113.

Eikelboom D. H. and van Buijsen H. J. J. (1981). Microscopic sludge investigation manual. Report A94a. IMG-TNO, Delft.

Eikelboom D. H. and van Buijsen H. J. J. (1983). Microscopic Sludge Investigation Manual. TNO Delft, The Netherlands.

Eikelboom D. H. and van Buijsen H. J. J. (2002). Handbuch für die mikroskopische Schlammuntersuchung, 4. Auflage, Hirthammer Verlag 1983; ISBN: 3-88721-013-1.

Ekama G. A. (2010). The role and control of sludge age in biological nutrient removal systems. *Water Science and Technology*, **61**(7), 1645–1652.

Ekama G. A. (2015). Review: Recent developments in biological nutrient removal. *Water SA*, **41**(4), 515–524.

Ekama G. A. and Wentzel M. C. (1999). Difficulties and developments in biological nutrient removal technology and modelling. *Water Science and Technology*, **39**(6), 1–11.

Ekama G. A., Wentzel M. C., Casey T. G. and Marais G. v. R. (1996a). Filamentous organism bulking in nutrient removal activated sludge systems. Paper 6: Review, evaluation and consolidation of results. *Water SA*, **22**(2), 147–152.

Ekama G. A., Wentzel M. C., Casey T. G. and Marais G. v. R. (1996b). Filamentous organism bulking in nutrient removal activated sludge systems. Paper 3: Stimulation of the selector effect under anoxic conditions. *Water SA*, **22**(2), 119–126.

Ekster A., Jenkins D. Nocardia Foaming Control in a BNR Plant. Proceedings of the Water Environment Federation, 72nd Annual Conference & Exposition, Water Environment Federation, Alexandria, VA, CD-ROM 1999.

Erhart R., Bradford D., Seviour R. J., Amann R. and Blackall L. L. (1997). Development and use of fluorescent in situ hybridization probes for the detection and identification of 'M. parvicella' in activated sludge. *Systematic and Applied Microbiology*, **20**, 310–318.

European Environment Agency (2012). Map of catchment-sensitive type and percentage distribution of type of treatment in Spain. http://www.eea.europa.eu/data-and-maps

Fainsod A., Pagilla K. R., Jenkins D., Pitt P. A. and Mamais D. (1996). The Effect of Anaerobic Selectors on Nocardia Growth in Activated Sludge. Proceedings of the Water Environment Federation 69st Annual Conference & Exposition. Volume 1, Wastewater Treatment Research and Municipal Wastewater Treatment. Water Environment Federation, Alexandria, VA, 1996, 189–200.

Fainsod A., Pagilla K. R., Jenkins D., Pitt P. A. and Mamais D. (1999). The Effect of Selectors on Nocardioform Organism Growth in Activated Sludge. *Water Environ Research*, **71**, 1151–1157.

Ferrer C. (2005). Eliminación biológica de espumas en reactores biológicos. Transferencia de Tecnología sobre ejercicios interlaboratorios en fangos activos como sistema de control de calidad en la EDAR. *Tecnología del Agua*, **260**, 82–92.

Gabb D. M. D., Ekama G. A., Jenkins D., Wentzel M. C., Casey T. G. and Marais G. v. R. (1996a). Filamentous organism bulking in nutrient removal activated sludge systems. Paper 4: System configurations and operating conditions to develop low F/M filament bulking sludges at laboratory-scale. *Water SA*, **22**(2), 127–137.

Gabb D. M. D., Ekama G. A., Jenkins D., Wentzel M. C., Casey T. G. and Marais G. v. R. (1996b). Filamentous organism bulking in nutrient removal activated sludge systems. Paper 5: Experimental examination of aerobic selectors in anoxic-aerobic systems. *Water SA*, **22**(2), 127–137.

Gaval G. and Pernelle J.-J. (2003). Impact of the repetition of oxygen deficiencies on the filamentous bacteria proliferation in activated sludge. *Water Research*, **37**, 1991–2000.

Gich F., Garcia-Gil J. and Overmann J. (2001). Previously unknown and phylogenetically diverse members of the green nonsulfur bacteria are indigenous to freshwater lakes. *Archives of Microbiology*, **177**, 1–10.

Golla G., Hocquet C., Rakedjian B. and Morin E. (2010). Bilan 2008 de l'assainissement en France (No. 28). Ministère de l'Écologie, de l'Énergie, du Développement Durable et de la Mer & Office National de l'Eau et des Milieux Aquatiques & Office internationale de l'eau, Paris, France.

Goodfellow M. and Williams S. (1983). Ecology of actinomycetes. *Annual Reviews in Microbiology*, **37**(1), 189–216.

Graveleau L., Cotteux E. and Duchene P. (2005). Bulking and foaming in France: The 1999–2001 Survey. *Acta Hydrochimica et Hydrobiologica*, **33**, 223–231.

Guo J., Peng Y., Wang S., Yang X., Wang Z. and Zhu A. (2012). Stable limited filamentous bulking through keeping the competition between the floc-formers and filaments in balance. *Bioresource Technology*, **103**, 7–15.

Gray N. F. (2010). Water Technology: An Introduction for Environmental Scientists and Engineers. IWA Publishing, United Kingdom.

Greek Ministry of Environment (2016). Greek National Database for the Monitoring of Performance of Wastewater Treatment Plants, http://astikalimata.ypeka.gr/

Hamlin E. J. and Wilson H. (1951). The history, development, present, past and future trends of sewage disposal in South Africa. *Journal and Proceedings of the Institute for Sewage Purification*, **3**, 235–256.

Hattori S., Watanabe M., Osono H., Togii H. and Sasaki K. (2001). Effects of an external magnetic field on the flock size and sedimentation of activated sludge. *World Journal of Microbiology and Biotechnology*, **17**, 833–838.

Higgins M. J. (1995). The Interaction of Cations Proteins and Poly-Saccharides in the Settling and Dewatering of Activated Sludge. PhD, Dissertation, Environmental Engineering, Virginia Polytechnic and State University, August 1995.

Higgins M. J. and Novak J. T. (1997a). The Effect of Cations on the Settling and Dewatering of Activated Sludges: Laboratory Results. *Water Environmental Research*, **69**, 215–224.

Higgins M. J. and Novak J. T. (1997b). Dewatering and settling of activated sludges: the case for using cation analysis. *Water Environmental Research*, **69**, 225–232.

Higgins M. J. and Novak J. T. (1997c). Characterization of Exocellular Protein and Its Role in Bioflocculation. *Journal of Environmental Engineering*, **123**, 479–485.

Higgins M. J. and Novak J. T. (1995). The Effect of Cations on the Settling and Dewatering of Activated Sludges. Proceedings of the Water Environment Federation, 68st Annual Conference & Exposition, Volume 1, Wastewater Treatment Research and Municipal Wastewater Treatment, Water Environment Federation, Alexandria, VA, pp. 225–236.

Higgins M. J., Miller L. A. and Sobeck D. (1999). Improving Industrial Wastewater Treatment Using Divalent Cation Addition. Proceedings of the Water Environment Federation 72nd Annual Conference & Exposition, Water Environment Federation, Alexandria, VA, CD-ROM 1999.

Hu B., Qi R., An W., Xu M., Zhang Y., Bai X., Bao H., Wen Y., Gu J. and Yang M. (2013). Dynamics of the microfauna community in a full-scale municipal wastewater treatment plant experiencing sludge bulking. *European Journal of Protistology*, **49**(4), 491–499.

Hu Z., Sötemann S., Moodley R., Wentzel M. C. and Ekama G. A. (2003). Experimemntal investigation of the external nitrification biological nutrient removal activated sludge system. *Biotechnology and Bioengeneering*, **83**(3), 260–273.

Infante P. (2012). Optimización del rendimiento depurativo de una planta de tratamiento biológico de efluentes de la industria química. VIII Jornadas de Transferencia de Tecnología sobre Microbiología del Fango Activo (Sesión on line). ISBN: 978-84-695-8114-8.

Infante P. and Rodríguez E. (2010). Control microbiológico de una EDAR de Refinería. Póster. Octubre 2010. VII Jornadas de Transferencia de tecnología sobre Microbiología del Fango Activo. ISBN: 978-84-614-7529-2

Infante P. and Rodríguez E. (2011). Control Microbiológico en una EDAr de refinería. Técnica y gestión del agua. N° 216.

Infante P., Rodríguez E., Moral A., Tijero A., Hernández M. D. and Linares M. R. (2012a) Depurative performance optimization of a biological effluents treatment plant in the chemical industry. (Póster). Water Integral Management. ANQUE 2012, Sevilla, 24–27 de Junio de 2012.

Infante P., Rodríguez E., Moral A., Tijero A. and Hernández M. D. (2012b). Microbiologic study in paper industries water treatment plants. (Comunicación oral). Paradigms for the pulp and paper Industry in the XXI Century: Opportunities for a Sustainable Future. ANQUE 2012, Sevilla, 24–27 de Junio de 2012. ISBN: 978-84-695-8114-8.

Istat (2014). Censimento delle acque per uso civile. Statistiche, Report, 30 Giugno 2014.

Jenkins D., Richard M. G. and Daigger G. T. (1993) Manual on the Causes and Control of Activated Sludge Bulking and Foaming, 2nd edn, Lewis Publishers, USA.

Jenkins D., Richard M. G. and Daigger G. T. (2003). Manual on the Causes and Control of Activated Sludge Bulking and Foaming, 3rd edn. Lewis publishers, Michigan.

Jenkins D., Richard M. G. and Daigger G. T. (2004). Manual on the Causes and Control of Activated Sludge Bulking, Foaming, and Other Solids Separation Problems. IWA Publishing, United Kingdom.

Ji Y., Wang Y., Sun J., Yan T., Li J., Zhao T., Yin X. and Sun C. (2010). Enhancement of biological treatment of wastewater by magnetic field. *Bioresource Technology*, **101**, 8535–8540.

Jiménez C., Fernández N., Rodríguez E., Isac L., Salas M. D. and Gómez E. (2001). Sistema rápido de estimación de los rendimientos en depuración de una EDAR en función de las características macroscópicas y microscópicas del fango activado (Rapid system of estimation of wastewater treatment plants performance based on macroscopic and microscopic characteristics of activated sludge). *Tecnología del Agua*, **216**, 40–44.

Jin L., Zhang G. and Tian H. (2014). Current state of sewage treatment in China. *Water Research*, **66**, 85–98.

Juang D.-F. (2005). Effects of synthetic polymer on the filamentous bacteria in activated sludge. *Bioresource Technology*, **96**, 31–40.

Kaetzke A., Jentzsch D. and Eschrich K. (2005). Quantification of M. parvicella in activated sludge bacterial communities by real-time PCR. *Letter in Applied Microbiology*, **40**, 207–211.

Kaewpipat K. and Grady Jr. C. P. L. (2002). Population dynamics in laboratory-scale activated sludge reactors. *Water Science and Technology*, **46**(1–2), 19–27.

Kanagawa T., Kamagata Y., Aruga S., Kohno T., Horn M. and Wagner M. (2000). Phylogenetic analysis of and oligonucleotide probe development for Eikelboom type 021N filamentous bacteria isolated from bulking activated sludge. *Applied and Environmental Microbiology*, **66**, 5043–5052.

Kappeler J. and Gujer W. (1994). Influences of wastewater composition and operating conditions on activated sludge bulking and scum formation. *Water Science and Technology*, **30**, 181–189.

Keegan A., van den Akker B., Luong J., Hallas G. Dalgleish V. and Krampe J. (2012). Seasonal changes in filamentous bacteria in activated sludge plants in south Australian WWTPs. Australian Water Association Proceedings, Ozwater, Sydney, Australia.

Kohno T., Sei K. and Mori K. (2002). Characterization of type 1851 organism isolated from activated sludge samples. *Water Science and Technology*, **46**(1–2), 111–114.

Kos M., Wanner J., Sorm I. and Grau P. (1992). R-D-N Activated Sludge Process. *Water Science and Technology*, **25**(4–5), 151–160.

Knoop S. and Kunst S. (1998). Influence of temperature and sludge loading on activated sludge settling, especially on M. *parvicella. Water Science and Technology*, **46**(4–5), 487–490.

Kragelund C., Levantesi C., Borger A., Thelen K., Eikelboom D., Tandoi V., Kong Y., Waarde J. van der, Krooneman J., Rossetti S., Thomsen T. R. and Nielsen P. H. (2007). Identity, abundance and ecophysiology of filamentous Chloroflexi species present in activated sludge treatment plants. *FEMS Microbiology Ecology*, **59**, 671–682.

Kragelund C., Levantesi C., Borger A., Thelen K., Eikelboom D., Tandoi V., Kong Y., Krooneman J., Larsen P., Thomsen T. R. and Nielsen P. H. (2008). Identity, abundance and ecophysiology of filamentous bacteria belonging to the Bacteroidetes present in activated sludge plants. *Microbiology*, **154**, 886–894.

Kragelund C., Nilsson B., Eskilsson K., Bøgh A. M. and Nielsen P. H. (2010). Full-scale control of Mycolata foam by FEX-120 addition. *Water Science & Technology*, **61**, 2443.

Kragelund C., Thomsen T. R., Mielczarek A. T. and Nielsen P. H. (2011). Eikelboom's morphotype 0803 in activated sludge belongs to the genus Caldilinea in the phylum Chloroflexi. *FEMS Microbiology Ecology*, **76**(3), 451–462.

Krhutkova O., Ruzickova I. and Wanner J. (2001). Microbial evaluation of activated sludge and filamentous population at eight Czech nutrient removal activated sludge plants during year 2000. Proc. of the 3rd IWA International Specialised Conference on Microorganisms in Activated Sludge and Biofilm Processes, 13–15 June, V. Tandoi, R. Passino and M. C. Blundo (eds), Rome, Italy.

Kristensen G. H., Jørgensen P. E. and Nielsen P. H. (1994). Settling characteristics for activated sludge in Danish treatment plants with biological nutrient removal. *Water Science & Technology*, **29**(7), 157–165.

Kristiansen R., Nguyen H. T. T., Saunders A. M., Nielsen J. L., Wimmer R., Le V. Q., McIlroy S. J., Petrovski S., Seviour R. J. and Calteau A. (2013). A metabolic model for members of the genus *Tetrasphaera* involved in enhanced biological phosphorus removal. *ISME Journal*, **7**(3), 543–554.

Kumari S. K. S., Marrengane Z. and Bux F. (2009). Application of quantitative RT-PCR to determine the distribution of M. parvicella in full-scale activated sludge treatment systems. *Applied Microbiology and Biotechnology*, **83**, 1135–1141.

Lacko N., Bux F. and Kasan H. C. (1999). Survey of filamentous bacteria in activated sludge plants in KwaZulu-Natal. *Water SA*, **25**, 63–68.

Lakay M. T., Hulsman A., Ketley D., Warburton C., de Villiers M., Casey T. G., Wentzel M. C. and Ekama G. A. (1999). Filamentous organism bulking in nutrient removal activated sludge systems Paper 7: Exploratory experimental investigations. *Water SA*, **25**(4), 338–396.

Lakay M. T., Wentzel M. C., Ekama G. A. and Marais G. v. R. (1988). Bulking control with chlorination in a nutrient removal activated sludge system. *Water SA*, **14**, 35–42.

Łebkowska M., Rutkowska-Narożniak A., Pajor E. and Pochanke Z. (2011). Effect of a static magnetic field on formaldehyde biodegradation in wastewater by activated sludge. *Bioresource Technology*, **102**(19), 8777–8782.

Łebkowska M., Narożniak-Rutkowska A. and Pajor E. (2013). Effect of a static magnetic field of 7 mT on formaldehyde biodegradation in industrial wastewater from urea-formaldehyde resin production by activated sludge. *Bioresource Technology*, **132**, 78–83.

Levantesi C., Rossetti S., Thelen K., Kragelund C., Krooneman J., Eikelboom D., Nielsen P. H. and Tandoi V. (2006). Phylogeny, physiology and distribution of '*Candidatus* Microthrix calida', a new Microthrix species isolated from industrial activated sludge wastewater treatment plants. *Environmental Microbiology*, **8**(9), 1552–1563.

Liu J., Yang M., Qi R., An W. and Zhou J. (2008). Comparative study of protozoan communities in full-scale MWTPs in Beijing related to treatment processes. *Water Research*, **42**(8–9), 1907–18.

Liu J. R. and Seviour R. J. (2001). Design and application of oligonucleotide probes for fluorescent in situ identification of the filamentous bacterial morphotype Nostocoida limicola in activated sludge. *Environmental Microbiology*, **3**, 551–560.

Liu S., Yang F., Meng F., Chen H. and Gong Z. (2008). Enhanced anammox consortium activity for nitrogen removal: Impacts of static magnetic field. *Journal ofBiotechnology*, **138**(3–4), 96–102.

Liu Y.-Q, Liu Y. and Tay J.-H. (2004). The effects of extracelular polymeric substances on the formation and stability of biogranules. *Applicated Microbiological Biotechnology*, **65**, 143–148.

Lou I. C. and De los Reyes III F. L. (2008). Clarifying the roles of kinetics and diffusion in activated sludge filamentous bulking. *Biotechnology and Bioengineering*, **101**(2), 327–336.

Madoni P. (1994). A sludge biotic index (SBI) for the evaluation of the biological performance of activated sludge plants based on the microfauna analysis. *Water Research*, **2**, 67–75.

Madoni P. (2011). Protozoa in wastewater treatment processes: a minireview. *Italian Journal of Zoology*, **78**, 3–11.

Madoni P., Davoli D., Guglielmi L. and Stefanini C. (1996). Guida all'analisi microscopica del fango attivo. Atlante Fotografico, AGAC Reggio Emilia and Università degli Studi di Parma Eds.

Madoni P., Davoli D. and Gibin G. (2000). Survey of filamentous microorganisms from bulking and foaming activated sludge plants in Italy. *Water Research*, **34**, 1767–1772.

Mamais D., Marneri M. and Noutsopoulos C. (2012). Causes and control practices of filamentous foaming in wastewater treatment plants, *Water Practice and Technology*, **7**(3), doi:10.2166/wpt.2012.049.

Marrengane Z., Kumari S. K., Pillay L. and Bux F. (2011). Rapid quantification and analysis of genetic diversity among *Gordonia* populations in foaming activated sludge plants. *Journal of Basic Microbiology*, **51**, 415–423.

Martens W. L. and Daigger G. T. (1995). Full-Scale Evaluation of Factors Affecting the Performance of Anoxic Selectors. Proceedings of the Water Environment Federation 68st Annual Conference & Exposition. Volume 1, Wastewater Treatment Research and Municipal Wastewater Treatment, Water Environment Federation, Alexandria, VA, pp. 669–680.

Martens W. L. and Daigger G. T. (1997). Full-scale evaluation of factors affecting the performance of anoxic selectors. *Water Environmental Research*, **69**, 1272–1281.

Martens W. L. and Daigger G. T. (1998). Closure to full-scale evaluation of factors affecting the performance of anoxic selectors. *Water Environmental Research*, **70**, 1229–1231.

Martins A. M. P., Pagilla K., Heijnen J. J. and van Loosdrecht M. C. M. (2004a). Filamentous bulking sludge – A review. *Water Research*, **38**, 793–817.

Martins A. M. P., Picioreanu C., Heijnen J. J. and van Loosdrecht M. C. M. (2004b). Three-dimensional dual morphotype species modeling of activated sludge flocs. *Environmental Science and Technology*, **38**(21), 5632–5641.

Martins A. M. P., Heijnen J. J. and Van Loosdrecht M. C. M. (2004c). Bulking Sludge in Biological Nutrient Removal Systems. *Biotechnology and Bioengineering*, **86**(2), 125–135.

Martins A. M. P., Karahan Ö. and van Loosdrecht M. C. M. (2011). Effect of polymeric substrate on sludge settleability. *Water Research*, **45**(1), 263–273.

McIlroy S. J., Nittami T., Seviour E. M. and Seviour R. J. (2010). Filamentous members of cluster III *Defluviicoccus* have the in situ phenotype of a glycogen-accumulating organism. *FEMS Microbiology Ecology*, **74**, 248–256.

McIlroy S. J., Kristiansen R., Albertsen M., Karst S. M., Rossetti S., Nielsen J. L., Tandoi V., Seviour R. J. and Nielsen P. H. (2013). Metabolic model for the filamentous '*Candidatus* M. parvicella'based on genomic and metagenomic analyses. *ISME Journal*, **7**(6), 1161–1172.

McIlroy S. J., Karst S. M., Nierychlo M., Dueholm M. S., Albertsen M., Kirkegaard R. H., Seviour R. J. and Nielsen P. H. (2016). Genomic and in situ investigations of the novel uncultured Chloroflexi associated with 0092 morphotype filamentous bulking in activated sludge. *ISME Journal*. doi:10.1038/ismej.2016.14.

McIlroy S. J., Saunders A. M., Albertsen M., Nierychlo M., McIlroy B., Hansen A. A., Karst S. M., Nielsen J. L. and Nielsen P. H. (2015). MiDAS: the field guide to the microbes of activated sludge. *Database*, 2015, bav062.

Miana P., Grando L., Caravello G. and Fabris M. (2002). M. parvicella foaming at the Fusina WWTP. *Water Science and Technology*, **46**(1–2), 499–502.

Mielczarek A. T., Kragelund C., Eriksen P. S. and Nielsen P. H. (2012). Population dynamics of filamentous bacteria in Danish wastewater treatment plants with nutrient removal. *Water Research*, **46**, 3781–3795.

Mielczarek A. T., Saunders A. M., Larsen P., Albertsen M., Stevenson M., Nielsen J. L. and Nielsen P. H. (2013). The Microbial Database for Danish wastewater treatment plants with nutrient removal (MiDas-DK) – a tool for understanding activated sludge population dynamics and community stability. *Water Science & Technology*, **67**, 2519–2526.

Ministerio de Agricultura, Alimentación y Medio Ambiente: Report on authorizations of discharges, 31 December 2014.

Ministry of Environment and Food of Denmark (2015). Punktkilder 2014 (Report: Point sources 2014). www.nst.dk ISBN: 978-87-7091-616-5.

Ministry of Housing and Urban-Rural Development of the People's Republic of China (2015). The Notification about the Construction and Operation of National Urban Sewage Treatment Facilities in the Second Quarter of 2015. http://www.mohurd.gov.cn/wjfb/201508/t20150817_223298.html

Molinari M. and Canziani R. (1992). Analisi Microscopica dei fanghi attivati. In: Rimozione Biologica dell'azoto: impianti per reflui industriali, R. Canziani, M. Dubini and M. Molinari (eds), Istituto per l'Ambiente, Milano, pp. 193–228.

Montoya T., Zornoza A., Granell P., Fayos G., Fajardo V., Zorrilla F., Alonso J. L., Morenilla J. J., Bernacer I. and Martínez F. (2009). Viabilidad celular como herramienta para el control del bulking mediante cloración. VI Jornadas de Transferencia de tecnología sobre Microbiología del Fango Activo. ISBN: 978-84-613-7691-9

Murakami C. and Babcock J. (1998). Effect of anoxic selector detention time and mixing rate on denitrification rate and control of Sphaerotilus natans bulking. Proceedings of the Water Environment Federation, 71st Annual Conference & Exposition, Wastewater Treatment Research and Municipal Wastewater Treatment, Volume 1, Alexandria, VA, USA, pp. 89–98.

Murthy S. N. and Novak J. T. (1998). Influence of cations on activated sludge effluent quality. Proceedings of the Water Environment Federation, 71st Annual Conference & Exposition, Volume 1. Wastewater Treatment Research and Municipal Wastewater Treatment. Alexandria, VA, USA, pp. 309–324.

Murthy S. N., Novak J. T. and De Haas R. D. (1998). Monitoring cations to predict and improve activated sludge settling and dewatering properties of industrial waters. *Water Science & Technology*, **38**(3), 119–126.

Musvoto E. V., Lakay M. T., Casey T. G., Wentzel M. C. and Ekama G. A. (1999). Filamentous organism bulking in nutrient removal activated sludge systems, Paper 8: The effect of nitrate and nitrite. *Water SA*, **25**(4), 397–408.

Nakhla G. and Lugowski A. (1999). Full-scale comparative performance assessment of selectors and intermittent aeration for filament control. Proceedings of the Water Environment Federation, 72nd Annual Conference & Exposition, CD-ROM, Alexandria (VA), USA.

Neef A. (1997). Anwendung der in situ Einzelzell-Identifizierung von Bakterien zur Populationsanalyse in komplexen mikrobiellen Biozonosen. PhD dissertation, Technical University of Munich, Germany.

Ng W. J., Ong S. L. and Hossain F. (2000). An algorithmic approach for system-specific modeling of activated sludge bulking in an SBR. *Environmental and Modelling Software*, **15**, 199–210.

Nielsen J. L. and Nielsen P. H. (2005). Advances in microscopy: microautoradiography of single cells. *Methods in Enzymology*, **397**, 237–256.

Nielsen P. H., Kragelund C., Nielsen J. L., Tiro S., Lebek M., Rosenwinkel K.-H. and Gessesse A. (2005). Control of M. parvicella in activated sludge plants by dosage of polyaluminium salts: possible mechanisms. *Acta hydrochim. Hydrobiologica*, **33**, 255–261.

Nielsen P. H., Kragelund C., Seviour R. J. and Nielsen J. L. (2009). Identity and ecophysiology of filamentous bacteria in activated sludge. *FEMS Microbiology Reviews*, **33**(6), 969–998.

Nittami T., McIlroy S., Seviour E. M., Schroeder S. and Seviour R. J. (2009). *Candidatus* Monilibacter spp., common bulking filaments in activated sludge, are members of Cluster III *Defluviicoccus*. *Systematic and Applied Microbiology*, **32**(7), 480–489.

Noutsopoulos C., Mamais D. and Andreadakis A. (2002). The effect of the reactor configuration and the operational mode on *M. parvicella* bulking and foaming in nutrient removal activated sludge systems. *Water Science & Technology*, **46**(1–2), 61–64.

Noutsopoulos C., Mamais D. and Andreadakis A. (2006). Effect of solids retention time on M. parvicella growth. *Water SA*, **32**, 315–321.

Noutsopoulos C., Andreadakis A., Mamais D. and Gavalakis E. (2007). Identification of type and causes of filamentous bulking under Mediterranean conditions. *Environmental Technology*, **28**, 115–122.

Noutsopoulos C., Mamais D. and Andreadakis A. (2012). A hypothesis on M. parvicella proliferation in biological nutrient removal activated sludge systems with selector tanks. *FEMS Microbiology Ecology*, **80**, 380–389.

Novak J. T. (1997). The effect of nitrification on activated sludge settling properties. Proceedings of the Water Environment Federation, 70st Annual Conference & Exposition, Volume 1, Research and Municipal Wastewater Treatment. Alexandria, VA, USA, pp. 377–384.

Novak J. T., Love N. G., Smith M. L. and Wheeler E. R. (1998). The effect of cationic salt addition on the settling and dewatering properties of an industrial activated sludge. *Water Environment Research*, **70**, 984–996.

Osborn D. W. (1988). Sewage purification in South Africa – past and present. *Water SA*, **3**, 139–151.

Pagilla K. R., Jenkins D. and Kido W. H. (1996). *Nocardia* control in activated sludge by classifying selectors. *Water Environment Research*, **68**, 235–239.

Parker D., Ekama G. A., Melcer H. and van Niekirk A. (1998). Discussion of full-scale evaluation of factors affecting the performance of anoxic selectors, by Marten W.L. and Daigger G.T. *Water Environment Research*, **70**, 1225–1228.

Pasinetti E. and Emondi V. (2005). Survey of activated sludge filamentous microrganisms from several north Italian wastewater treatment plants. Proceedings, 6th EWA International Conference and Exhibition, Teplice (Czech Republic), 10–12 May 2005.

Pedrazzani R., Menoni L., Nembrini S., Manili L. and Bertanza G. (2016). Suitability of Sludge Biotic Index (S.B.I.), Sludge Index (S.I.) and filamentous bacteria analysis for assessing activated sludge process performance: the case of piggery slaughterhouse wastewater. *Journal of Industrial Microbiology & Biotechnology*, **43**(7), 953–964.

Peng Z. X. and He Z. G. (2015). Optimal control of anoxic selector based on activated sludge with different phosphorus removal ability. In: Energy, Environment and Green Building Materials, Sheng A. (ed.), Taylor and Francis Group, London, pp. 29–33.

Petrovski S., Dyson Z. A., Quill E. S., McIlroy S. J., Tillett D. and Seviour R. J. (2011). An examination of the mechanisms for stable foam formation in activated sludge systems. *Water Research*, **45**(5), 2146–2154.

Petrovski S., Tillett D. and Seviour R. J. (2012). Isolation and complete genome sequence of a bacteriophage lysing *Tetrasphaera jenkinsii*, a filamentous bacteria responsible for bulking in activated sludge. *Virus Genes*, **45**(2), 380–388.

Pitman A. R. (1994). Process Design for Nutrient Removal Facicilitities In: Nutrient Removal from Wastewater, N. J. Horan (ed.), Technomic Publishing Company, USA, pp. 167–72.

Pitman A. R. (1998). Nutrient removal: experiences in Johannesburg. *Water and Environmental Journal*, **12**, 338–443.

Pujol R. and Boutin P. (1989). Control of activated sludge bulking: From lab to the plant. *Water Science and Technology*, **21**(6/7), 717–726.

Pujol R. and Arnaud P. (2015). Plaidoyer pour l'optimisation technique et énergétique des stations d'épuration. *Techniques Sciences Méthodes*, **1–2**, 32–38.

Pujol R., Duchene P., Schetrite S. and Canler J. P. (1991). Biological foams in activated-sludge plants – characterization and situation. *Water Research*, **25**, 1399–1404.

Qi R., Yu T., Li Z., Li D., Mino T., Shoji T., Fujie K. and Yang M. (2012). Comparison of conventional and inverted A2/O processes: Phosphorus release and uptake behaviors. *Journal of Environmental Sciences*, **24**(4), 571–578.

Quast C., Pruesse E., Yilmaz P., Gerken J., Schweer T., Yarza P., Peplies J. and Glöckner F. O. (2013). The SILVA ribosomal RNA gene database project: improved data processing and web-based tools. *Nucleic Acids Research*, **41**, D590–D596.

Ramírez G., Dacal E., Van Rossum F., Ferrer C., Castell C., Martínez F., Bernácer I. and Morenilla J. J. (2009). Aplicación de ozono para la eliminación de espumas en la EDAR de Castellón. VI Jornadas de Transferencia de tecnología sobre Microbiología del Fango Activo. ISBN: 978-84-613-7691-9

Rein D. A. and Paulson S. (1997). Selectors for high strength industrial wastewater treatment. Proceedings of the Water Environment Federation 70th Annual Conference & Exposition. Volume 3. Remediation of Soil & Groundwater and Industrial Issues & Treatment Technology, Alexandria, VA, USA, pp. 415–424.

Richard M. G. (ed.) (1989). Activated Sludge Microbiology, Water Pollution Control Federation, Alexandria.

Richard M. G., Hao O. J., Shimiju G. and Jenkins D. (1981). The isolation and characterization of filamentous microorganisms from activated sludge bulking. National Science Foundation Report, December.

Rodríguez E. (2009). Control o descontrol de Galo. VI Jornadas de Transferencia de tecnología sobre Microbiología del Fango Activo. ISBN: 978-84-613-7691-9

Rodríguez E. and Reina E. (2014). Los ejercicios interlaboratorios de fangos activos como sistema de chequeo del personal técnico. X Jornadas de Transferencia de Tecnología sobre Microbiología del Fango Activo. ISBN: 978-84-697-1426-3. DEPÓSITO LEGAL: SE 1854–2014.

Rodríguez E., Isac L., Salas M. D., Fernández N., Zornoza A., Perez-Uz B., Serrano S., Arregui L., Calvo P., Guinea A. and Estévez F. (2008). Manual práctico para el estudio de grupos bioindicadores en fangos activos (Working manual for biological analysis in activated sludge). Ed Reed Business Information-Tecnología del Agua, Barcelona.

Roels T., Dauwe F., Van Damme S., De Wilde K. and Roelandt F. (2002). The influence of PAX-14 on activated sludge systems and in particular on M. parvicella. *Water Science and Technology*, **46**(1–2), 487–490.

Rossetti S. and Tandoi V. (1999). Microbiologia dei fanghi attivati, in Il problema del bulking filamentoso e delle schiume biologiche negli impianti a fanghi attivati, irsa cnr, Ramadori Ed., Quaderni, 110.

Rossetti S., Carucci A. and Rolle E. (1994). Survey on the occurrence of filamentous organisms in municipal wastewater treatment plants relating to their operating conditions. *Water Science and Technology*, **29**(7), 305–308.

Rossetti S., Tomei M. C., Nielsen P. H. and Tandoi V. (2005). "M. parvicella", a filamentous bacterium causing bulking and foaming in activated sludge systems: A review of current knowledge. *FEMS Microbiology Reviews*, **29**(1), 49–64.

Roux A., Munganga G. and Mali-Bolo S. (2014). Market Intelligent Report: Water. South Africa.

Saayman G. B., Schutte C. F. and van Leeuwen J. (1996). The effect of chemical bulking control on biological nutrient removal in a full scale activated sludge plant. *Water Science and Technology*, **34**(3), 275–282.

Saayman G. B., Schutte C. F. and van Leeuwen J. (1999). Chemical control of filamentous sludge bulking in a full-scale biological nutrient removal activated sludge plant. *Ozone: Science and Engineering*, **20**, 1–15.

Seder-Colomina M., Goubet A., Lacroix S., Morin G., Ona-Nguema G., Esposito G., Van Hullebusch E. D. and Pernelle J.-J. (2015). Moderate oxygen depletion as a factor favouring the filamentous growth of *Sphaerotilus natans. Antonie Van Leeuwenhoek*, **107**(5), 1135–1144.

Serrano S. and Arregui L. (2014). Evolución de los parámetros biológicos indicadores en el control de las depuradoras de aguas residuales con sistemas de fango activos. X Jornadas de Transferencia de Tecnología sobre Microbiología del Fango Activo. ISBN: 978-84-697-1426-3. DEPÓSITO LEGAL: SE 1854–2014. Editorial: GBS.

Serrano S., Arregui L., Pérez-Uz B. and Guinea A. (2006). Simbiosis Universidad-Empresa, una relación beneficiosa para nuevas propuestas. Descripción de nuevas especies. III Jornadas Técnicas de Transferencia de Tecnología sobre Microbiología del Fango Activo. Sevilla (España). 26 Octubre, 2006. ISBN 978-84-611-5208-7.

Seviour E., Williams C., Seviour R., Soddell J. and Lindrea K. (1990). A survey of filamentous bacterial populations from foaming activated sludge plants in eastern states of Australia. *Water Research*, **24**(4), 493–498.

Seviour R. and Nielsen P. (2010). Microbial Ecology of Activated Sludge, IWA Publishing, London.

Shao Y. J., Starr M., Kaporis K., Kim H. S. and Jenkins D. (1997). Polymer addition as a solutions to *Nocardia* foaming Problems. *Water Environmental Research*, **69**, 25–27.

Sladka A. and Ottova V. (1973). Filamentous Organisms in Activated Sludge. *Hydrobiologia*, **43**, 285–299.

Soukup B., Rosenbergová R., Todt V. and Chudoba P. (2015). Methods of Microthrix parvicell growth control in practice. Proc. of the 20th CzWA Seminar on New Methods and Procedures in the Operation of Wastewater Treatment Plants, Moravská Třebová, 14–15 April 2015, Czech Rrepublic.

Speirs L., Nittami T., McIlroy S., Schroeder S. and Seviour R. J. (2009). Filamentous bacterium Eikelboom type 0092 in activated sludge plants in Australia is a member of the phylum Chloroflexi. *Applied and Environmental Microbiology*, **75**(8), 2446–2452.

Speirs L., McIlroy S. J., Petrovski S. and Seviour R. J. (2011). The activated sludge bulking filament Eikelboom morphotype 0914 is a member of the Chloroflexi. *Environmental Microbiology Reports*, **3**(2), 159–165.

Speirs L. B. M., Tucci J. and Seviour R. J. (2015). The activated sludge bulking filament Eikelboom morphotype 0803 embraces more than one member of the Chloroflexi. *FEMS Microbiology Ecology*, **91**(9).

Starr M. A., Shao Y. J., Kaporis K., Kim H. S. and Jenkins D. (1996). Polymer addition: a solution to Nocardia foaming problems at the Terminal Island treatment plant, Los Angles, California. Proceedings of the Water Environment Federation, 69st Annual Conference & Exposition, Wastewater Treatment Research and Municipal Wastewater Treatment, Volume 1. Alexandria, VA, USA, pp. 183–188.

Still D. A., Ekama G. A., Wentzel M. C., Casey T. G. and Marais C. v. R. (1996). Filamentous organisms bulking in nutrient removal activated sludge systems. Paper 2: Stimulation of the selector effect under aerobic conditions, *Water SA*, **22**, 97–118.

Stratton H., Seviour R., Soddell J., Blackall L. and Muir D. (1996). The opportunistic pathogen *Nocardia farcinica* is a foam-producing bacterium in activated sludge plants. *Letters in Applied Microbiology*, **22**(5), 342–346.

Strom P. F. and Jenkins D. (1984). Identification and significance of filamentous microorganisms in activated sludge. *Journal of the Water Pollution Control Federation*, **56**, 449–459.

Suarez J. (2014). Prueba e instalación de selectores biológicos en varias EDAR de Madrid. X Jornadas de Transferencia de Tecnología sobre Microbiología del Fango Activo. ISBN: 978-84-697-1426-3. DEPÓSITO LEGAL: SE 1854–2014. Editorial: GBS.

Tandoi V., Jenkins D. and Wanner J. (2006). IWA Scientific Technical Report, 16, Activated Sludge Separation Problems: Theory, Control Measures, Practical Experiences, IWA Publishing, London.

Thirion N. (1983). Chlorination as a control measure for bulking sludge at the Daspoort sewage treatment plant, Pretoria. SA Water Bulletin 13, Water Research Commision, Pretoria, South Africa.

Thompson G. and Forster C. (2003). Bulking in activated sludge plants treating paper mill wastewaters. *Water Research*, **37**, 2636–2644.

Tsai M. W., Wentzel M. C. and Ekama G. A. (2004). The effect of residual ammonia concentration under aerobic conditions on the growth of M. parvicella in bilogocal nutrient removal plants. Proceedings of the 2004 Water Institute of Southern Africa (WISA) Biennial Conference, Cape Town, South Africa.

Vaiopoulou E., Melidis P. and Aivasidis A. (2007). Growth of filamentous bacteria in an enhanced biological phosphorus removal system. *Desalination*, **213**, 288–296.

Váňa M. (1996). Evaluation of the appearance of filamentous microorganisms at the WWTP Liberec and verification of the control of M. parvicella filaments by dosing aluminium coagulants in a pilot-plant model. Report for the Czech grant agency, project No. 206/94/1183 The Control of Activated Sludge Separation Properties in Nutrient Removal Activated Sludge Systems. Inst. Of Chemical Technology, Prague.

Váňa M. and Wanner J. (1996). Report on The Liberec and Jablonec Wastewater Treatment Plant, Newsletters of IAWQ specialist group on Design and Operation of large Wastewater Treatment Plant, March 1996.

Váňa M. and Wanner J. (1998). The effect of aluminium sulphate dosing on the population of activated sludge at the WWTP Sedlec. Proc. Conf. Progressive Methods of Wastewater Treatment, Slovak Association of Wastewater Experts, Tatranská Štrba, 2–5 June 1998, Slovak Republic.

Van den Akker B., Beard H., Kaeding U., Giglio S. and Short M. D. (2010). Exploring the relationship between viscous bulking and ammonia-oxidiser abundance in activated sludge: A comparison on conventional and IFAS systems. *Water Research*, **44**(9), 2919–2929.

Vanysacker L., Denis C., Roels J., Verhaeghe K. and Vankelecom I. F. J. (2014). Development and evaluation of a TaqMan duplex real-time PCR quantification method for reliable enumeration of *Candidatus* Microthrix. *Journal of Microbiological Methods*, **97**, 6–17.

van Leeuwen J. (1988). Bulking control with ozonation in a nutrient removal activated sludge system. *Water SA*, **14**, 119–124.

van Leeuwen J. (1992). A review of the potential application of non-specific activated sludge bulking. *Water SA*, **18**, 101–106.

van Loosdrecht M. C., Nielsen P. H., Lopez-Vazquez C. M. and Brdjanovic D. (2016). Experimental Methods in Wastewater Treatment. IWA Publishing, London.

Vismara R. (1998). Problemi di sedimentabilità dei fanghi attivi. *Ingegneria Ambientale*, **11**, 414–427.

Wagner M., Nielsen P. H., Loy A., Nielsen J. L. and Daims H. (2006). Linking microbial community structure with function: fluorescence in situ hybridization-microautoradiography and isotope arrays. *Current Opinion in Biotechnology*, **17**(1), 83–91.

Wágner D. S., Ramin E., Szabo P., Dechesne A. and Plósz B. G. (2015). Microthrix parvicella abundance associates with activated sludge settling velocity and rheology – Quantifying and modelling filamentous bulking. *Water Research*, **78**, 121–132.

Wang J., Li Q., Qi R., Tandoi V. and Yang M. (2014). Sludge bulking impact on relevant bacterial populations in a full-scale municipal wastewater treatment plant. *Process Biochemistry*, **49**(12), 2258–2265.

Wanner J. (1994a). The Implementation of Bulking Control in the Design of Activated Sludge Plants. *Water Science & Technology*, **29**(7), 193–202.

Wanner J. (1994b). Activated Sludge Bulking and Foaming Control. Technomic Publishing, Co Lancaster, PA, USA.

Wanner J. and Grau P. (1989). Identification of filamentous microorganisms from activated sludge: a compromise between wishes needs and possibilities. *Water Research*, **23**, 883–891.

Wanner J. and Jobbágy A. (2014). Activated sludg solids separation. Chapter 10. In: Activated Sludge – 100 Years and Counting, D. Jenkins and J. Wanner (eds), IWA Publishing, London, pp. 171–194. ISBN 9781780404936

Wanner J., Chudoba J., Kucman K. and Proske L. (1987). Control of activated sludge filamentous bulking – VII. Effect of anoxic conditions. *Water Research*, **21**(12), 1447–1451.

Wanner J., Kucman K. and Grau P. (1988). Activated-sludge process combined with biofilm cultivation. *Water Research*, **22**(2), 207–215.

Wanner J., Růzickova I., Jetmarová P., Krhutkova O. and Paraniakova J. (1998). A national survey of activated sludge separation problems in the Czech Republic: Filaments, floc characteristics and activated sludge metabolic properties. *Water Science & Technology*, **37**(4–5), 271–279.

Wanner J., Ruzickova I., Krhutkova O. and Pribyl M. (2000). Activated sludge population dynamics and wastewater treatment plant design and operation. *Water Science & Technology*, **41**(9), 217–225.

Ward D. B., Oldham K. and Abraham (1999). Resolution of capacity constraints and performance variations in a biological nutrient removal process. Proceedings of the Water Environment Federation, 72nd Annual Conference & Exposition, CD-Rom, Alexandria, VA, USA.

Welz P. J., Esterhuysen A., Vulindlu M. and Bezuidenhout C. (2014). Filament identification and dominance of Eikelboom Type 0092 in activated sludge from wastewater treatment facilities in Cape Town, South Africa. *Water SA*, **40**, 649.

Wiechers H. N. S. (1983). Sludge bulking in Heidelberg, SA Water Bulletin 11, Water Research Commision, Pretoria, South Africa.

Wu Y. C., Peng Y. Z., Wang S. Y., Liu Y., Peng Y. and Li X. L. (2008). Study on Activated Sludge Bulking and Sludge Character istics in A2 /O Process. *China Water and Wastewater*, **24**(23), 33–35.

Xie B., Dai X.-C. and Xu Y.-T. (2007). Cause and pre-alarm control of bulking and foaming by M. parvicella. A case study in triple oxidation ditch at a wastewater treatment plant. *Journal of Hazardous Materials*, **143**(1–2), 184–191.

Xu Y. B., Duan X. J., Yan J. N. and Sun S. Y. (2009). Influence of magnetic field on Cr (VI) adsorption capability of given anaerobic sludge. *Biodegradation*, **21**(1), 1–10.

Yavuz H. and Çelebi S. S. (2000). Effect of magnetic field on activity of activated sludge in wastewater treatment. *Enzyme Microbial Technology*, **26**, 22–27.

Zaidi N. S., Sohaili J., Muda K. and Sillanpää M. (2014). Magnetic field application and its potential in water and wastewater treatment systems: A review. *Separation Purification Reviews*, **43**, 206–240.

Zhang S., Wei W., Zhang J., Mao Y. and Liu S. (2002). Effect of static magnetic field on growth of *Escherichia coli* and relative response model of series piezoelectric quartz crystal. *Analyst*, **127**, 373–377.

Index

Note: Page numbers in *italics* refer to Figures while those in **bold** denote Tables.